Bolyai Society Mathematical Studies

Volume 30

Editor-in-Chief

János Pach, HUN-REN Alfréd Rényi Institute of Mathematics, Budapest, Hungary

Series Editor

Gergely Ambrus, HUN-REN Alfréd Rényi Institute of Mathematics, Budapest, Hungary

Editorial Board Members

Zoltán Buczolich, ELTE Eötvös Loránd University, Budapest, Hungary

Lajos Hajdu, University of Debrecen, Debrecen, Hungary

Lajos Molnár, University of Szeged, Szeged, Hungary

László Lovász, HUN-REN Alfréd Rényi Institute of Mathematics, Budapest, Hungary

Péter Pál Pálfy, HUN-REN Alfréd Rényi Institute of Mathematics, Budapest, Hungary

András Stipsicz, HUN-REN Alfréd Rényi Institute of Mathematics, Budapest, Hungary

Bálint Tóth, HUN-REN Alfréd Rényi Institute of Mathematics, Budapest, Hungary

This series will typically consist of survey volumes. It may also include conference proceedings and monographs.

Javier Fernández de Bobadilla • Marco Marengon •
András Némethi • András Stipsicz
Editors

Singularities and Low Dimensional Topology

BOLYAI SOCIETY

Springer

Editors

Javier Fernández de Bobadilla
Basque Foundation for Science
IKERBASQUE
Bilbao, Spain

András Némethi
Rényi Institute of Mathematics
Budapest, Hungary

Marco Marengon
Rényi Institute of Mathematics
Budapest, Hungary

András Stipsicz
Rényi Institute of Mathematics
Budapest, Hungary

ISSN 1217-4696 ISSN 2947-9460 (electronic)
Bolyai Society Mathematical Studies
ISBN 978-3-031-56613-4 ISBN 978-3-031-56611-0 (eBook)
https://doi.org/10.1007/978-3-031-56611-0

Mathematics Subject Classification: 14H20, 14J17, 57K10, 57K41, 57K18

© The Editor(s) (if applicable) and The Author(s), under exclusive license to Springer Nature Switzerland AG 2024

This work is subject to copyright. All rights are solely and exclusively licensed by the Publisher, whether the whole or part of the material is concerned, specifically the rights of translation, reprinting, reuse of illustrations, recitation, broadcasting, reproduction on microfilms or in any other physical way, and transmission or information storage and retrieval, electronic adaptation, computer software, or by similar or dissimilar methodology now known or hereafter developed.
The use of general descriptive names, registered names, trademarks, service marks, etc. in this publication does not imply, even in the absence of a specific statement, that such names are exempt from the relevant protective laws and regulations and therefore free for general use.
The publisher, the authors and the editors are safe to assume that the advice and information in this book are believed to be true and accurate at the date of publication. Neither the publisher nor the authors or the editors give a warranty, expressed or implied, with respect to the material contained herein or for any errors or omissions that may have been made. The publisher remains neutral with regard to jurisdictional claims in published maps and institutional affiliations.

Cover illustration: The image on the cover features the iconic Chain Bridge of Budapest rendered as a surface with singularities and compressing circles. Symbolizing both its architectural significance and its connection to objects studied in low-dimensional topology, the bridge serves as a testament to the themes explored within this volume. The figure was designed by Marco Marengon.

This Springer imprint is published by the registered company Springer Nature Switzerland AG
The registered company address is: Gewerbestrasse 11, 6330 Cham, Switzerland

If disposing of this product, please recycle the paper.

Preface

The Spring semester of 2023 at the Erdős Center (Rényi Institute, Budapest) was devoted to the study of the natural interactions between singularity theory and low-dimensional topology.

A fundamental example of a connection between singularities and low-dimensional topology is the concept of the *link* of a singularity: given a complex analytic surface together with an isolated singular point, the intersection of the variety with a small sphere centred at the singular point is an oriented compact three-manifold. For example, the link of the singularity defined by the equation $x^2 + y^3 + z^5 = 0$ (where the singular point is the origin) is the famous Poincaré homology sphere, a manifold well-known to low-dimensional topologists. In general, the complex algebro-geometric behaviour of a singularity imposes severe restrictions on the differential topology of its linking three-manifold. An example of this close connection is provided by the fact that the singularity is rational (a property which demands the vanishing of certain cohomology groups) if and only if the link has simple gauge-theoretic invariants (in technical terms: it is an L-space). Of course, the connection between the two subjects is much deeper and goes way beyond this equivalence.

Another example of the natural ties between singularities and low-dimensional topology is that of complex analytic plane curve singularities, which are characterised up to local homeomorphism by (algebraic) knots and links in the three-sphere. In such a case, a deep and challenging question is to compare the *topological* invariants of the algebraic link in the three-sphere with the *analytic* invariants of the singular germ, for example, the Jones polynomial of the link with the Hilbert schemes of the germ.

Due to the strong connections between singularity theory and low-dimensional topology, results in one of these two neighbouring fields can have strong impacts on the other, but unfortunately the difference in techniques and language used in these two research areas can sometimes discourage researchers from investing time in learning about recent developments in the other subject. To address this problem, in this volume we collected six expository contributions, all of which aims to provide a gentle introduction to some aspects of these fields. These manuscripts are based on notes taken during six lecture series given during the special semester at the Erdős Center. The first four notes cover the topics of the lecture series given at the Winter School in January 2023, whose aim was to introduce the main themes of the special semester to doctoral students and advanced master students. Throughout the semester, several 'focused lecture series' were organised to discuss details of more advanced, novel results and topics in low-dimensional topology and singularity theory. The lecture notes of two of these focused lecture series make up the last two chapters of this volume. Below we give a quick overview of the topics discussed in the individual notes.

The first lecture series (given by Michael Wong, with notes taken by Sarah Zampa) introduces Heegaard Floer homology through its (technically) simplest reincarnation, namely grid homology. This version of the general 'Heegaard Floer package' displays all the important features of the theory, but avoids the sometimes daunting analytic details in the definitions and proofs of the basic features. Nevertheless, the general theory is also sketched in the lectures.

The second lecture series (by Kyle Hayden, with notes taken by Márton Beke) discusses the basics of four-dimensional manifold topology, and connects this rich subject to knot theory on one side, and to contact, symplectic and complex analysis (through Stein surfaces) on the other side of the spectrum. The lectures cover parts of Kirby calculus (providing pictorial presentations of handle decompositions), and also discuss embedded surfaces in four-manifolds.

The third lecture series, presented by Duco van Straten (notes taken by László Koltai), discusses invariants of complex plane curve singularities. It starts with the delta-invariant and certain deformations of germs. The 'Severi strata' are introduced as the parameter spaces of adjacencies with prescribed number of nodes. Then a mysterious coincidence of Oblomkov and Shende (now a theorem of Maulik) is discussed: this connects the multiplicities of the Severi strata of a plane curve singularity with certain coefficients of the HOMFLY polynomial of the corresponding algebraic link.

Lattice cohomology plays the main role in the fourth lecture series (given by András Némethi, with notes taken by Tamás Ágoston). This cohomology theory provides a strong connection between the complex geometry of an isolated surface singularity and the topology of its link (which is an oriented three-manifold): while the cohomology theory is defined using algebro-geometric considerations, the resulting groups and modules turn out to depend only on the topology of the linking three-manifold, and are in fact isomorphic to some versions of its Heegaard Floer homology. In addition, the algebraic framework of lattice cohomology allows for generalisations to higher dimensions, and can be modified to study curve singularities and their delta-invariants as well.

The last two lecture series focus on the recent concept of *spacification*. The introduction of Khovanov homology and knot Floer homology provided 'categorifications' of famous knot invariants (the Jones and the Alexander polynomials): in this procedure the polynomials are enhanced to homology theories, so that the graded Euler characteristics of the homologies give back the polynomial invariants. 'Spacification' takes this procedure one step further, by constructing spaces (more precisely spectra) which have the aforementioned homological invariants as their *singular homologies*, and the spectra themselves are invariants of the objects of study (knots, links, or three-manifolds). This idea is relatively new, and we are at the beginning of understanding its full potential for the subject.

The fifth lecture series (given by Matthew Stoffregen) addresses the foundational aspects of Monopole Floer spectra. After introducing the Seiberg-Witten equations and the Bauer-Furuta invariants, and outlining the proof of Furuta's $10/8$ theorem, a connection between these spectra and lattice homology is also discussed.

In the concluding lecture series, Sucharit Sarkar (notes by Marco Marengon and András Stipsicz) describes the spacification of Khovanov homology constructed by himself and Robert Lipshitz, and his work with Ciprian Manolescu on the spacification of knot Floer homology; once again, to avoid some of the complications, the knot Floer spectrum is constructed using the approach provided by grid homology.

This set of notes contains several interesting and important aspects of a very dynamically changing field, and we hope that the lecture series will serve as a help and guidance in understanding these aspects on their own, and also shed light on the numerous connections among the various approaches.

This volume is a combined effort of several people. We would like to thank the staff of the Erdős Center, in particular Ildikó Bogáti and Annamária Tárczy, for their excellent work in organising the special semester. Special thanks go to Márton Beke, who organised the files into a coherent volume.

Bilbao, Spain — Javier Fernández de Bobadilla
Budapest, Hungary — Marco Marengon
Budapest, Hungary — András Némethi
Budapest, Hungary — András Stipsicz

Contents

1 **Heegaard Floer Homology** 1
C.-M. Michael Wong and Sarah Zampa

2 **A Knot-Theoretic Tour of Dimension Four** 45
Kyle Hayden and Márton Beke

3 **Severi Strata of Plane Curve Singularities** 77
Duco van Straten and László Koltai

4 **Lattice Cohomology** 107
András Némethi and Tamás Ágoston

5 **A Tour of Monopole Floer Spectra** 139
Matthew Stoffregen

6 **Spectra in Khovanov and Knot Floer Theories** 187
Sucharit Sarkar, Marco Marengon, and András Stipsicz

Chapter 1
Heegaard Floer Homology

C.-M. Michael Wong and Sarah Zampa

Contents

1.1 Grid Homology 3
1.1.1 Definition and $\partial^2 = 0$ 3
1.1.2 Invariance 7
1.1.3 Maslov and Alexander Gradings 13
1.1.4 Graded Isomorphism Types 14
1.1.5 Other Flavors 16
1.2 "Naive" Heegaard Floer Homology 17
1.2.1 Definition 17
1.2.2 Admissibility 20
1.2.3 $\partial^2 \equiv 0$ 20
1.3 Lagrangian Floer Homology 22
1.3.1 Setup 22
1.3.2 Morse Homology: An Interlude 23
1.3.3 The Differential 24
1.3.4 Technicalities and Additional Structures 28
1.4 Heegaard Floer Homology 30
1.4.1 Definition 30
1.4.2 Unraveling the Definition 31
1.4.3 Spin^c-Decomposition, Gradings, Admissibility, and Invariance 33
1.4.4 Some Computational Tools 36
1.4.5 Knot Floer Homology: Definition and Features 38
1.4.6 Multi-Pointed Heegaard and Knot Floer Homologies: Definition and Relationship with Grid Homology 41
References 42

C.-M.M. Wong
University of Ottawa, Ottawa, ON, Canada
e-mail: mike.wong@uottawa.ca

S. Zampa (✉)
Budapest University of Technology and Economics, Budapest, Hungary
e-mail: zampa.sarah@renyi.hu

© The Author(s), under exclusive license to Springer Nature Switzerland AG 2024
J. Fernández de Bobadilla et al. (eds.), *Singularities and Low Dimensional Topology*,
Bolyai Society Mathematical Studies 30,
https://doi.org/10.1007/978-3-031-56611-0_1

Heegaard Floer homology has grown to be a vast subject in low-dimensional topology since its inception in the early 2000s. It is thus impossible to cover in four lectures its numerous features, variants, applications, and connections to other theories. These lecture notes are aimed at students and researchers who are not yet acquainted with the theory, and represent my attempt at providing the context of how various ideas fit in it, in a short amount of time. As a result, we will inevitably be vague at times and omit details that the expert will find important. We will also have to omit or glide over many significant contributions by members of the Floer theory community, and the reader is encouraged to fill in these gaps by consulting other references.

Overview

Floer homology is a kind of Morse homology on an infinite-dimensional space. A particular example is Heegaard Floer homology, a 3-manifold invariant defined by Ozsváth and Szabó in the early 2000s. The following is a simplified overview of one way to construct it, as well as some properties.

Given a 3-manifold Y, we take a Heegaard diagram $\mathcal{H}$ for Y, and use $\mathcal{H}$ to construct an infinite-dimensional manifold $\mathcal{P}(\mathcal{H})$ with a functional $\mathcal{A}$. With some further choices, we then apply "Morse homology" to the pair $(\mathcal{P}(\mathcal{H}), \mathcal{A})$ to get a chain complex $\mathrm{CF}^\circ(\mathcal{H})$ over a nice ring $\mathcal{R}$ (for example, the field $\mathbb{F} = \mathbb{F}_2$ of two elements), whose homotopy type turns out to depend only on Y—under further constraints on $\mathcal{H}$ and the choices along the way. Consequently, we obtain invariants $\mathrm{CF}^\circ(Y)$ and $\mathrm{HF}^\circ(Y)$, where $\mathrm{HF}^\circ(Y)$ is the homology of $\mathrm{CF}^\circ(Y)$. There are also "relative" invariants $\mathrm{CFK}^\circ(Y, K)$ and $\mathrm{HFK}^\circ(Y, K)$ for knots $K \subset Y$, defined similarly; note, however, that $\mathrm{CFK}^\circ(Y, K)$ is often a filtered complex and $\mathrm{HFK}^\circ(Y, K)$ is the *associated graded homology* of $\mathrm{CFK}^\circ(Y, K)$ rather than its honest homology.

Here are some basic structural properties of the Heegaard Floer complex:

- It admits a direct sum decomposition by elements of $\mathrm{Spin}^c(Y)$, the set of Spin^c-structures on Y;
- It has a homological $\mathbb{Z}/m$-grading, known as the *Maslov grading*. In many cases (e.g., when $b_1(Y) = 0$), we have $m = 0$, and so we get a $\mathbb{Z}$-graded chain complex;
- There are many different flavors: $\widehat{\mathrm{CF}}$, CF^-, CF^+, CF^∞, $\widetilde{\mathrm{CF}}$,[1] which are defined over different rings $\mathcal{R}$. (The notation CF° denotes any of these flavors.)

In my experience, in learning Heegaard Floer theory, it can be somewhat overwhelming to try to absorb the large amount of geometry and analysis required in its definition while keeping track of the relevant algebraic structures. Thus, in

[1] As we shall see later, the *tilde* version is almost but not quite an invariant.

most of the first two lectures, we will separate these two aspects, focusing on the algebra first via a combinatorial model of $\mathrm{CFK}^\circ(S^3, K)$ known as *grid homology*.

1.1 Grid Homology

1.1.1 Definition and $\partial^2 = 0$

Grid homology was first defined in [MOS09], and the first combinatorial proof of its invariance appeared in [MOSzT07]. A comprehensive treatise can be found in [OSSz15], which we will mostly be following. (However, we will be omitting a lot of details and proofs, and the reader is encouraged to consult [OSSz15] for a fuller understanding.)

Definition 1.1.1 A (toroidal) *grid diagram* $\mathcal{G} = (n, \mathbb{O}, \mathbb{X})$ of an oriented knot $K \subset S^3$ is a diagram of an $n \times n$ grid, with a set $\mathbb{O}$ of O's and a set $\mathbb{X}$ of X's, such that there is exactly one O and one X in each row and in each column. (There is often an additional condition that any square is not occupied by an O and an X at the same time.) The diagram is to be thought of as lying on a torus, with the leftmost and rightmost (resp. topmost and bottommost) lines identified into circles. See Fig. 1.1 for an example.

The diagram $\mathcal{G}$ encodes a knot K as follows: Connect the X's to the O's vertically, and the O's to the X's horizontally, with the vertical strands crossing over the horizontal strands; this gives a rectilinear projection of an oriented knot K. We say that $\mathcal{G}$ is a *grid diagram of* K. It is not hard to see that any knot $K \subset S^3$ (and indeed any link in S^3) can be presented this way.

We now define a chain complex associated to a grid diagram.

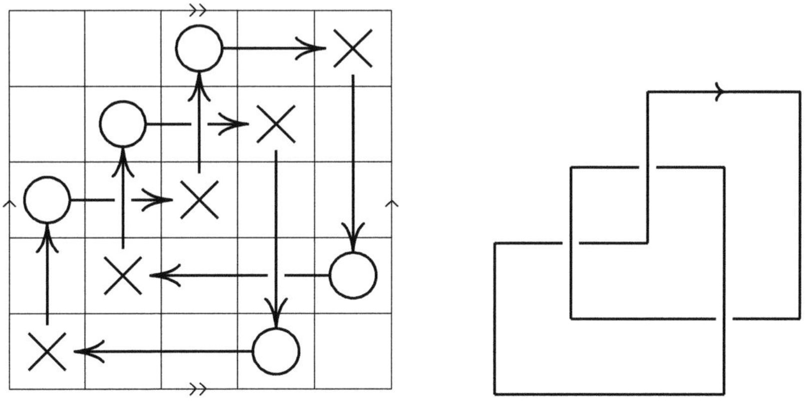

Fig. 1.1 A grid diagram of the left-handed trefoil

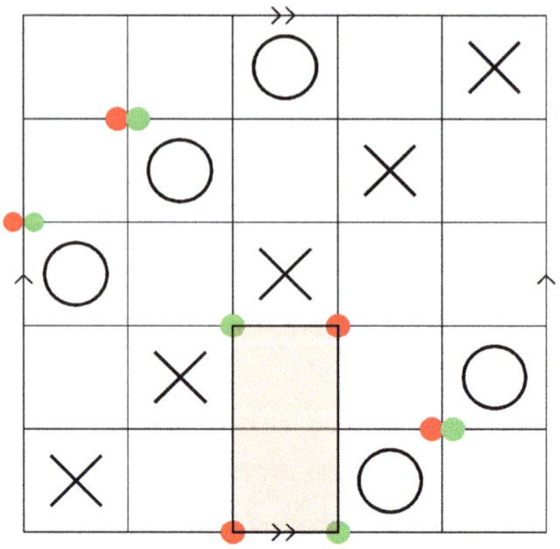

Fig. 1.2 Two generators of $S(\mathcal{G})$ on a diagram $\mathcal{G}$ of the left-handed trefoil

Definition 1.1.2 Given a grid diagram $\mathcal{G}$, we define the generating set

$$\mathcal{S}(\mathcal{G}) = \{\text{matchings between vertical and horizontal circles in } \mathcal{G}\},$$

where a matching is a one-to-one correspondence between the vertical and horizontal circles. A matching is denoted on the grid diagram by dots on the intersections between vertical and horizontal circles. Figure 1.2 illustrates two different matchings, in red and in green respectively.

The *tilde grid chain complex* is now defined as the $\mathbb{F}$-module

$$\widetilde{\mathrm{GC}}(\mathcal{G}) = \mathbb{F}\langle \mathcal{S}(\mathcal{G})\rangle,$$

with boundary homomorphism $\widetilde{\partial}\colon \widetilde{\mathrm{GC}}(\mathcal{G}) \to \widetilde{\mathrm{GC}}(\mathcal{G})$ given by

$$\widetilde{\partial}(\mathbf{x}) = \sum_{\mathbf{y}\in\mathcal{S}(\mathcal{G})} \sum_{\substack{r\in\mathrm{Rect}^\circ(\mathbf{x},\mathbf{y}) \\ r\cap(\mathbb{O}\cup\mathbb{X})=\emptyset}} \mathbf{y},$$

where $\mathrm{Rect}^\circ(\mathbf{x}, \mathbf{y})$ denotes the space of empty rectangles from $\mathbf{x}$ to $\mathbf{y}$. A *rectangle r from $\mathbf{x}$ to $\mathbf{y}$* is a 2-chain on the (toroidal) grid diagram, such that ∂r consists exactly of two arcs on horizontal circles from a point in $\mathbf{x}$ to a point in $\mathbf{y}$ and two arcs on vertical circles from a point in $\mathbf{y}$ to a point in $\mathbf{x}$, in its induced orientation. A succinct way to say the above is that ∂r consists of 4 arcs, and

$$\partial((\partial r) \cap (\text{horizontal circles})) = \mathbf{y} - \mathbf{x}.$$

Such a rectangle exists only if $\mathbf{x}$ and $\mathbf{y}$ coincide everywhere except at two points each. A rectangle is *empty* if it does not contain any points in $\mathbf{x}$ (or equivalently $\mathbf{y}$) in its interior.

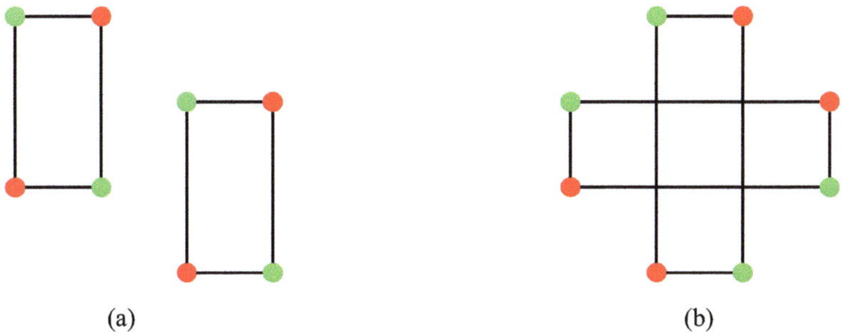

Fig. 1.3 Two rectangles with no shared vertex. (**a**) Disjoint rectangles. (**b**) Intersecting rectangles with no shared vertex

In other words, the differential counts rectangles r whose northeast and southwest corners are in $\mathbf{x}$, whose northwest and southeast corners are in y, and which do not contain any other points in $\mathbf{x}$ or $\mathbf{y}$. In this *tilde* flavor, the contributing rectangles r are further required to not contain any O's or X's in its interior, expressed in the requirement $r \cap (\mathbb{O} \cup \mathbb{X}) = \emptyset$.

Example 1.1.3 Consider the diagram of the left-handed trefoil again.

The generators $\mathbf{x}$ (in red) and $\mathbf{y}$ (in green) coincide everywhere, except on the vertices of the empty rectangle (shaded in orange) from $\mathbf{x}$ to $\mathbf{y}$.

Of course, one needs to show that $\widetilde{\mathrm{GC}}$ is indeed a chain complex.

Proposition 1.1.4 *$\widetilde{\partial}$ is a boundary map; i.e., $\widetilde{\partial} \circ \widetilde{\partial} = 0$.*

Proof Suppose that $\mathbf{z}$ is a generator that appears as a term in $\widetilde{\partial} \circ \widetilde{\partial}(\mathbf{x})$. We will show that there are always an even number of ways to connect $\mathbf{x}$ to $\mathbf{z}$ with two empty rectangles in order. Since $\widetilde{\mathrm{GC}}$ is defined over $\mathbb{F} = \mathbb{F}_2$, this will complete the proof.

Marking $\mathbf{x}$ in red and $\mathbf{z}$ in green in the diagrams below, there are three different possible configurations of the rectangles. The first two configurations are where the rectangles do not share a vertex, as in Fig. 1.3.

In these two cases, there are two distinct ways to count the rectangles: either counting r_1 first then r_2—connecting $\mathbf{x}$ to some $\mathbf{y}$ and then to $\mathbf{z}$—or counting r_2 first then r_1—connecting $\mathbf{x}$ to some $\mathbf{w}$ and then to $\mathbf{z}$.[2]

The third case consists of L-shaped domains that arise from juxtaposing two rectangles sharing a single vertex, as in Fig. 1.4, and its reflections and rotations. In this case, with the same configuration for $\mathbf{x}$ (in red) and $\mathbf{z}$ (in green), there are two ways of decomposing such an L-shaped domain into rectangles.

As in the left of Fig. 1.4, we can first count the bottom left rectangle (shaded in orange), arriving at the intermediate generator $\mathbf{y}$ (in blue), before counting the right rectangle (in red), ending in $\mathbf{z}$. Or, as in the right of the figure, we can first count the

[2] Technically, using r_i to denote the rectangles in different orders is a slight abuse of notation, since, e.g., the first r_1 belongs to $\mathrm{Rect}^\circ(\mathbf{x}, \mathbf{y})$, while the second r_1 belongs to $\mathrm{Rect}^\circ(\mathbf{w}, \mathbf{z})$.

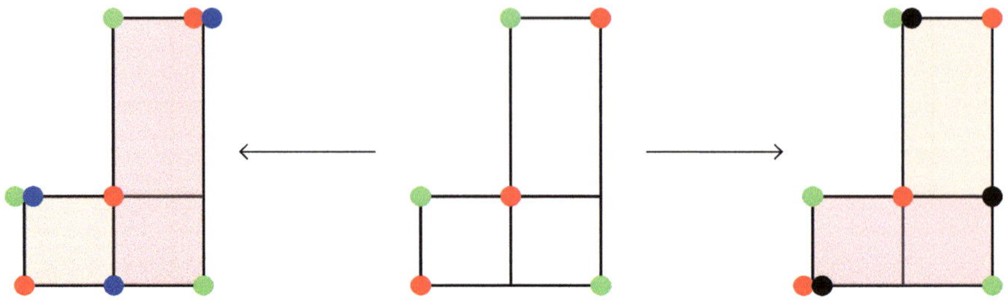

Fig. 1.4 Intersecting rectangles with a shared vertex

top right rectangle (shaded in red), giving the intermediate generator **w** (in black), before counting the bottom rectangle (in orange), again ending in **z**. This shows that such decompositions always come in pairs, concluding the argument. □

Remark 1.1.5 One can similarly define $\widetilde{\mathrm{GC}}(\mathcal{G})$ over $\mathbb{Z}$, but it requires adding signs to all definitions and proofs, which represents significantly more work.

In a similar fashion, we can define the *minus* flavor of the grid chain complex.

Definition 1.1.6 The *minus grid chain complex* is the $\mathbb{F}[U_1, \ldots, U_n]$-module

$$\mathrm{GC}^-(\mathcal{G}) = \mathbb{F}[U_1, \ldots, U_n]\langle \mathcal{S}(\mathcal{G})\rangle,$$

with boundary homomorphism $\partial^-\colon \mathrm{GC}^-(\mathcal{G}) \to \mathrm{GC}^-(\mathcal{G})$ given by

$$\partial^-(\mathbf{x}) = \sum_{\mathbf{y}\in\mathcal{S}(\mathcal{G})} \sum_{\substack{r\in\mathrm{Rect}^\circ(\mathbf{x},\mathbf{y})\\ r\cap\mathbb{X}=\emptyset}} U_1^{O_1(r)} \cdots U_n^{O_n(r)}\mathbf{y},$$

where $O_i(r)$ is the number of times O_i appears in r, which is either 0 or 1. Here, we have fixed an (arbitrary) indexing of the elements of $\mathbb{O}$, from O_1 to O_n.

In other words, in the *minus* flavor, we allow O's in the rectangles that are counted in the boundary homomorphism, and record them by formal variables U_i. We continue to block all X's as before.

Proposition 1.1.7 *∂^- is also a boundary map, i.e., $\partial^- \circ \partial^- = 0$.*

Proof The proof is identical to that for $\widetilde{\partial}$, *mutatis mutandis.* □

Since $\widetilde{\mathrm{GC}}(\mathcal{G})$ and $\mathrm{GC}^-(\mathcal{G})$ are chain complexes, we can compute their homologies, denoted $\widetilde{\mathrm{GH}}(\mathcal{G})$ and $\mathrm{GH}^-(\mathcal{G})$ respectively.

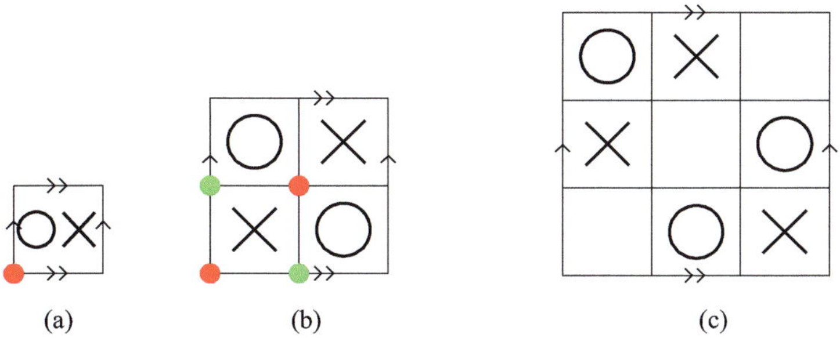

Fig. 1.5 Grid diagrams of the unknot of different sizes. (**a**) $\mathcal{G}_1$. (**b**) $\mathcal{G}_2$. (**c**) $\mathcal{G}_3$

1.1.2 Invariance

The natural question to ask is if $\widetilde{\mathrm{GH}}(\mathcal{G})$ and $\mathrm{GH}^-(\mathcal{G})$ are invariants of the knot K that $\mathcal{G}$ encodes. The following example shows that $\widetilde{\mathrm{GH}}(\mathcal{G})$ is not an invariant.

Example 1.1.8 Figure 1.5 shows three grid diagrams for the unknot. (The leftmost 1×1 grid does not satisfy the additional condition in Definition 1.1.1 but is otherwise a fine grid diagram.)

We can compute:

- $\widetilde{\mathrm{GH}}(\mathcal{G}_1) \cong \widetilde{\mathrm{GC}}(\mathcal{G}_1) \cong \mathbb{F}$, because there is only one generator (in red in Fig. 1.5a), and $\widetilde{\partial} \equiv 0$ as there are no rectangles;
- $\widetilde{\mathrm{GH}}(\mathcal{G}_2) \cong \widetilde{\mathrm{GC}}(\mathcal{G}_2) \cong \mathbb{F} \oplus \mathbb{F}$, since there are two matchings, and again $\widetilde{\partial} \equiv 0$ as each of the four empty rectangles contains an O or an X;
- $\widetilde{\mathrm{GH}}(\mathcal{G}_3) \cong \mathbb{F}^{\oplus 4}$, left as an exercise for the reader. (There are $3! = 6$ generators in $\widetilde{\mathrm{GC}}(\mathcal{G}_3)$.)

What about $\mathrm{GH}^-(\mathcal{G})$? It turns out that it is an invariant of K, but in a slightly subtle way. In particular, note that if $\mathcal{G}_1$ and $\mathcal{G}_2$ are grid diagrams of K of different sizes, $\mathrm{GC}^-(\mathcal{G}_1)$ and $\mathrm{GC}^-(\mathcal{G}_2)$ are *a priori* modules over different rings.

How might one prove the invariance of grid homology? To do so, one can use the following theorem by Cromwell, that may be thought of as the "Reidemeister Theorem" for grid diagrams.

Theorem 1.1.9 (Cromwell [Cro95]) *Two grid diagrams $\mathcal{G}_1$ and $\mathcal{G}_2$ represent the same knot $K \subset S^3$ if and only if they are related by a finite sequence of row and column commutations, and (de)stabilizations.*

A *column commutation* swaps two adjacent columns when either the two corresponding vertical strands project to disjoint vertical intervals, or one projected vertical interval is contained in the interior of the other. See Fig. 1.6 for examples. A *row commutation* is similar but swaps adjacent rows.

Given a square in the grid diagram that is occupied by either an X or an O, a *stabilization* splits the row and column of the square the symbol is in into two rows

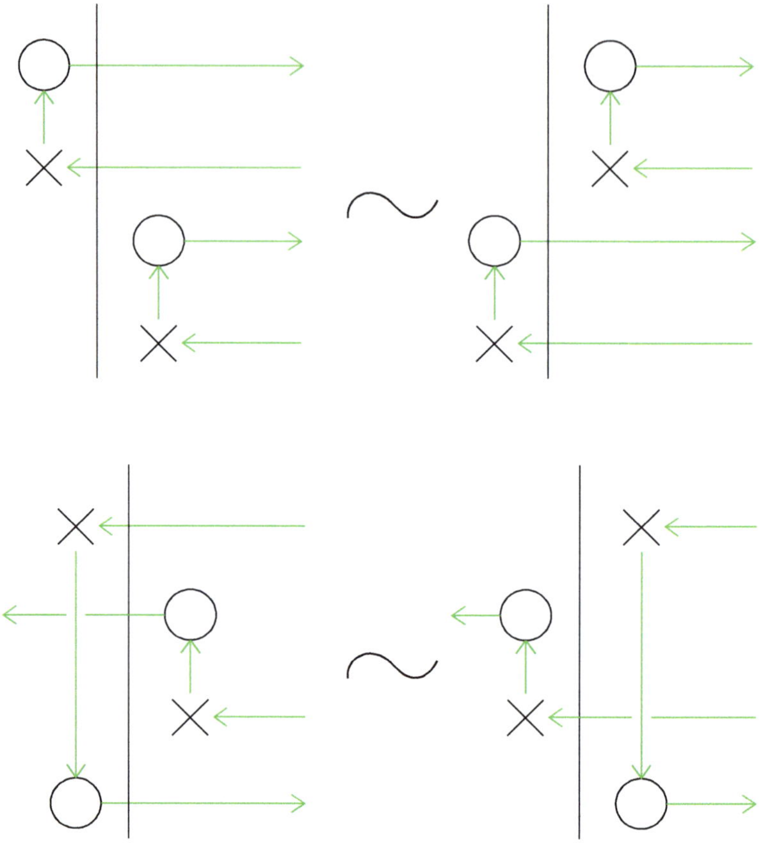

Fig. 1.6 Two column commutations

and two columns, replacing the 1×1 marked square with a 2×2 grid of which three squares are marked. For a given marker type (say, an X), there are four distinct ways to achieve this, corresponding to which square in the 2×2 grid is unmarked. The rest of the diagram is left unchanged. See Fig. 1.7 for examples. *Destabilization* is the process inverse to stabilization.

Proposition 1.1.10 *The isomorphism types of* $\widetilde{\mathrm{GH}}$ *and* GH^- *are invariant under a row or column commutation.*

Proof Suppose that $\mathcal{G}_1$ and $\mathcal{G}_2$ are related by a column commutation. We can combine $\mathcal{G}_1$ and $\mathcal{G}_2$ into a single diagram, as in the right of Fig. 1.8. Here, if we focus on the blue circle and erase the green circle, we obtain (a diagram isotopic to) $\mathcal{G}_1$; similarly, erasing the blue circle, we obtain $\mathcal{G}_2$. Note that we position the blue and green curves so as to avoid triple intersections with horizontal circles. Label the intersection points between and blue and green circles by a and b as shown.

Fig. 1.7 Stabilization; the diagram shows two possibilities out of the four

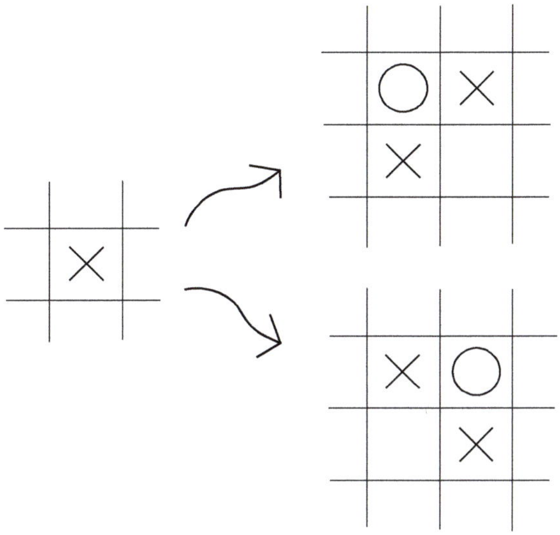

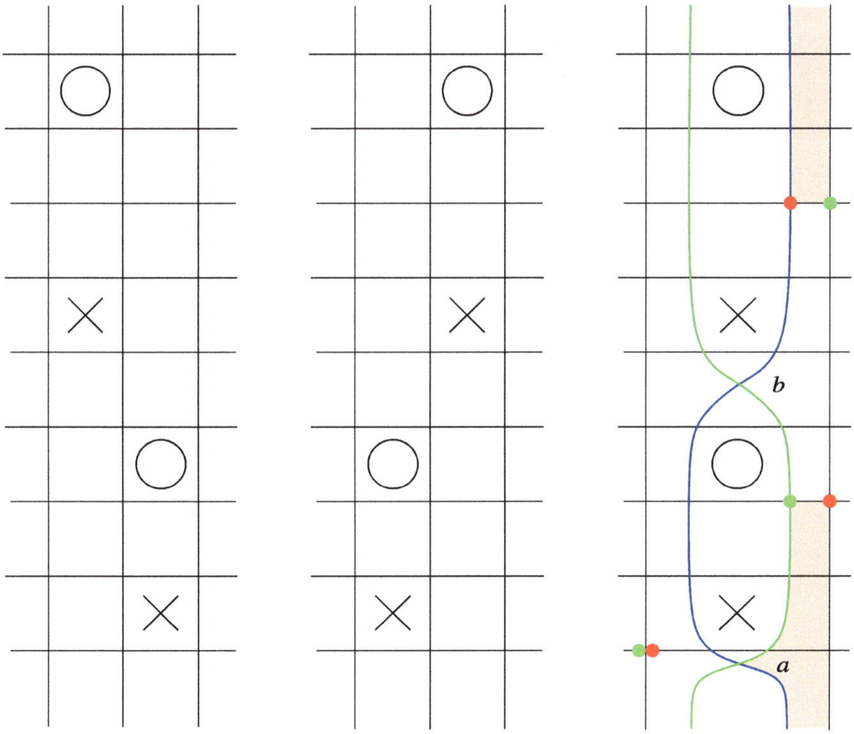

Fig. 1.8 Combining two grid diagrams related by a column commutation into one diagram

The idea now is to use this combined diagram to define a chain homotopy equivalence $P_1\colon \mathrm{GC}^-(\mathcal{G}_1) \to \mathrm{GC}^-(\mathcal{G}_2)$, which counts empty pentagons. Instead of defining empty pentagons rigorously, we will just refer to Fig. 1.8, and say that for $\mathbf{x} \in \mathcal{S}(\mathcal{G}_1)$ and $\mathbf{y} \in \mathcal{S}(\mathcal{G}_2)$, an element of $\mathrm{Pent}^\circ(\mathbf{x}, \mathbf{y})$ is an embedded pentagon whose sides lie on the curves in the combined diagram, with vertices on $\mathbf{x}$, $\mathbf{y}$, and the distinguished point a, and leave the details to the reader. (Note that pentagons can exist to the left or the right of the column in question.) Then P_1 is defined by

$$P_1(\mathbf{x}) = \sum_{\mathbf{y}\in\mathcal{S}(\mathcal{G}_2)} \sum_{\substack{p\in\mathrm{Pent}^\circ(\mathbf{x},\mathbf{y})\\ p\cap\mathbb{X}=\emptyset}} U_1^{O_1(p)} \cdots U_n^{O_n(p)} \mathbf{y}.$$

First, we need to show that P_1 is a chain map, i.e.,

$$\partial^-_{\mathrm{GC}^-(\mathcal{G}_2)} \circ P_1 + P_1 \circ \partial^-_{\mathrm{GC}^-(\mathcal{G}_1)} = 0.$$

(Recall that we are in characteristic 2.) The proof is similar to the proof that $\partial^- \circ \partial^- = 0$, but we juxtapose a pentagon with a rectangle (in either order), instead of two rectangles.

Next, we need to show that P_1 has a homotopy inverse $P_2\colon \mathrm{GC}^-(\mathcal{G}_2) \to \mathrm{GC}^-(\mathcal{G}_1)$. The definition of P_2 is *verbatim* identical to that of P_1, but this time we count empty pentagons with a vertex on the distinguished point b instead of a. The homotopies $H_i\colon \mathrm{GC}^-(\mathcal{G}_i) \to \mathrm{GC}^-(\mathcal{G}_i)$ are defined by counting empty hexagons that have both a and b as vertices. The claim is then that

$$P_2 \circ P_1 + \partial^-_{\mathrm{GC}^-(\mathcal{G}_1)} \circ H_1 + H_1 \circ \partial^-_{\mathrm{GC}^-(\mathcal{G}_1)} = \mathrm{Id}_{\mathrm{GC}^-(\mathcal{G}_1)},$$

$$P_1 \circ P_2 + \partial^-_{\mathrm{GC}^-(\mathcal{G}_2)} \circ H_2 + H_2 \circ \partial^-_{\mathrm{GC}^-(\mathcal{G}_2)} = \mathrm{Id}_{\mathrm{GC}^-(\mathcal{G}_2)}.$$

Indeed, most terms that appear on the left-hand side of these equations cancel out in pairs as in the proof that $\partial^- \circ \partial^- = 0$, with the exception of domains that are topologically annuli, as in Fig. 1.9, which cancel with the identity maps on the right-hand side.

The paragraphs above show that $(P_1)_*\colon \mathrm{GH}^-(\mathcal{G}_1) \to \mathrm{GH}^-(\mathcal{G}_2)$ and $(P_2)_*\colon \mathrm{GH}^-(\mathcal{G}_2) \to \mathrm{GH}^-(\mathcal{G}_1)$ are inverses of each other, and are thus isomorphisms. The proof for row commutations is similar, and the proof for $\widetilde{\mathrm{GH}}$ is obtained by modifying the proof above to block all O markers in all maps. □

The invariance of GH^- under (de)stabilization is quite a bit more tedious, and so we refer the reader to [OSSz15, Section 5.2] for full details. We simply give a brief outline here:

- First, we claim that the actions by the formal variables U_i and U_j on $\mathrm{GC}^-(\mathcal{G})$ are chain homotopic for all $i, j \in \{1, \ldots, n\}$. The proof proceeds by defining a

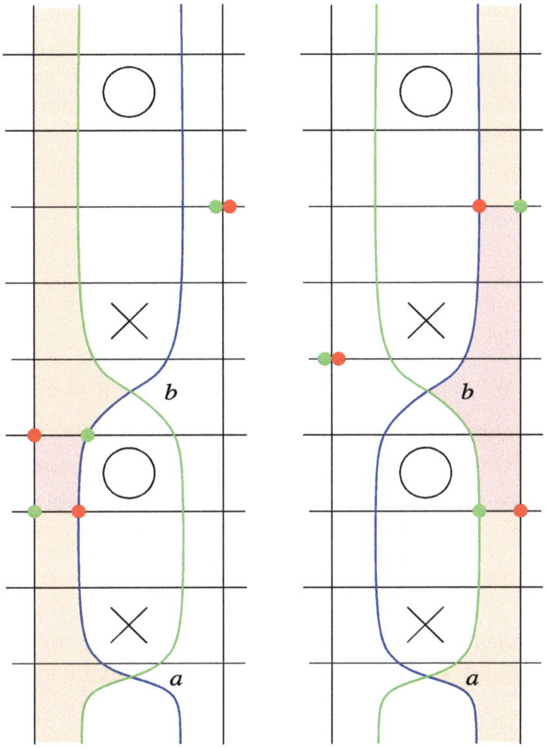

Fig. 1.9 Two annuli, one representing a term in $\partial^-_{\mathrm{GC}^-(\mathcal{G}_1)} \circ H_1(\mathbf{x})$ (left) and one representing a term in $P_2 \circ P_1(\mathbf{x}')$ (right), that cancel with the identity. Here $\mathbf{x}$ and $\mathbf{x}'$ (both in red) are two distinct generators in $\mathcal{S}(\mathcal{G}_1)$. Note that there are other possibilities of such annular domains. Depending on $\mathbf{x}$, there is always exactly one term that results in such an annular domain

homotopy $\phi\colon \mathrm{GC}^-(\mathcal{G}) \to \mathrm{GC}^-(\mathcal{G})$ that counts rectangles containing a single X marker X_i in it. One can show that

$$\partial^- \circ \phi + \phi \circ \partial^- = U_i + U_{i+1},$$

where the corresponding markers O_i and O_{i+1} are in the same row and same column as X_i respectively; this proof is again similar to the proof that $\partial^- \circ \partial^- = 0$.

- The upshot is that $\mathrm{GH}^-(\mathcal{G})$ can be thought of as an $\mathbb{F}[U]$-module, where U acts by multiplication by any of the U_i's.
- One then proceeds to prove that the isomorphism type of GH^- is invariant under (de)stabilization, by considering certain mapping cones. Combined with invariance under commutation, we see that the isomorphism type of $\mathrm{GH}^-(\mathcal{G})$ is a knot invariant, which we denote by $\mathrm{GH}^-(K)$.
- One may define the *hat grid chain complex* by setting one of the U_i's to be zero, i.e.,

$$\widehat{\mathrm{GC}}(\mathcal{G}) = \mathrm{GC}^-(\mathcal{G}) \otimes_{\mathbb{F}[U_1,\dots,U_n]} \left(\mathbb{F}[U_1,\dots,U_n]/_{U_1}\right).$$

This has the effect of setting the action of U to be zero, and so we get an $\mathbb{F}$-module $\widehat{\mathrm{GC}}(\mathcal{G})$. It is clear from the above that the isomorphism type of the $\mathbb{F}$-module $\widehat{\mathrm{GH}}(\mathcal{G})$ is also an invariant, which we denote by $\widehat{\mathrm{GH}}(K)$.

- Finally, we may place $\widetilde{\mathrm{GH}}(\mathcal{G})$ in this context also. Note that $\widetilde{\mathrm{GC}}(\mathcal{G})$ is obtained from $\widehat{\mathrm{GC}}(\mathcal{G})$ by also setting all the other U_i's to zero. In other words, we have set $n-1$ "too many" U_i's to be zero, and some homological algebra shows that

$$\widetilde{\mathrm{GH}}(\mathcal{G}) \cong \widehat{\mathrm{GH}}(\mathcal{G}) \otimes_{\mathbb{F}} \mathbb{F}^{\otimes 2^{n-1}},$$

and so the isomorphism type of $\widetilde{\mathrm{GH}}(\mathcal{G})$ is "almost" a knot invariant and can be denoted $\widetilde{\mathrm{GH}}(K, n)$.

We may summarize the invariance of grid homology here.

Theorem 1.1.11 (Manolescu et al. [MOS09, Theorem 1.1] and Manolescu et al. [MOSzT07, Theorem 1.2], cf. [OSSz15, Theorem 5.3.1]) *Let $\mathcal{G}$ be a grid diagram of $K \subset S^3$. The graded isomorphism types of* $\mathrm{GH}^-(\mathcal{G})$ *and* $\widehat{\mathrm{GH}}(\mathcal{G})$ *are invariants of the knot K. Thus, we may write* $\mathrm{GH}^-(K)$ *and* $\widehat{\mathrm{GH}}(K)$ *for these homologies.*

Remark 1.1.12 Instead of the homology, one may in fact state the invariance of the grid complex (and, later, of the Heegaard Floer complex) on the chain level: The *graded chain homotopy type* of the chain complex is an invariant. This is important, as many applications rely on invariants extracted from the chain complex.

In this lecture, we have chosen to exhibit in detail certain proofs while hiding others entirely. The point of showing the proof that $\partial^-_{\mathrm{GC}^-(\mathcal{G})} \circ \partial^-_{\mathrm{GC}^-(\mathcal{G})} = 0$ in detail is that it is structurally the same as the proof that $\partial^-_{\mathrm{CF}^-(\mathcal{H})} \circ \partial^-_{\mathrm{CF}^-(\mathcal{H})} = 0$ later, when we define the Heegaard Floer complex $\mathrm{CF}^-(\mathcal{H})$. There, the boundary homomorphism is defined by counting certain (pseudo)holomorphic bigons, and so the analysis becomes much more intricate; however, one will still consider a generator $\mathbf{z}$ that appears in $\partial^- \circ \partial^-(\mathbf{x})$, juxtapose two holomorphic bigons (analogous to rectangles here), and prove that there is an even number of ways that the resulting holomorphic curve can be decomposed into two holomorphic bigons that are counted.

Similarly, the proof that GH^- is invariant under commutation is structurally the same as the proof that HF^-, the homology of CF^-, is invariant under handleslides. There, one also combines multiple Heegaard diagrams (analogous to grid diagrams) into a single diagram, define chain maps by counting holomorphic triangles (analogous to pentagons here), and define chain homotopies by counting holomorphic quadrilaterals (analogous to hexagons here).

Limited by time, we will not be able to delve into the details of these proofs in later lectures. However, it is my hope that the analogies above will provide some insight into them.

1.1.3 Maslov and Alexander Gradings

Next, we move on to the homological grading of grid homology, known as the *Maslov grading*. This will allow us to discuss the graded isomorphism types of $\mathrm{GH}^-(K)$. While the definitions of gradings on Heegaard Floer homology are somewhat different in nature, the hope is that this will help us gain some insight into those gradings also. We continue to follow [OSSz15].

Definition 1.1.13 Let A and B be two finite sets in $\mathbb{R}^2$. We define the following quantities:

$$\mathcal{I}(A, B) = \#\{(a, b) \in A \times B \mid a = (a_1, a_2),\ b = (b_1, b_2),\ a_1 < b_1,\ a_2 < b_2\},$$

$$\mathcal{J}(A, B) = \frac{1}{2}(\mathcal{I}(A, B) + \mathcal{I}(B, A)).$$

In other words, $\mathcal{I}(A, B)$ counts the number of pairs of points $(a, b) \in A \times B$ such that a is strictly to the southwest of b—such a pair is known as a *northeast–southwest pair*—and $\mathcal{J}$ symmetrizes $\mathcal{I}$.

Given $\mathbf{x} \in \mathcal{S}(\mathcal{G})$, the *Maslov grading* $M(\mathbf{x}) = M_{\mathbb{O}}(\mathbf{x})$ of $\mathbf{x}$ is defined by

$$M(\mathbf{x}) = M_{\mathbb{O}}(\mathbf{x}) = \mathcal{J}(\mathbf{x}, \mathbf{x}) - 2\mathcal{J}(\mathbf{x}, \mathbb{O}) + \mathcal{J}(\mathbb{O}, \mathbb{O}) + 1.$$

Proposition 1.1.14 *If there is a rectangle r from $\mathbf{x}$ to $\mathbf{y}$, then*

$$M(\mathbf{x}) - M(\mathbf{y}) = 1 - 2\#(r \cap \mathbb{O}).$$

Proof If there are no O's inside the rectangle r, then $\mathcal{J}(\mathbf{x}, \mathbb{O}) = \mathcal{J}(\mathbf{y}, \mathbb{O})$ and $\mathcal{J}(\mathbf{x}, \mathbf{x}) - \mathcal{J}(\mathbf{y}, \mathbf{y}) = 1$, because r "destroys" a northeast–southwest pair in $\mathbf{x} \times \mathbf{x}$ to create a northwest–southeast pair in $\mathbf{y}$, and $\mathbf{x}$ and $\mathbf{y}$ coincide everywhere else.

If $r \cap \mathbb{O} \neq \emptyset$, then for each O inside the rectangle r, both $\mathcal{I}(\mathbf{x}, \mathbb{O})$ and $\mathcal{I}(\mathbb{O}, \mathbf{x})$ are increased by 1; thus, $\mathcal{J}(\mathbf{x}, \mathbb{O}) - \mathcal{J}(\mathbf{y}, \mathbb{O})$ counts the number of O's inside r. □

This implies that M induces a homological $\mathbb{Z}$-grading on $\widetilde{\mathrm{GC}}(\mathcal{G})$, since $r \cap \mathbb{O} = \emptyset$ in all relevant rectangles r in the boundary homomorphism $\widetilde{\partial}$.

For $\mathrm{GC}^-(\mathcal{G})$, note that $\#(r \cap \mathbb{O})$ may be a positive integer m. This would mean that $U^m\mathbf{y}$ appears as a term in $\partial^-(\mathbf{x})$, where by U^m we mean a product of m distinct U_i's. Since $M(\mathbf{x}) - M(\mathbf{y}) = 1 - 2m$, if we set $M(U_i) = -2$ for all U_i's, then

$$M(U^m\mathbf{y}) = M(U^m) + M(\mathbf{y}) = -2m + (M(\mathbf{x}) - 1 + 2m) = M(\mathbf{x}) - 1,$$

and we have a homological $\mathbb{Z}$-grading on $\mathrm{GC}^-(\mathcal{G})$.

Remark 1.1.15 The reason that we have a homological $\mathbb{Z}$-grading on $\mathrm{GH}^-(K)$ is because, as we shall see, it is a version of knot Floer homology in the three-sphere, $\mathrm{HFK}^-(S^3, K)$. For a general 3-manifold Y, neither $\mathrm{HF}^-(Y)$ nor $\mathrm{HFK}^-(Y, K)$ is necessarily $\mathbb{Z}$-graded; in fact, one may get different $\mathbb{Z}/m$ gradings for different Spin^c-structures on Y. These gradings are also not necessarily *absolute*, meaning that they may be well-defined only up to a constant shift. We will briefly discuss the Spin^c-decomposition and gradings in the last lecture.

In fact, $\widetilde{\mathrm{GC}}(\mathbb{G})$ and $\mathrm{GC}^-(\mathbb{G})$ are equipped with an internal grading A (meaning that the boundary homomorphism preserves A), known as the *Alexander grading*.

Definition 1.1.16 Given $\mathbf{x} \in \mathbf{S}(\mathbb{G})$, the *Alexander grading* $A(\mathbf{x})$ of $\mathbf{x}$ is defined by

$$A(\mathbf{x}) = \frac{1}{2}(M_{\mathbb{O}}(\mathbf{x}) - M_{\mathbb{X}}(\mathbf{x})) - \frac{n-1}{2},$$

where $M_{\mathbb{X}}$ is defined in the same way as $M_{\mathbb{O}}$ but with $\mathbb{O}$ replaced by $\mathbb{X}$ in its definition.

The existence of A is a feature of knot Floer invariants (e.g., CFK^-) and can be understood in terms of a "relative" Spin^c-decomposition. We will not discuss it further until the final lecture, but will point out that A plays an important role in many applications; for example, see Remark 1.1.18.

1.1.4 Graded Isomorphism Types

From now until the final lecture, we will focus only on the Maslov grading and ignore the Alexander grading.

Let us now look at the possible graded isomorphism types that the various flavors of grid homology can have, starting with $\widetilde{\mathrm{GH}}(K, n)$ and $\widehat{\mathrm{GH}}(K)$. Since they are finitely generated graded $\mathbb{F}$-modules (that is to say, $\mathbb{F}$-vector spaces), they must have graded isomorphism types

$$\bigoplus_{m \in \mathbb{Z}} \mathbb{F}_{(m)}^{\oplus a_m} \quad \text{with} \quad \sum_m a_m < \infty,$$

where $\mathbb{F}_{(m)}$ denotes a summand in Maslov grading m.

As for GH^-, it is an $\mathbb{F}[U]$-module. Since $\mathbb{F}[U]$ is a principal ideal domain, the Fundamental Theorem for Finitely Generated Modules over a Principal Ideal Domain implies that $\mathrm{GH}^-(K)$ is a direct sum of cyclic modules, i.e., it must have ungraded isomorphism type

$$\mathrm{GH}^-(K) \cong \mathbb{F}[U]^{\oplus a} \oplus \left(\mathbb{F}[U]/(f_1(U))\right) \oplus \cdots \oplus \left(\mathbb{F}[U]/(f_k(U))\right),$$

where the f_i's are polynomials. Moreover, since U strictly decreases grading, the f_i's must actually be monomials! Hence, we have

$$\mathrm{GH}^-(K) \cong \mathbb{F}[U]^{\oplus a} \oplus \left(\mathbb{F}[U]/U^{b_1}\right) \oplus \cdots \oplus \left(\mathbb{F}[U]/U^{b_k}\right).$$

Remark 1.1.17 This last assertion is no longer true if the grading is a $\mathbb{Z}/m$-grading instead of a $\mathbb{Z}$-grading, because $U^{m/2}\mathbf{x}$ can now have the same grading as $\mathbf{x}$. This may occur for the Heegaard Floer homology of a general 3-manifold.

One can, of course, write a graded version of this statement as well, which is left to the reader.

It turns out that, because $\mathrm{GH}^-(K)$ is a version of $\mathrm{HFK}^-(S^3, K)$, i.e., the ambient 3-manifold is S^3, there is only one $\mathbb{F}[U]$-summand, meaning that $a = 1$.

Remark 1.1.18 The invariant $\tau(K)$ is defined as $-A$, where A is the (top) Alexander grading of the generator of the $\mathbb{F}[U]$-summand of $\mathrm{HFK}^-(K)$. τ provides a very useful concordance homomorphism.

One feature that is often overlooked is that the graded isomorphism type of $\widehat{\mathrm{GH}}(K)$ is completely determined by that of $\mathrm{GH}^-(K)$. Indeed, note that

$$\widehat{\mathrm{GC}}(\mathbb{G}) \cong \mathrm{GC}^-(\mathbb{G}) \otimes_{\mathbb{F}[U]} \left(\mathbb{F}[U]/U\right).$$

Thus, the Universal Coefficient Theorem implies that

$$\widehat{\mathrm{GH}}(K) \cong \mathrm{GH}^-(K) \otimes_{\mathbb{F}[U]} \left(\mathbb{F}[U]/U\right) \oplus \mathrm{Tor}_1^{\mathbb{F}[U]}\left(\mathrm{GH}^-(K), \mathbb{F}[U]/U\right).$$

Example 1.1.19 With some work, one can work out the *minus* grid homology of the left-handed trefoil to be

$$\mathrm{GH}^-(\overline{T}_{2,3}) \cong (\mathbb{F}[U])_{(2)} \oplus \left(\mathbb{F}[U]/U\right)_{(1)},$$

and so we can also read off the *hat* grid homology to be

$$\begin{aligned}\widehat{\mathrm{GH}}(\overline{T}_{2,3}) &\cong \left(\mathbb{F}[U]/U\right)_{(2)} \oplus \left(\mathbb{F}[U]/U\right)_{(1)} \oplus \left(\mathbb{F}[U]/U\right)_{(0)} \\ &\cong \mathbb{F}_{(2)} \oplus \mathbb{F}_{(1)} \oplus \mathbb{F}_{(0)}.\end{aligned}$$

Exercise 1.1.20 Compute Tor_1 here. Bonus: Compute it with gradings.

1.1.5 Other Flavors

Finally, we mention some more flavors of grid homology. First, we have the *infinity* and *plus* flavors.

Definition 1.1.21 The *infinity* and *plus grid chain complexes* are defined respectively by

$$\mathrm{GC}^{\infty}(\mathbb{G}) = \mathrm{GC}^{-}(\mathbb{G}) \otimes_{\mathbb{F}[U]} \mathbb{F}[U, U^{-1}],$$

$$\mathrm{GC}^{+}(\mathbb{G}) = \mathrm{GC}^{\infty}(\mathbb{G}) \otimes_{\mathbb{F}[U]} \left({}^{\mathbb{F}[U, U^{-1}]}\!\big/_{U} \right),$$

both thought of as (bi)graded chain complexes of $\mathbb{F}[U]$-modules.

One can recover $\mathrm{GC}^{-}(\mathbb{G})$ from $\mathrm{GC}^{\infty}(\mathbb{G})$ as a subcomplex, and $\mathrm{GC}^{+}(\mathbb{G})$ from $\mathrm{GC}^{\infty}(\mathbb{G})$ as a quotient complex, and so these three flavors contain roughly the same amount of information.

Exercise 1.1.22 Work out some obvious mapping cones among the *minus*, *plus*, *infinity*, and *hat* flavors of grid chain complexes.

All of these flavors are considered in the literature, as it is often easier to formulate a certain statement in one flavor versus another. They also all appear in general Heegaard Floer homology, as $\mathrm{CF}^{-}(\mathcal{H})$, $\mathrm{CF}^{+}(\mathcal{H})$, $\mathrm{CF}^{\infty}(\mathcal{H})$, and $\widehat{\mathrm{CF}}(\mathcal{H})$. In each of these cases, the graded isomorphism type of the homology (or in fact, the graded chain homotopy type of the chain complex) is an invariant.

Remark 1.1.23 Again by the Universal Coefficient Theorem, one can compute

$$\mathrm{GH}^{\infty}(K) \cong \mathbb{F}[U, U^{-1}],$$

and so the *infinity* homology is not very interesting. However, the *infinity* chain complex $\mathrm{GC}^{\infty}(K)$ enjoys a rich structure, and many invariants, such as $\varepsilon(K)$ [Hom14] and $\Upsilon(K)$ [OSSz17a], can be extracted from it.

There are more flavors that are worth noting:

- If we allow rectangles to contain X's but do not record them, the Alexander grading is no longer an internal grading; instead, it can be lowered by the boundary homomorphism, meaning that it becomes only a *filtration*. (The Maslov grading remains a homological grading.) The filtered, graded homotopy type of this filtered chain complex, denoted $\mathcal{GC}^{-}(\mathbb{G})$, is a knot invariant. To recover $\mathrm{GC}^{-}(K)$, one takes the *associated graded object* of $\mathcal{GC}^{-}(K)$, which exactly blocks all rectangles that contain X's. The object in general Heegaard Floer theory analogous to $\mathcal{GC}^{-}(K)$ is $\mathrm{CFK}^{-}(Y, K)$, which is a filtered, graded chain complex, and $\mathrm{HFK}^{-}(Y, K)$, which is analogous to $\mathrm{GH}^{-}(K)$, is the homology of the associated graded object of $\mathrm{CFK}^{-}(Y, K)$.

- An approach alternative to filtered complexes is to introduce formal variables V_i for X_i, and define

$$\partial^-_{\mathbb{OX}}(\mathbf{x}) = \sum_{\mathbf{y}\in\mathcal{S}(\mathcal{G})} \sum_{r\in\mathrm{Rect}^\circ(\mathbf{x},\mathbf{y})} U_1^{O_1(r)} \cdots U_n^{O_n(r)} V_1^{X_1(r)} \cdots V_n^{X_n(r)} \mathbf{y},$$

  This is the approach taken in [Zem19b, Zem19a] and some other recent work. However, one must grapple with the fact that $\partial^-_{\mathbb{OX}} \circ \partial^-_{\mathbb{OX}} \equiv c \neq 0$, where $c \in \mathcal{R}$ is the *curvature* of a *curved chain complex.*
- We can also set $U_1 = \cdots = U_n = V_1 = \cdots = V_n$ and consider a chain complex with

$$\partial'(\mathbf{x}) = \sum_{\mathbf{y}\in\mathcal{S}(\mathcal{G})} \sum_{r\in\mathrm{Rect}^\circ(\mathbf{x},\mathbf{y})} U^{O(r)+X(r)} \mathbf{y}.$$

  Since O's and X's are treated equally in the boundary homomorphism, we lose the information of the orientation of the knot. This is called *unoriented grid homology* [OSSz17b].

1.2 "Naive" Heegaard Floer Homology

1.2.1 Definition

Having caught a glimpse of some of the features of Heegaard Floer homology as exhibited in grid homology, the rest of this lecture will be devoted to discussing a "naive" definition of Heegaard Floer homology, where the boundary homomorphism counts domains on a Heegaard diagram, just as the grid boundary homomorphism counts rectangles on a grid diagram. For someone unacquainted with Heegaard Floer homology attending a research talk on the subject, this may seem like what is going on when the speaker shades certain domains on a Heegaard diagram. However, this is not the actual definition of Heegaard Floer homology, as it may not even produce a legitimate chain complex. And yet, I think there is some merit in investigating this "naive" definition.

This exercise serves several purposes. First, not yet burdened with the analysis involved in Morse or Floer theory, we can transfer some of the intuition we gained from grid homology to help us make sense of some definitions. Second, by looking at some of the simplest examples, for which the "naive" definition seems to make sense, we can get a taste of some of the issues that must be addressed in the actual theory. Third, by looking at an example where $\partial^2 \not\equiv 0$, we will get to see why the "naive" definition *cannot* make sense.

We assume familiarity with Heegaard diagrams of 3-manifolds. In Heegaard Floer homology, we consider *pointed 3-manifolds* (Y, z), where $z \in Y$. Correspondingly, we consider *pointed Heegaard diagrams* $\mathcal{H} = (\Sigma, \boldsymbol{\alpha}, \boldsymbol{\beta}, z)$. As usual, if Σ

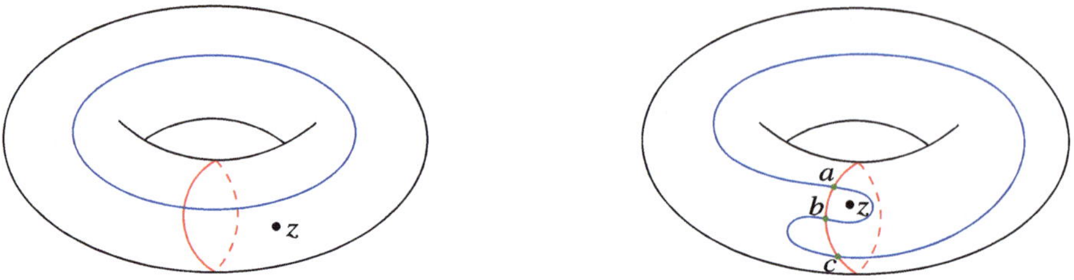

(a) $\mathcal{H}_1$ (b) $\mathcal{H}_2$

Fig. 1.10 Two Heegaard diagrams of (S^3, z). (**a**) $\mathcal{H}_1$. (**b**) $\mathcal{H}_2$

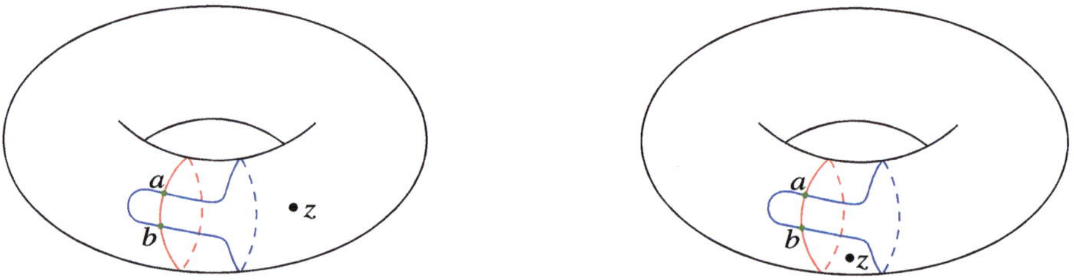

(a) $\mathcal{H}_3$ (b) $\mathcal{H}_4$

Fig. 1.11 Two Heegaard diagrams of $(S^1 \times S^2, z)$. (**a**) $\mathcal{H}_3$. (**b**) $\mathcal{H}_4$

has genus g, then $\boldsymbol{\alpha}$ is a set of g mutually disjoint circles $\{\alpha_1, \ldots, \alpha_g\}$ on Σ, and $\boldsymbol{\beta}$ is a set of g mutually disjoint circles $\{\beta_1, \ldots, \beta_g\}$. We often abuse notation and use $\boldsymbol{\alpha}$ to denote $\alpha_1 \cup \cdots \cup \alpha_g$ and $\boldsymbol{\beta}$ to denote $\beta_1 \cup \cdots \cup \beta_g$. We assume that z lies on Σ away from all α- and β-curves.

Example 1.2.1 Figure 1.10 shows two pointed Heegaard diagrams of (S^3, z), while Fig. 1.11 shows two diagrams of $(S^1 \times S^2, z)$. As is customary, α-curves are colored red, while β-curves are colored blue.

Inspired by the grid chain complexes, here is a "naive" definition of the Heegaard Floer chain complex associated to a pointed Heegaard diagram.

Definition 1.2.2 Given a pointed Heegaard diagram $\mathcal{H} = (\Sigma, \boldsymbol{\alpha}, \boldsymbol{\beta}, z)$, we define the generating set

$$\mathcal{S}(\mathcal{H}) = \{\text{matchings between } \boldsymbol{\alpha} \text{ and } \boldsymbol{\beta} \text{ in } \mathcal{H}\},$$

where a matching is a one-to-one correspondence between the α- and β-circles.

We define the *hat* and *minus "naive" Heegaard Floer chain complexes* respectively as the $\mathbb{F}$-module and the $\mathbb{F}[U]$-module

$$\widehat{\mathrm{CF}}(\mathcal{H}) = \mathbb{F}\langle \mathcal{S}(\mathcal{H})\rangle, \qquad \mathrm{CF}^-(\mathcal{H}) = \mathbb{F}[U]\langle \mathcal{S}(\mathcal{H})\rangle,$$

with boundary homomorphisms $\widehat{\partial}\colon \widehat{\mathrm{CF}}(\mathcal{H}) \to \widehat{\mathrm{CF}}(\mathcal{H})$ and $\partial^-\colon \mathrm{CF}^-(\mathcal{H}) \to \mathrm{CF}^-(\mathcal{H})$ given by

$$\widehat{\partial}(\mathbf{x}) = \sum_{\mathbf{y}\in\mathcal{S}(\mathcal{H})} \sum_{\substack{p\in\pi_2(\mathbf{x},\mathbf{y})\\ p\cap\{z\}=\emptyset}} \mathbf{y}, \qquad \partial^-(\mathbf{x}) = \sum_{\mathbf{y}\in\mathcal{S}(\mathcal{H})} \sum_{p\in\pi_2(\mathbf{x},\mathbf{y})} U^{n_z(p)}\mathbf{y},$$

where $n_z(p)$ is the multiplicity of z in p, and $\pi_2(\mathbf{x}, \mathbf{y})$ denotes the space of domains from $\mathbf{x}$ to $\mathbf{y}$. Here, a *region* is a connected component of $\Sigma \setminus (\boldsymbol{\alpha} \cup \boldsymbol{\beta})$, and a *domain* p *from* $\mathbf{x}$ *to* $\mathbf{y}$ is a 2-chain on Σ that is a linear combination of regions with non-negative coefficients, such that

$$\partial((\partial p) \cap \boldsymbol{\alpha}) = \mathbf{y} - \mathbf{x}$$

in the induced orientation on ∂p.

Something seems immediately suspicious in this definition. Indeed, if we concatenate a domain from $\mathbf{x}$ to $\mathbf{y}$ and a domain from $\mathbf{y}$ to $\mathbf{z}$, we will get a domain from $\mathbf{x}$ to $\mathbf{z}$; but clearly we do not want a scenario where $\mathbf{y}$ appears as a term in $\partial(\mathbf{x})$, and $\mathbf{z}$ appears as a term in both $\partial(\mathbf{y})$ and $\partial(\mathbf{x})$. In grid homology, concatenating two rectangles always results in a domain that is not a rectangle; here, we need to instead rely on a *Maslov index* on elements of $\pi_2(\mathbf{x}, \mathbf{y})$. We will not go into details here, except that such an index exists and can be computed [Lip06, Corollary 4.10], and gives rise to the homological *Maslov grading* on our chain complex. The idea, then, is that we will only count domains that are of index 1; concatenating two such domains will give a domain of index 2, which is not counted. Applying this idea back to the grid chain complex, this would eliminate

- domains that have reflex angles in them; and
- domains that contain points in $\mathbf{x}$ or $\mathbf{y}$ in them.

This is why only empty rectangles are counted in grid homology. In any case, this represents the first modification we would like to make.

When $g(\Sigma) = 1$, the condition that a domain must be of Maslov index 1 will imply that it must in fact be an immersed bigon with non-reflex angles at its corners. (A bigon is a disk with two corners.)

Example 1.2.3 We can compute the chain complexes associated to the two diagrams of (S^3, z) in Example 1.2.1. For $\mathcal{H}_1$, let x be the unique intersection point of the α and β curves. Since there are no domains, it follows that

$$\widehat{\mathrm{CF}}(\mathcal{H}_1) \cong \mathbb{F}\langle x\rangle, \qquad \mathrm{CF}^-(\mathcal{H}_1) \cong \mathbb{F}[U]\langle x\rangle, \qquad \widehat{\partial} \equiv 0 \equiv \partial^-,$$

$$\widehat{\mathrm{HF}}(\mathcal{H}_1) \cong \mathbb{F}, \qquad \mathrm{HF}^-(\mathcal{H}_1) \cong \mathbb{F}[U].$$

For $\mathcal{H}_2$, labeling the intersection points by a, b, and c as in Fig. 1.10b, there are two domains, one from a to b containing z, and one from c to b not containing z. Thus,

$$\widehat{\mathrm{CF}}(\mathcal{H}_2) \cong \mathbb{F}\langle a, b, c\rangle, \qquad \mathrm{CF}^-(\mathcal{H}_2) \cong \mathbb{F}[U]\langle a, b, c\rangle,$$
$$\widehat{\partial}(a) = \widehat{\partial}(b) = 0, \quad \widehat{\partial}(c) = b, \qquad \partial^-(a) = Ub, \quad \partial^-(b) = 0, \quad \partial^-(c) = b,$$
$$\widehat{\mathrm{HF}}(\mathcal{H}_2) \cong \mathbb{F}\langle a, b\rangle/(b) \cong \mathbb{F}, \qquad \mathrm{HF}^-(\mathcal{H}_2) \cong \mathbb{F}[U]\langle a + Uc, b\rangle/(b) \cong \mathbb{F}[U].$$

1.2.2 Admissibility

So far, we have not run into any major problems. Let us look an example that seems more mysterious:

Example 1.2.4 We now look at the two diagrams of $(S^1 \times S^2, z)$ in Example 1.2.1. One can compute that

$$\widehat{\mathrm{HF}}(\mathcal{H}_3) \cong \mathbb{F} \oplus \mathbb{F}, \qquad \widehat{\mathrm{HF}}(\mathcal{H}_4) \cong 0,$$

and

$$\mathrm{HF}^-(\mathcal{H}_3) \cong \mathbb{F}[U] \oplus \mathbb{F}[U], \qquad \mathrm{HF}^-(\mathcal{H}_4) \cong {}^{\mathbb{F}[U]}\!/_{(1+U)}.$$

This seems to suggest that $\widehat{\mathrm{HF}}$ and HF^- are not invariants of 3-manifolds! One may naturally suspect that the culprit is the "naive" definition, but in fact, it is not—we are indeed computing the correct chain complexes here.

So what gives? It turns out that this has to do with a requirement on the Heegaard diagrams called *admissibility*, a condition that guarantees the diagram to be suitable for computation. In other words, we cannot simply choose any Heegaard diagram and expect it to give us the correct answer. This is closely related to the Spin^c-decomposition of Heegaard Floer homology, and we will briefly touch upon this issue again in the last lecture.

1.2.3 $\partial^2 \equiv 0$

There is, in fact, a more serious problem that has been lurking in our "naive" approach to the Heegaard Floer complex: We have not proved $\partial^2 \equiv 0$ for either the *hat* or the *minus* flavor; in other words, we do not know that we have defined chain complexes.

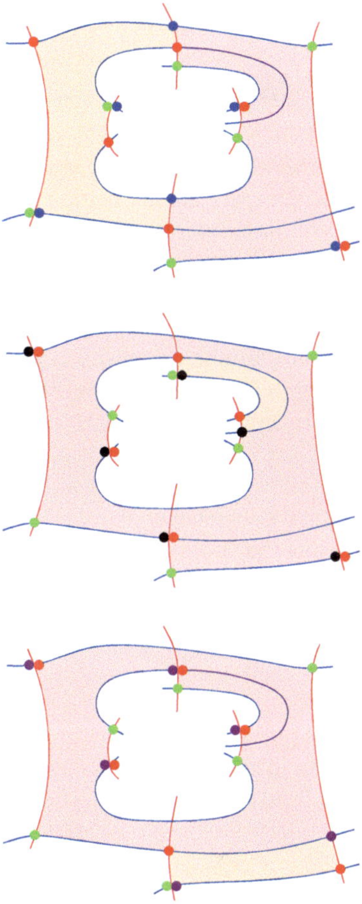

Fig. 1.12 A composite domain that has three decompositions into domains that are counted in the differential

This turns out to be a real problem. See Fig. 1.12; the shaded domain has three decompositions into domains, each of which has the correct index. This means that $\mathbf{z}$ (in green) appears as a term in $\widehat{\partial} \circ \widehat{\partial}(\mathbf{x})$ that is not canceled by any other term.

This illustrates a key point, upon understanding which, we graduate from the "naive" approach: The Heegaard Floer chain complex is not supposed to be defined by counting polygons on a Heegaard diagram. Instead, it is defined by counting (pseudo)holomorphic curves on a space *associated* to the Heegaard diagram, distinct from the Heegaard diagram itself unless $g(\Sigma) = 1$. One can think of the domains we have been counting as "shadows" of such holomorphic curves; given a domain of the right index, there may be 0, 1, or any number of holomorphic curves with that shadow. It is the count of holomorphic curves, rather than the count of domains, that satisfies $\partial^2 \equiv 0$.

1.3 Lagrangian Floer Homology

1.3.1 Setup

So now we understand that the boundary homomorphism in the Heegaard Floer chain complex counts certain holomorphic curves in some yet mysterious space. One of the questions most often asked is "why": What is the motivation behind the Heegaard Floer complex? In this lecture, we attempt to address this by providing an introduction to Lagrangian Floer homology; in short, Heegaard Floer homology can be defined via Lagrangian Floer homology, which in turn can be understood as a kind of infinite-dimensional Morse homology.

Lagrangian intersection Floer homology (or simply *Lagrangian Floer homology*) is an invariant of (M, ω, L_0, L_1), where (M, ω) is a symplectic manifold, and L_0, L_1 are two transversely intersecting Lagrangian submanifolds of (M, ω). The following is a quick survey that mostly follows the exposition in [Ped18]. We will make whichever assumptions we need to simplify the exposition, regardless of whether they hold in Heegaard Floer theory. First, we will assume that $\pi_2(M) = 0$—or less restrictively,

$$\int_{S^2} f^*\omega = 0 \tag{1.1}$$

for all $[f] \in \pi_2(M)$—and $\pi_1(L_i) = 0$ for each $i = 0, 1$.

The idea is to perform "Morse homology" on an infinite-dimensional manifold associated to (M, ω, L_0, L_1) defined as follows. Consider first

$$\mathcal{P} = \mathcal{P}(L_0, L_1) = \{\gamma : [0, 1] \to M \mid \gamma \text{ is smooth}, \ \gamma(0) \in L_0, \ \gamma(1) \in L_1\},$$

the set of smooth paths from L_0 to L_1, with the C^∞ topology. Note that $\mathcal{P}$ is not necessarily connected. Choose an element $\widehat{\gamma} \in \mathcal{P}$ (and by doing so, a connected component of $\mathcal{P}$), and consider the universal cover $\widetilde{\mathcal{P}}$ based at $\widehat{\gamma}$.

This means that we get a set of paths of paths:

$$\begin{aligned}\widetilde{\mathcal{P}} = \widetilde{\mathcal{P}}(L_0, L_1; \widehat{\gamma}) = \{(\gamma, w) \mid \gamma \in \mathcal{P}, \ & w \text{ is a path from } \widehat{\gamma} \text{ to } \gamma, \ \textit{i.e.} \\ & w : [0, 1] \times [0, 1] \to M \text{ with } w(\cdot, r) \in \mathcal{P} \text{ for all } r \in [0, 1], \\ & w(s, 0) = \widehat{\gamma}(s), \ w(s, 1) = \gamma(s) \text{ for all } s \in [0, 1]\}/\sim,\end{aligned}$$

where $(\gamma, w) \sim (\gamma, w')$ if w, w' are homotopic relative to $(\widehat{\gamma}, \gamma)$; see Fig. 1.13.

It is on this space that we will apply "Morse homology".

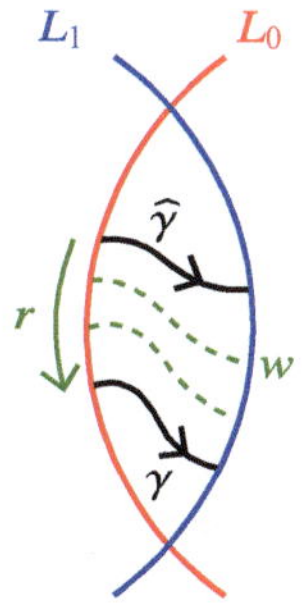

Fig. 1.13 A path w from $\widehat{\gamma}$ to γ, where each value of r gives a path $w(\cdot, r)$ from L_0 to L_1

1.3.2 Morse Homology: An Interlude

We briefly recall here the construction of Morse homology. Let M be a closed, smooth manifold. Recall that a smooth function $f : M \to \mathbb{R}$ is *Morse* if at each of its critical points (where in local coordinates all first derivatives $\partial f/\partial x_i$ vanish), the Hessian matrix $H = (\partial^2 f/\partial x_i \partial x_j)$ is not degenerate (i.e., has non-zero determinant). By the Morse Lemma, all such critical points must be isolated. The *Morse index* $\mathrm{Ind}(x)$ of a non-degenerate critical point x is the dimension of the largest subspace of the tangent space that is negative definite. Classical Morse theory says that Morse functions are generic, and that, given a Morse function f, M is homotopy equivalent to a CW complex with an n-cell for each critical point of f of index n.

In fact, one can even reconstruct a CW chain complex of M using f and auxiliary data. Since the critical points of f correspond to cells in a CW complex, let $\mathrm{CM}(M; f)$ be the $\mathbb{F}$-module generated by the critical points of f. To define a boundary homomorphism the idea is to count the number of negative gradient flow lines from a critical point of index n to one of index $n - 1$. The notion of the gradient requires the choice of a Riemannian metric g on M, meaning that we will have a chain complex $\mathrm{CM}(M; f, g) = \mathrm{CM}(M; f)$, with boundary homomorphism $\partial : \mathrm{CM}(M; f, g) \to \mathrm{CM}(M; f, g)$. Concretely, we consider the moduli space $\mathcal{M}_g(x, y)$ of curves $u : \mathbb{R} \to M$ satisfying

- $\dfrac{du}{dt} = -\operatorname{grad} f$;
- $\lim_{t\to-\infty} u(t) = x$; and
- $\lim_{t\to\infty} u(t) = y$.

To ensure that $\mathcal{M}_g(x, y)$ is a manifold of dimension $\mathrm{Ind}(x) - \mathrm{Ind}(y)$, the pair (f, g) needs to satisfy the *Morse–Smale* condition, which is luckily also generic. (The desired property that $\mathcal{M}_g(x, y)$ is be a manifold of the right dimension, is known as *transversality*.) There is a natural action of $\mathbb{R}$ on $\mathcal{M}_g(x, y)$ by shifting $t \in \mathbb{R}$; to consider flow lines rather than flows, one takes the quotient $\mathcal{M}_g(x, y)/\mathbb{R}$, which has dimension $\mathrm{Ind}(x) - \mathrm{Ind}(y) - 1$. The manifold $\mathcal{M}_g(x, y)/\mathbb{R}$ can be compactified to a compact manifold $\widehat{\mathcal{M}}_g(x, y)$ with boundary by including the limits of sequences of elements, which are broken gradient flow lines. (The desired property that $\mathcal{M}_g(x, y)$ can be compactified to a compact manifold with boundary is known as *compactness*.) When the index difference between x and y is 1, the moduli space $\widehat{\mathcal{M}}_g(x, y)$ is a compact manifold of dimension 0. Thus, one can define the matrix element $\langle \partial x, y \rangle$ as follows: If $\mathrm{Ind}(x) - \mathrm{Ind}(y) = 1$, it is the number of elements in $\widehat{\mathcal{M}}_g(x, y)$ modulo 2; otherwise, it is 0.

To show that $\partial^2 \equiv 0$, we consider $\widehat{\mathcal{M}}_g(x, z)$ where $\mathrm{Ind}(x) - \mathrm{Ind}(z) = 2$. This is a compact 1-dimensional manifold with boundary, and so there is an even number of boundary points; since these correspond to broken flow lines from x to z that appear in $\partial \circ \partial(x)$, this means that all terms cancel in pairs.

The main result is then that the *Morse homology*

$$\mathrm{HM}(M; f, g) = H_*(\mathrm{CM}(M; f, g), \partial)$$

is isomorphic to the CW or singular homology $H_*(M; \mathbb{Z}/2)$ of M. One could deduce from this that $\mathrm{HM}(M; f, g)$ is invariant under a different choice of f or g. Absent the isomorphism with singular homology, one would need to define chain homotopy equivalences between $\mathrm{CM}(M; f, g)$ and $\mathrm{CM}(M; f', g')$. This is important, since in the context of Floer homology, there is often no classical invariant to rely on, and this is the only way to prove invariance.

1.3.3 The Differential

Turning back to Lagrangian Floer homology, we would like to follow the steps above in this setting. The role of the manifold M will be played by the space $\mathcal{P}$; what we need now is a "Morse function".

Definition 1.3.1 The *action functional* $\mathcal{A} \colon \widetilde{\mathcal{P}} \to \mathbb{R}$ is defined by

$$\mathcal{A}([\gamma, w]) = \int_{[0,1]^2} w^* \omega.$$

(Here and below, we abbreviate $[(\gamma, w)]$ to $[\gamma, w]$.) We need to check that it does not depend on the choice of w, i.e., that we have

$$(\gamma, w) \sim (\gamma, w') \implies \mathcal{A}([\gamma, w]) = \mathcal{A}([\gamma, w']). \tag{1.2}$$

This follows from Eq. (1.1).

Exercise 1.3.2 Verify Eq. (1.2).

In local coordinates, we have

$$w^* \omega \left(\frac{\partial}{\partial s}, \frac{\partial}{\partial r} \right) = \omega \left(\frac{\partial w}{\partial s}, \frac{\partial w}{\partial r} \right),$$

and so

$$\mathcal{A}([\gamma, w]) = \int_0^1 \int_0^1 \omega\left(\frac{\partial w}{\partial s}, \frac{\partial w}{\partial r}\right) dr\, ds.$$

Now, we want to run "Morse homology" on $(\widetilde{\mathcal{P}}, \mathcal{A})$, which is structurally similar to the finite-dimensional case but has a lot of technicalities that come with infinite-dimensional spaces.

Let us compute grad $\mathcal{A}$. We need a Riemannian metric on $\widetilde{\mathcal{P}}$, which we will get by integrating a Riemannian metric on M. This is possible because a tangent vector $\xi \in T_{[\gamma,w]}\widetilde{\mathcal{P}}$ can be regarded, roughly, as an infinitesimal change to $\gamma(s)$ for every $s \in [0, 1]$; in other words, it can be viewed as a collection of tangent vectors $\xi(s) \in T_{\gamma(s)}M$.

We will find a natural metric on M using ω and a further choice. Recall that the following structures are related.

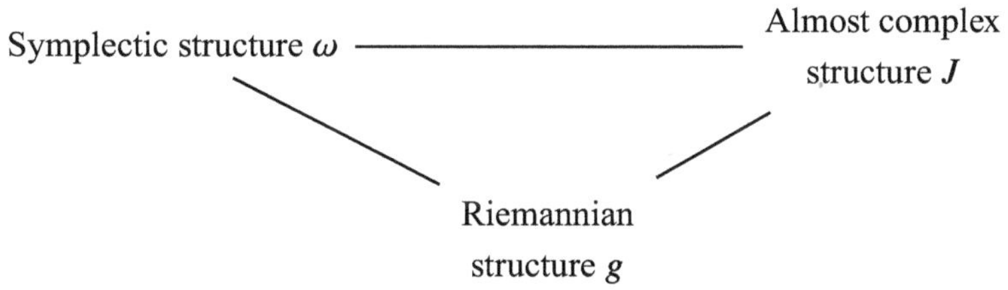

Here, a choice of any two of these structures that are compatible determines the third one. This is related to the structure groups of these structures, which are subgroups of $GL_{2n}(\mathbb{R})$.

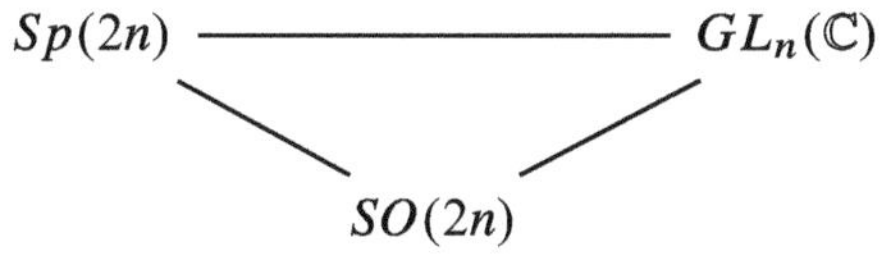

The intersection of any two of these is $U(n)$. The notion of compatibility mentioned above is specified in the following definition.

Definition 1.3.3 An almost complex structure J is *compatible* with ω if

$$\omega(Jv, Jv') = \omega(v, v') \qquad \text{for all } v, v' \in T_pM, \tag{1.3}$$

$$\omega(v, Jv) > 0 \qquad \text{for all } v \neq 0 \in T_pM. \tag{1.4}$$

It can be shown that compatible almost complex structures on symplectic manifolds are abundant.

Given ω and a compatible J, we can now define the metric g_J by the formula

$$g_J(v, v') = \omega(v, Jv'). \tag{1.5}$$

Exercise 1.3.4 Check that g_J is a Riemannian structure.

With a metric g_J on M, we can define a metric on $\widetilde{\mathcal{P}}$ as follows: Given $\xi_1, \xi_2 \in T_{[\gamma,w]}\widetilde{\mathcal{P}}$, we can define

$$\langle\langle \xi_1, \xi_2 \rangle\rangle = \int_0^1 g_J(\xi_1(s), \xi_2(s))\, ds.$$

But in fact, to allow sufficient flexibility to achieve genericity, we will allow the almost complex structure J to vary for different values of s in the integral. This means that we will make a choice of a family

$$\mathcal{J} = \{J_s\}_{0 \le s \le 1}$$

of ω-compatible almost complex structures on M, and define

$$\langle\langle \xi_1, \xi_2 \rangle\rangle = \int_0^1 g_{J_s}(\xi_1(s), \xi_2(s))\, ds. \tag{1.6}$$

We now determine $\operatorname{grad}_{\langle\langle \cdot,\cdot \rangle\rangle} \mathcal{A}([\gamma, w])$, by first computing

$$\begin{aligned}
d\mathcal{A}_{[\gamma,w]}(\xi) &= \int_0^1 \omega\left(\frac{\partial \gamma}{\partial s}, \xi(s)\right) ds \\
&= \int_0^1 \omega\left(J_s\frac{\partial \gamma}{\partial s}, J_s\xi(s)\right) ds \\
&= \int_0^1 g_{J_s}\left(J_s\frac{\partial \gamma}{\partial s}, \xi(s)\right) ds \\
&= \left\langle\!\!\left\langle J_s\frac{\partial \gamma}{\partial s}, \xi \right\rangle\!\!\right\rangle,
\end{aligned}$$

where the second, third, and fourth equalities follow from Eqs. (1.3), (1.5), and (1.6) respectively. This means that

$$\operatorname{grad}_{\langle\langle \cdot,\cdot \rangle\rangle} \mathcal{A}([\gamma, w])(s) = J_s \frac{\partial \gamma}{\partial s},$$

which is depicted in Fig. 1.14.

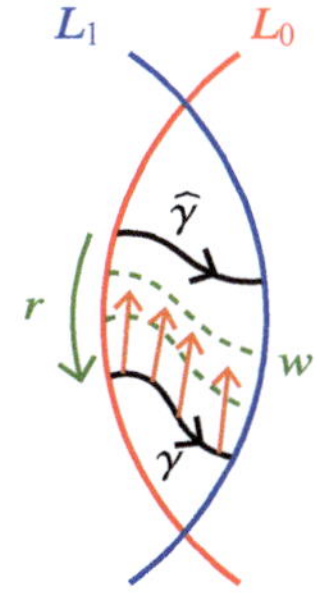

Fig. 1.14 The gradient of $\mathcal{A}$ represented (in brown) as a collection of tangent vectors "rotated" counterclockwise by J_s

We will now determine the critical points of $\mathcal{A}$. Since the almost complex structure J_s is an automorphism of $T_{\gamma(s)}M$, we have

$$\operatorname{grad}\mathcal{A}([\gamma, w]) = 0 \iff \frac{\partial \gamma}{\partial s} = 0 \quad \text{for all } s \in [0, 1],$$

i.e., γ is constant. As γ is a path from L_0 to L_1, it must be a constant path with $\gamma(0) = \gamma(1) \in L_0 \cap L_1$. This means that there is a bijection between the set of critical points of $\mathcal{A}$ and the intersection points $L_0 \cap L_1$.

Next, we move on to gradient flows, which are maps $u\colon \mathbb{R} \to \widetilde{\mathcal{P}}$ such that

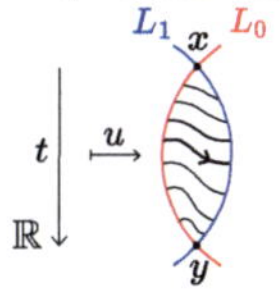

- $\dfrac{du}{dt} = -\operatorname{grad}\mathcal{A} \implies \left.\dfrac{du}{dt}\right|_{t=t_0} = -J_s\dfrac{\partial u(t_0)}{\partial s}$,
- $\lim_{t\to-\infty} u(t) = x$,
- $\lim_{t\to+\infty} u(t) = y$.

Equivalently, we can regard u as a function $u\colon [0, 1] \times \mathbb{R} \to M$, with parameters $s \in [0, 1]$ and $t \in \mathbb{R}$, and translate the first condition above to the condition

$$\frac{\partial u}{\partial t} + J_s\frac{\partial u}{\partial s} = 0.$$

Of course, this is nothing but the Cauchy–Riemann equation; in other words, u is a *pseudoholomorphic curve*. It is also called a *(pseudo)holomorphic disk* or a *(psuedo)holomorphic bigon*, as $[0, 1]\times\mathbb{R}$ is conformally equivalent to $\mathbb{D}^2\setminus\{i, -i\} \subset \mathbb{C}$, a disk with two corners (punctures) on its boundary.

Therefore, to do "Morse homology", we define a chain complex

$$\mathrm{CF}(M, \omega, L_0, L_1; \widehat{\gamma}, \mathcal{J}) = \mathbb{F}\langle L_0 \cap L_1\rangle.$$

Given $x, y \in L_0 \cap L_1$, consider the moduli spaces

$$\mathcal{M}_{\mathcal{J}}(x, y) = \left\{ u\colon [0,1] \times \mathbb{R} \to M \;\middle|\; \frac{\partial u}{\partial t} + J_s \frac{\partial u}{\partial s} = 0,\ u(\cdot, t) = \gamma(\cdot) \in \mathcal{P}\ \forall t, \right.$$
$$\left. \lim_{t \to +\infty} u(s,t) = x,\ \lim_{t \to -\infty} u(s,t) = y\ \ \forall s \right\}$$

Often, we want one more "finite energy" condition:

$$\int_{[0,1]\times\mathbb{R}} u^*\omega < \infty.$$

Hopefully, we will have

1. (Transversality) $\mathcal{M}_{\mathcal{J}}(x, y)$ is a manifold with an $\mathbb{R}$-action, of dimension $\mathrm{Ind}(u)$ for some notion of *index* Ind of u; and
2. (Compactness) $\mathcal{M}_{\mathcal{J}}(x, y)/\mathbb{R}$ can be compactified to a manifold $\widehat{\mathcal{M}}_{\mathcal{J}}(x, y)$ with boundary.

Then we can define the boundary homomorphism $\partial\colon \mathrm{CF}(M, \omega, L_0, L_1; \widehat{\gamma}, \mathcal{J}) \to \mathrm{CF}(M, \omega, L_0, L_1; \widehat{\gamma}, \mathcal{J})$ by

$$\partial(x) = \sum_{y \in L_0 \cap L_1} \#\widehat{\mathcal{M}}_{\mathcal{J}}^{\mathrm{Ind}(u)=1}(x, y) \cdot y.$$

Transversality is necessary to ensure that the moduli space is 0-dimensional and can be counted. Compactness is necessary to prove that $\partial(x)$ is a finite sum. Both are necessary for proving $\partial^2 \equiv 0$.

1.3.4 Technicalities and Additional Structures

Below, we discuss some technicalities that arise in the definitions above.

First, $\mathrm{Ind}(u)$ in fact depends on the class $B = [u] \in \pi_2(M, L_0 \cup L_1)$. Therefore, we need to write

$$\mathcal{M}_{\mathcal{J}}(x, y) = \bigsqcup_{B \in \pi_2(M, L_0 \cap L_1)} \mathcal{M}_{\mathcal{J}}^B(x, y).$$

In other words, different components of $\mathcal{M}_{\mathcal{J}}(x, y)$ might be of different dimensions. We then really have

$$\partial(x) = \sum_{y \in L_0 \cap L_1} \sum_{\substack{B \in \pi_2(M, L_0 \cup L_1) \\ \mathrm{Ind}(B)=1}} \#\widehat{\mathcal{M}}_{\mathcal{J}}^B(x, y) \cdot y$$

Second, the issue of transversality is way more technical than in finite-dimensional Morse homology, requiring one to do analysis on Sobolev spaces.

This is the step that requires us to consider families of almost complex structures on M, rather than a single almost complex structure, in our earlier definitions. The lack of appropriate regularity and transversality results is the reason that some versions of Floer homology are not yet well defined.

Third, the desired compactness result is provided by the following, known as *Gromov compactness*.

Theorem 1.3.5 (Floer [Flo88], Oh [Oh93], cf. Gromov [Gro85], et al.) *For a dense subset* $\{\mathcal{J}\}$ *of* 1*-parameter families of almost complex structures on* M, $\mathcal{M}^B_{\mathcal{J}}(x, y)/\mathbb{R}$ *can be compactified to a compact manifold* $\widehat{\mathcal{M}}^B_{\mathcal{J}}(x, y)$ *with boundary, of dimension* $\operatorname{Ind}(B)$.

Here, $\partial\widehat{\mathcal{M}}^B_{\mathcal{J}}(x, y)$ consists of holomorphic buildings (i.e., broken flow lines and bubbles). Extra work then goes into dealing with bubbling.

When $\operatorname{Ind}(B) = 1$, $\widehat{\mathcal{M}}^B_{\mathcal{J}}(x, y)$ is a compact 0-dimensional manifold, which guarantees that $\partial(x)$ is a finite sum. When $\operatorname{Ind}(B) = 2$, under favorable conditions (when bubbles are absent or appear in a controlled manner), $\widehat{\mathcal{M}}^B_{\mathcal{J}}(x, y)$ is a compact 1-dimensional manifold, whose boundary points correspond to broken flow lines. As in Morse homology, this allows one to prove $\partial \circ \partial \equiv 0$. The pictures in this proof are reminiscent of the pictures of $\partial^2 \equiv 0$ in grid homology. Fortunately, bubbling is in general not an issue for Heegaard Floer homology.

Finally, as mentioned before, unlike Morse homology on a finite-dimensional manifold, there is no homology we could relate our Lagrangian Floer homology to, that could provide a proof of invariance. In the present context, it turns out that the chain homotopy type of

$$\operatorname{CF}(M, \omega, L_0, L_1; \widehat{\gamma}, \mathcal{J})$$

is an invariant of the quadruple (M, ω, L_0, L_1). To prove this, one must define chain homotopy equivalences between complexes defined using different auxiliary data $(\widehat{\gamma}, \mathcal{J})$. In particular, given two families of almost complex structures $\mathcal{J}_1$ and $\mathcal{J}_2$, we choose a family of almost complex structures $\mathcal{J}$ that "connects" the two, and use $\mathcal{J}$ to define a chain homotopy equivalence between the chain complexes.

We end this discussion with a couple of features in this theory:

- Given a divisor $D^{2n-2} \subset M^{2n}$, we can modify the definition of $\mathcal{M}^B_{\mathcal{J}}(x, y)$ to block all holomorphic curves u that intersect D. This would result in a chain complex

$$\operatorname{CF}(M, \omega, L_0, L_1, D; \widehat{\gamma}, \mathcal{J}).$$

- For brevity, we will write $\operatorname{CF}(L_0, L_1)$ to mean $\operatorname{CF}(M, \omega, L_0, L_1)$. Given three mutually transversely intersecting Lagrangian submanifolds L_0, L_1, and L_2, there is a map

$$\mu_2\colon \operatorname{CF}(L_0, L_1) \otimes \operatorname{CF}(L_1, L_2) \to \operatorname{CF}(L_0, L_2),$$

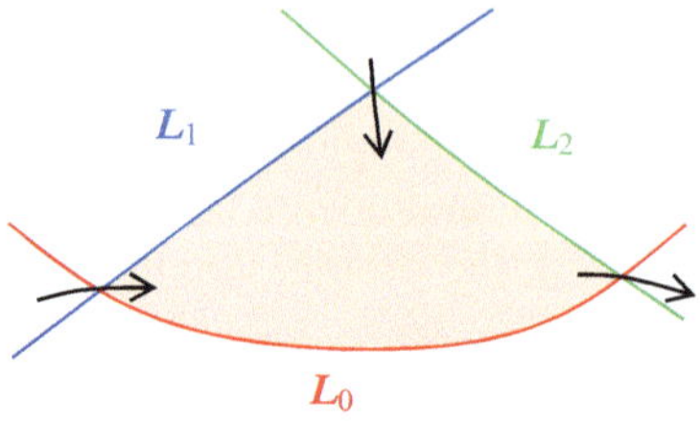

Fig. 1.15 A holomorphic triangle with two inputs and one output

defined by counting (pseudo)holomorphic triangles. A holomorphic triangle is much like a holomorphic bigon, but it has three instead of two punctures on the boundary of its domain (e.g., $\mathbb{D}^2 \setminus \{1, \zeta, \zeta^2\} \subset \mathbb{C}$); see Fig. 1.15.

The map μ_2 is like a multiplication, in that it satisfies the Leibniz rule with the boundary homomorphisms ∂. However, it is not necessarily associative; instead, it is associative only up to homotopy, given by a map

$$\mu_3\colon \mathrm{CF}(L_0, L_1) \otimes \mathrm{CF}(L_1, L_2) \otimes \mathrm{CF}(L_2, L_3) \to \mathrm{CF}(L_0, L_3),$$

defined by counting holomorphic quadrilaterals. Writing μ_1 for ∂, there is in fact a collection $\{\mu_n\}_{n=1}^{\infty}$ of maps, each defined by counting holomorphic $(n+1)$-gons, which satisfy a compatibility condition. This is known as an A_∞*-structure*. Given (M, ω), the collection of Lagrangian submanifolds, the Lagrangian Floer complexes—assuming they are well defined—and the A_∞-structure on them, together form the *Fukaya category of* (M, ω).

1.4 Heegaard Floer Homology

1.4.1 Definition

Finally, we are ready to give a definition of the Heegaard Floer chain complex. Let $\mathcal{H} = (\Sigma_g, \boldsymbol{\alpha}, \boldsymbol{\beta}, z)$ be a pointed Heegaard diagram, and consider $M = \mathrm{Sym}^g(\Sigma) = (\Sigma \times \cdots \times \Sigma)/S_g$, where we quotient the product of g copies of Σ by the group S_g of permutations on g letters. One can prove that M is a manifold of dimension $2g$, by a standard (but non-trivial) argument.

Next, choose a symplectic structure ω_Σ on Σ; this induces a symplectic structure $\omega_{\Sigma^{\times g}}$ on $\Sigma^{\times g} = \Sigma \times \cdots \times \Sigma$. Since the α_i's and β_i's are Lagrangian submanifolds of Σ, it follows that $\mathbb{T}_{\boldsymbol{\alpha}} = \alpha_1 \times \cdots \times \alpha_g$ and $\mathbb{T}_{\boldsymbol{\beta}} = \beta \times \cdots \times \beta_g$ are Lagrangian submanifolds of $\Sigma^{\times g}$ also. By a result of Perutz [Per08a], there is a symplectic form $\omega_{\mathrm{Sym}^g(\Sigma)}$ on $\mathrm{Sym}^g(\Sigma)$ that coincides, outside of a neighborhood of the diagonal, with the pushforward of $\omega_{\Sigma^{\times g}}$ under the quotient map. Since $\mathbb{T}_{\boldsymbol{\alpha}}$ and $\mathbb{T}_{\boldsymbol{\beta}}$ are disjoint from the diagonal, it follows that their images, which we denote by the same symbols, are also Lagrangian submanifolds of $(\mathrm{Sym}^g(\Sigma), \omega_{\mathrm{Sym}^g(\Sigma)})$.

Definition 1.4.1 (Ozsváth and Szabó [OSz04d], reformulation by Perutz [Per08a, Per08b]) Given $\mathcal{H} = (\Sigma, \boldsymbol{\alpha}, \boldsymbol{\beta}, z)$, where Σ is of genus g, we define the *hat Heegaard Floer chain complex* of $\mathcal{H}$ as the Lagrangian Floer complex of $M = \mathrm{Sym}^g(\Sigma)$ with the two submanifolds $L_0 = \mathbb{T}_{\boldsymbol{\alpha}}$ and $L_1 = \mathbb{T}_{\boldsymbol{\beta}}$,

$$\widehat{\mathrm{CF}}(\mathcal{H}) = \mathrm{CF}\left(\mathrm{Sym}^g(\Sigma), \omega_{\mathrm{Sym}^g(\Sigma)}, \mathbb{T}_{\boldsymbol{\alpha}}, \mathbb{T}_{\boldsymbol{\beta}}, \{z\} \times \mathrm{Sym}^{g-1}(\Sigma)\right),$$

blocking all holomorphic bigons that intersect the divisor $\{z\} \times \mathrm{Sym}^{g-1}(\Sigma) \subset \mathrm{Sym}^g(\Sigma)$. (Note that the divisor is also symmetrized.)

1.4.2 Unraveling the Definition

Ozsváth and Szabó [OSz04d] gave a more direct definition, however, given Perutz's result [Per08a], it differs from the standard Lagrangian Floer homology framework only in how it handles energy bounds. In any case, we unravel the definition here to make it more explicit.

First, an intersection point of $L_0 = \mathbb{T}_{\boldsymbol{\alpha}}$ and $L_1 = \mathbb{T}_{\boldsymbol{\beta}}$ corresponds exactly to a matching between the α- and β-circles. Second, given two generators $\mathbf{x}$ and $\mathbf{y}$, recall that the index of a holomorphic bigon $u \in \mathcal{M}_{\mathcal{J}}(\mathbf{x}, \mathbf{y})$ is determined by the class $B = [u] \in \pi_2(\mathrm{Sym}^g(\Sigma), \mathbb{T}_{\boldsymbol{\alpha}} \cup \mathbb{T}_{\boldsymbol{\beta}})$. It turns out that B specifies and is specified by $\mathbf{x}$, $\mathbf{y}$, and a *domain* on Σ, which is a formal linear combination of regions of Σ, i.e., connected components of $\Sigma \setminus (\boldsymbol{\alpha} \cup \boldsymbol{\beta})$. We may thus consider B to be an element of the space $\pi_2(\mathbf{x}, \mathbf{y})$ of domains from $\mathbf{x}$ to $\mathbf{y}$. Finally, the fact that we block holomorphic bigons that intersect the divisor $\{z\} \times \mathrm{Sym}^{g-1}(\Sigma)$ translates to blocking domains that contain z in its interior, a condition denoted by $n_z(B) = 0$, where n_z is the multiplicity of z.

Definition 1.4.2 (Ozsváth and Szabó [OSz04d]) Given a pointed Heegaard diagram $\mathcal{H} = (\Sigma, \boldsymbol{\alpha}, \boldsymbol{\beta}, z)$ where Σ is of genus g, and a choice of a family of almost complex structures $\mathcal{J} = \{J_s\}_{0 \le s \le 1}$, we define the *hat Heegaard Floer chain complex* of $(\mathcal{H}, \mathcal{J})$ as the $\mathbb{F}$-module

$$\widehat{\mathrm{CF}}(\mathcal{H}; \mathcal{J}) = \mathbb{F}\langle \mathcal{S}(\mathcal{H}) \rangle$$

with boundary homomorphism $\widehat{\partial} \colon \widehat{\mathrm{CF}}(\mathcal{H}; \mathcal{J}) \to \widehat{\mathrm{CF}}(\mathcal{H}; \mathcal{J})$ given by

$$\widehat{\partial}(\mathbf{x}) = \sum_{\mathbf{y} \in \mathcal{S}(\mathcal{H})} \sum_{\substack{B \in \pi_2(\mathbf{x}, \mathbf{y}) \\ \mathrm{Ind}(B) = 1 \\ n_z(B) = 0}} \#\widehat{\mathcal{M}}^B_{\mathcal{J}}(\mathbf{x}, \mathbf{y}) \cdot \mathbf{y}.$$

Similarly, we define the *minus Heegaard Floer chain complex* of $(\mathcal{H}, \mathcal{J})$ as the $\mathbb{F}[U]$-module

$$\mathrm{CF}^-(\mathcal{H}; \mathcal{J}) = \mathbb{F}[U]\langle \mathcal{S}(\mathcal{H}) \rangle$$

with boundary homomorphism $\partial^- : \mathrm{CF}^-(\mathcal{H}; \mathcal{J}) \to \mathrm{CF}^-(\mathcal{H}; \mathcal{J})$ given by

$$\partial^-(\mathbf{x}) = \sum_{\mathbf{y} \in \mathcal{S}(\mathcal{H})} \sum_{\substack{B \in \pi_2(\mathbf{x}, \mathbf{y}) \\ \mathrm{Ind}(B)=1}} \# \widehat{\mathcal{M}}^B_{\mathcal{J}}(\mathbf{x}, \mathbf{y}) \cdot U^{n_z(B)} \cdot \mathbf{y}.$$

Finally, we define the *infinity* and *plus Heegaard Floer chain complexes* of $(\mathcal{H}, \mathcal{J})$ respectively by

$$\mathrm{CF}^\infty(\mathcal{H}; \mathcal{J}) = \mathrm{CF}^-(\mathcal{H}; \mathcal{J}) \otimes_{\mathbb{F}[U]} \mathbb{F}[U, U^{-1}],$$

$$\mathrm{CF}^+(\mathcal{H}; \mathcal{J}) = \mathrm{CF}^\infty(\mathcal{H}; \mathcal{J}) \otimes_{\mathbb{F}[U]} \left(\mathbb{F}[U, U^{-1}] \big/ U \right),$$

viewed as chain complexes of $\mathbb{F}[U]$-modules. Note that, alternatively, these complexes can also be defined without reference to $\mathrm{CF}^-(\mathcal{H}; \mathcal{J})$, with the boundary homomorphism explicitly defined.[3]

A few comments are in order:

- As discussed before, both the proof that the boundary homomorphisms are finite sums and the proof that $\partial^2 \equiv 0$ require Gromov compactness.
- When $\mathrm{Ind}(\mathbf{x}) - \mathrm{Ind}(\mathbf{y}) = 1$, Gromov compactness guarantees that $\widehat{\mathcal{M}}^B_{\mathcal{J}}(\mathbf{x}, \mathbf{y})$ consists of a finite number of points. But to get that the boundary homomorphism is a finite sum, one would also need to know that there are only finitely many elements in $B \in \pi_2(\mathbf{x}, \mathbf{y})$ that contribute a non-zero term. We will need to address this later, when we discuss the admissibility of diagrams.
- As in general Lagrangian Floer homology, one can prove that the chain complex is independent of the choice of $\mathcal{J}$. Namely, given $\mathcal{J}_1$ and $\mathcal{J}_2$, one uses a "connecting" family of almost complex structures $\mathcal{J}$ to define a chain homotopy equivalence. This justifies writing $\mathrm{CF}^\circ(\mathcal{H})$.
- However, the goal of Heegaard Floer homology is not to create invariants of pointed Heegaard diagrams $\mathcal{H}$. Instead, it is to create invariants of pointed 3-manifolds (Y, z). This is indeed the case, but we defer the statement of invariance until after the discussion of Spin^c-decompositions.
- When the genus $g(\Sigma) = 1$, we have that $\mathrm{Sym}^g(\Sigma) = \Sigma$. This means that we are really counting holomorphic bigons in Σ. Given $B \in \pi_2(\mathbf{x}, \mathbf{y})$ with the right index, by the Riemann Mapping Theorem, there is a unique holomorphic disk

[3] As we shall see later, when $\mathcal{H}$ is weakly but not strongly admissible, it is possible that we cannot guarantee $\mathrm{CF}^-(\mathcal{H}; \mathcal{J})$ to be well defined, while we can guarantee $\mathrm{CF}^+(\mathcal{H}; \mathcal{J})$ to be well defined.

(up to translation by $\mathbb{R}$) in the moduli space $\widehat{\mathcal{M}}^B_{\mathcal{J}}(\mathbf{x}, \mathbf{y})$, which means that the count is necessarily 1 for each such B. This means that, when $g(\Sigma) = 1$, the complex $\mathrm{CF}^\circ(\mathcal{H})$ can be computed combinatorially rather than holomorphically; in other words, our "naive" definition of the Heegaard Floer chain complex, while problematic in general, is mostly fine when $g(\Sigma) = 1$.

- Like the grid chain complexes, the Heegaard Floer chain complexes can also be defined over $\mathbb{Z}$ or $\mathbb{Z}[U]$. However, even more work is required here, as one needs to orient the relevant moduli spaces consistently.

1.4.3 **Spin^c-*Decomposition, Gradings, Admissibility, and Invariance***

We now explain, without proof or details, the Spin^c-decomposition and the homological Maslov grading on Heegaard Floer homology. Let $\circ \in \{\frown, -, +, \infty\}$ denote any of the four flavors.

- First, there is a decomposition of $\mathrm{CF}^\circ(\mathcal{H})$ into a direct sum

$$\mathrm{CF}^\circ(\mathcal{H}) = \bigoplus_{\mathfrak{s} \in \mathrm{Spin}^c(Y)} \mathrm{CF}^\circ(\mathcal{H}, \mathfrak{s}),$$

with each summand corresponding to a Spin^c-structure of Y. We give a terse summary below, and encourage the reader to consult references on this topic. For example, other than the description contained in [OSz04d, Section 2.6], see also [Tur97, Section 1] for the equivalence between Spin^c-structures and Euler structures. For our purposes, it suffices to know that

  - The set $\mathrm{Spin}^c(Y)$ of Spin^c-structures on Y is an affine copy of $H^2(Y; \mathbb{Z})$, with an action by $H^2(Y; \mathbb{Z})$;
  - There is a one-to-one correspondence between Spin^c-structures and *Euler structures*, which are homology classes of nowhere-vanishing vector fields. In other words, each $\mathfrak{s} \in \mathrm{Spin}^c(Y)$ is represented non-uniquely by a nowhere vanishing vector field v on Y. Replacing v by $-v$ corresponds to replacing $\mathfrak{s}$ by its *conjugate* Spin^c*-structure* $\bar{\mathfrak{s}}$.
  - For $\mathfrak{s} \in \mathrm{Spin}^c(Y)$, one can define its *first Chern class* $c_1(\mathfrak{s}) \in H^2(Y; \mathbb{Z})$ as $\mathfrak{s} - \bar{\mathfrak{s}}$, interpreted as the element in $H^2(Y; \mathbb{Z})$ that sends $\bar{\mathfrak{s}}$ to $\mathfrak{s}$ by its action. Alternatively, it can be defined as the first Chern class of the orthogonal complement of v. Intuitively, c_1 acts like "multiplication by 2" on $\mathrm{Spin}^c(Y)$ (which is of course not well defined on an affine space).

The key point is then the following. Given $\mathbf{x} \in \mathcal{S}(\mathcal{H})$, observe that $\mathbf{x}$ pairs up the index-1 and index-2 critical points of a Morse function f corresponding to $\mathcal{H}$; in fact, it specifies a gradient flow line between each pair. Likewise, z specifies a gradient flow line between the index-0 and index-3 critical points. The gradient vector field of f does not vanish outside a neighborhood of these flow lines, and can be completed to a vector field that is nowhere vanishing on Y. In short, $\mathbf{x}$ and z determine some $\mathfrak{s}_z(\mathbf{x}) \in \mathrm{Spin}^c(Y)$. It can be proved that $\pi_2(\mathbf{x}, \mathbf{y}) \neq \emptyset$ if and only if $\mathfrak{s}_z(\mathbf{x}) = \mathfrak{s}_z(\mathbf{y})$, which gives the direct sum decomposition.

- Next, each summand $\mathrm{CF}^\circ(\mathcal{H}, \mathfrak{s})$ is relatively $\mathbb{Z}/m$-graded by $\mathrm{Ind}(B)$, where

$$m = \gcd_{\xi \in H_2(Y;\mathbb{Z})} \langle c_1(\mathfrak{s}), \xi \rangle.$$

Here, since $\mathrm{Ind}(B)$ only provides the grading difference between $\mathbf{x}$ and $\mathbf{y}$, the grading is *relative*, meaning that it is defined only up to a shift.

Note that if $c_1(\mathfrak{s}) \in H^2(Y; \mathbb{Z})$ is torsion, then the evaluation against ξ is always 0. In this case, $m = 0$, and $\mathrm{CF}^\circ(\mathcal{H}, \mathfrak{s})$ is relatively $\mathbb{Z}$-graded. (With further work [OSz06, OSz03a], it can be lifted to an absolute $\mathbb{Q}$-grading.) In particular, this applies when $H^2(Y; \mathbb{Z}) \cong H_1(Y; \mathbb{Z})$ is torsion, i.e., when $b_1(Y) = 0$.

- Given $\mathfrak{s} \in \mathrm{Spin}^c(Y)$, to compute the summand of the Heegaard Floer complex associated to $\mathfrak{s}$, we need to use a diagram $\mathcal{H}$ such that $\mathfrak{s}_z(\mathbf{x}) = \mathfrak{s}$ for some $\mathbf{x} \in \mathcal{S}(\mathcal{H})$. This condition is called *$\mathfrak{s}$-realizability.*

Furthermore, for a fixed $\mathfrak{s}$, we need to ensure that whenever $\mathfrak{s}_z(\mathbf{x}) = \mathfrak{s}_z(\mathbf{y}) = \mathfrak{s}$, there are only finitely many $B \in \pi_2(\mathbf{x}, \mathbf{y})$ that contribute a non-zero term to the differential. It turns out that the different flavors requires admissibility of different strengths. For a given $\mathfrak{s}$, a Heegaard diagram $\mathcal{H}$ may be *strongly $\mathfrak{s}$-admissible*, *weakly $\mathfrak{s}$-admissible*, or neither. We do not give the definitions here; see [OSz04d, Section 4.2.2].

  - If $\mathcal{H}$ is strongly $\mathfrak{s}$-admissible, then $\widehat{\partial}$, ∂^-, ∂^+, and ∂^∞ all give finite sums; while
  - If $\mathcal{H}$ is weakly $\mathfrak{s}$-admissible, then $\widehat{\partial}$ and ∂^+ give finite sums.

Notably, if $\mathcal{H}$ is strongly $\mathfrak{s}_0$-admissible for a torsion $\mathfrak{s}_0$, then it is weakly $\mathfrak{s}$-admissible for all $\mathfrak{s}$.

We are now ready to state the invariance of Heegaard Floer homology.

Theorem 1.4.3 (Ozsváth and Szabó [OSz04d, Theorem 11.1]) *Let $\mathcal{H}$ be a pointed Heegaard diagram of (Y, z). If $\mathcal{H}$ is strongly (resp. weakly) $\mathfrak{s}$-admissible, then the graded isomorphism type of* $\mathrm{HF}^\circ(\mathcal{H}, \mathfrak{s})$ *is an invariant of* $(Y, \mathfrak{s})$ *for* $\circ \in \{\frown, -, +, \infty\}$ *(resp. for* $\circ \in \{\frown, +\}$*). Thus, we may write* $\mathrm{HF}^\circ(Y, \mathfrak{s})$ *for these homologies.*

Remark 1.4.4 As in grid homology, in fact, the graded chain homotopy type of $\mathrm{CF}^\circ(\mathcal{H}, \mathfrak{s})$ is an invariant of $(Y, \mathfrak{s})$. Also, it may seem that the point z here does not play a role. However, if we wish to get concrete homology modules $\mathrm{HF}^\circ(Y, \mathfrak{s})$ rather than isomorphism types, we would need to know that the isomorphisms relating $\mathrm{HF}^\circ(\mathcal{H}; \mathcal{J})$ for different choices of $(\mathcal{H}, \mathcal{J})$ are canonical. This *naturality* property is proved in [JTZ21], and is required for functoriality as well as the definition of a variant called involutive Heegaard Floer homology [HM17]. It turns out that, taking naturality into account, Heegaard Floer theory is really a *pointed* theory, with concrete homology modules $\mathrm{HF}^\circ(Y, z, \mathfrak{s})$.

The proof of Theorem 1.4.3 proceeds as follows. Analogous to commutation and (de)stabilization for grid diagrams, two weakly or strongly $\mathfrak{s}$-admissible pointed Heegaard diagrams are related by a finite sequence of isotopies, handleslides, and (de)stabilizations.

Here, isotopies refer to isotopies of the sets of α- and β-curves, independent of each other, and relative to z (i.e., the curves are not allowed to cross z). If the α- and β-curves remain transverse to each other throughout the isotopy, this is equivalent to choosing a path of different $\mathcal{J}$'s, which has been dealt with. Otherwise, one defines the chain homotopy equivalence by counting holomorphic bigons with a (time t)–dependent constraint.

Handleslides refer to handlesliding α-circles over α-circles, and sliding β-circles over β-circles. Let us focus on the latter. We get two sets of β-curves, $\boldsymbol{\beta}_1$ and $\boldsymbol{\beta}_2$, corresponding to $\mathcal{H}_1$ and $\mathcal{H}_2$ respectively. As in grid homology, we may combine them with $\boldsymbol{\alpha}$ into a single *Heegaard triple diagram*

$$(\Sigma, \boldsymbol{\alpha}, \boldsymbol{\beta}_1, \boldsymbol{\beta}_2, z),$$

as in Fig. 1.16.

One may now define a map $\Phi\colon \mathrm{CF}^\circ(\mathcal{H}_1) \to \mathrm{CF}^\circ(\mathcal{H}_2)$ using the μ_2 map in the A_∞-structure on the Fukaya category,

$$\mu_2\colon \mathrm{CF}(\mathbb{T}_\alpha, \mathbb{T}_{\beta_1}) \otimes \mathrm{CF}(\mathbb{T}_{\beta_1}, \mathbb{T}_{\beta_2}) \to \mathrm{CF}(\mathbb{T}_\alpha, \mathbb{T}_{\beta_2}),$$

defined by counting holomorphic triangles. To define Φ, we choose a generator Θ in $\mathrm{CF}(\mathbb{T}_{\beta_1}, \mathbb{T}_{\beta_2})$ and define

$$\Phi(\mathbf{x}) = \mu_2(\mathbf{x} \otimes \Theta).$$

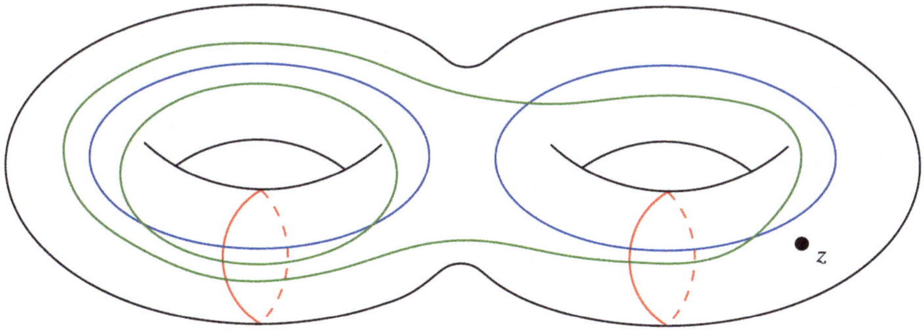

Fig. 1.16 A Heegaard triple diagram. Red circles represent $\boldsymbol{\alpha}$, blue circles represent $\boldsymbol{\beta}_1$, and green circles represent $\boldsymbol{\beta}_2$. Both $(\Sigma, \boldsymbol{\alpha}, \boldsymbol{\beta}_1, z)$ and $(\Sigma, \boldsymbol{\alpha}, \boldsymbol{\beta}_2, z)$ are Heegaard diagrams of (S^3, z)

One can then use a compatibility condition of the A_∞-structure to check that Φ is a chain map if Θ is a cycle. Then to define chain homotopies, one uses μ_3 and count holomorphic quadrilaterals.

Finally, stabilization refers to taking the connected sum with a toroidal Heegaard diagram (T^2, α, β) of S^3 where α and β meet exactly once, and destabilization is the inverse process. If $\mathcal{H}'$ is a stabilization of $\mathcal{H}$, there is a tautological identification of $\mathcal{S}(\mathcal{H}')$ with $\mathcal{S}(\mathcal{H})$; however, to identify the boundary homomorphisms, some work is required to identify the corresponding moduli spaces by gluing holomorphic curves. Combining all these proofs together would then conclude the proof of invariance.

Example 1.4.5 In Example 1.2.4, we looked at two Heegaard diagrams $\mathcal{H}_3$ and $\mathcal{H}_4$ of $(S^1 \times S^2, z)$. There, we used a "naive" definition of the Heegaard Floer chain complexes—which turns out to be fine in this case—to compute

$$\widehat{\mathrm{HF}}(\mathcal{H}_3) \cong \mathbb{F} \oplus \mathbb{F}, \qquad \widehat{\mathrm{HF}}(\mathcal{H}_4) \cong 0$$

$$\mathrm{HF}^-(\mathcal{H}_3) \cong \mathbb{F}[U] \oplus \mathbb{F}[U], \qquad \mathrm{HF}^-(\mathcal{H}_4) \cong {}^{\mathbb{F}[U]}\!/\!_{(1+U)}.$$

We can now explain this phenomenon. Observe that $H^2(S^1 \times S^2; \mathbb{Z}) \cong \mathbb{Z}$, and so $\mathrm{Spin}^c(S^1 \times S^2)$ is an affine copy of $\mathbb{Z}$. The word "affine" here is perhaps not necessary in this case, since there is in fact a unique self-conjugate, torsion Spin^c-structure $\mathfrak{s}_0$, satisfying $c_1(\mathfrak{s}_0) = 0$. Fixing an isomorphism $H^2(S^1 \times S^2; \mathbb{Z}) \cong \mathbb{Z}$, we let $\mathfrak{s}_i$ denote the Spin^c-structure satisfying $\mathfrak{s}_i = \mathfrak{s}_0 + i$. As it turns out, $\mathcal{H}_3$ is strongly $\mathfrak{s}_0$-admissible, implying that it is weakly $\mathfrak{s}_i$-admissible for all $i \in \mathbb{Z}$. On the other hand, $\mathcal{H}_4$ is strongly $\mathfrak{s}_1$-admissible; since $\mathfrak{s}_1$ is not torsion, one cannot conclude that $\mathcal{H}_4$ is weakly $\mathfrak{s}_i$-admissible for all $i \in \mathbb{Z}$. In fact, it is weakly $\mathfrak{s}_i$-admissible for all $i \neq 0$.

From the above, we may conclude that

$$\widehat{\mathrm{HF}}(S^1 \times S^2, \mathfrak{s}_0) \cong \mathbb{F} \oplus \mathbb{F}, \qquad \widehat{\mathrm{HF}}(S^1 \times S^2, \mathfrak{s}_1) \cong 0$$

$$\mathrm{HF}^-(S^1 \times S^2, \mathfrak{s}_0) \cong \mathbb{F}[U] \oplus \mathbb{F}[U], \qquad \mathrm{HF}^-(S^1 \times S^2, \mathfrak{s}_1) \cong {}^{\mathbb{F}[U]}\!/\!_{(1+U)}.$$

Exercise 1.4.6 Determine the relative gradings on these homologies.

1.4.4 Some Computational Tools

One of the biggest strengths of Heegaard Floer homology, compared to other Floer-theoretic invariants, is its computability. We now present a brief survey of some key computational tools that have been developed in the past several decades, for the reader to pursue further:

- **Cylindrical reformulation (Lipshitz [Lip06])**
Instead of counting holomorphic disks in $\mathrm{Sym}^g(\Sigma)$, one can count holomorphic curves in $\Sigma \times [0, 1] \times \mathbb{R}$ instead. The price is that one must now count holomorphic maps $u\colon S \to \Sigma \times [0, 1] \times \mathbb{R}$, whose domain S is any Riemann surface (possibly with multiple components). Lipshitz also provided an index formula to compute $\mathrm{Ind}(B)$. This is the formulation on which the following is built.
- **Bordered Heegaard Floer homology (Lipshitz et al. [LOT18])**
This theory introduces cut-and-paste methods to Heegaard Floer homology. It assigns a dg or A_∞-module to a 3-manifold Y with parametrized boundary, over an algebra associated with that boundary surface. If Y is obtained from gluing Y_1 and Y_2 along their common parametrized boundary, then $\widehat{\mathrm{CF}}(Y)$ can be reconstructed as a derived tensor product of the modules associated to Y_1 and Y_2.
- **Plumbed 3-manifolds and lattice homology (Ozsváth and Szabó [OSz03c], Némethi [Ném05, Ném08], Ozsváth et al. [OSSz14], Zemke [Zem23])**
Ozsváth and Szabó [OSz03c] first computed $\mathrm{HF}^-(Y)$ of certain plumbed 3-manifolds with negative-definite plumbing graphs. Némethi [Ném08] defined a combinatorial *lattice (co)homology* $\mathbb{H}$ for any negative-definite plumbing graph, connecting Heegaard Floer homology to singularity theory. For almost rational plumbing graphs (which includes the graphs that Ozsváth and Szabó considered), Némethi [Ném05] proved that $\mathbb{H}^-(Y)$ coincides with $\mathrm{HF}^-(Y)$ [Ném05] (but $\mathbb{H}^-(Y)$ enjoys more structure); this is later extended to a larger class of graphs by Ozsváth et al. [OSSz14], and to all negative-definite plumbing trees by Zemke [Zem23].
- **Nice diagrams (Sarkar and Wang [SW10])**
If every region, i.e., every component of $\Sigma \setminus (\boldsymbol{\alpha} \cup \boldsymbol{\beta})$, either is blocked by z (in the *hat* flavor), or is a bigon or a rectangle, then the count of holomorphic curves is in fact entirely combinatorial: Every domain that is a bigon or a rectangle has a unique holomorphic representative. This is of great importance, since this means that, by judiciously choosing a Heegaard diagram, the Heegaard Floer chain complex can be combinatorially computed without any reference to almost complex structures or solving any partial differential equations. Sarkar and Wang showed that any closed, oriented three-manifold admits such Heegaard diagram, which they called *nice*.
- **Surgery exact triangle (Ozsváth and Szabó [OSz04c])**
There is long exact sequence

$$\cdots \to \widehat{\mathrm{HF}}(Y) \to \widehat{\mathrm{HF}}(Y_r(K)) \to \widehat{\mathrm{HF}}(Y_{r+1}(K)) \to \cdots$$

where $Y_r(K)$ is the r-surgery on Y along the knot K. Knowing two of these terms often allows one to compute the third.
- **Large, integer, and rational surgery formulas (Ozsváth and Szabó [OSz04b, OSz08, OSz11], Rasmussen [Ras03], Manolescu and Ozsváth [MO22])**
For a $\mathbb{Z}$-homology sphere Y, the large surgery formula [OSz04b, Ras03] allows one to extract $\mathrm{CF}^\circ(Y_p(K))$,when $p \in \mathbb{Z}$ is large, from the *infinity knot Floer*

chain complex $\mathrm{CFK}^\infty(Y, K)$ as a subquotient. More complicated formulas allow one to extract $\mathrm{CF}^\circ(Y_p(K))$ for any $p \in \mathbb{Z}$ [OSz08], extended to $\mathbb{Z}$-surgery on links [MO22], and $\mathrm{CF}^\circ(Y_r(K))$ for $r \in \mathbb{Q}$ [OSz11].

1.4.5 Knot Floer Homology: Definition and Features

The knot Floer complex mentioned above is a variant of the Heegaard Floer complex, and may be viewed as a *relative* version of the theory.

We start by considering *doubly pointed Heegaard diagrams* for a pair (Y, K), where $K \subset Y$ is an oriented knot. (For ease of exposition, it is often assumed that Y is a $\mathbb{Z}$- or $\mathbb{Q}$-homology sphere, and K is null-homotopic.) Such diagrams are similar to the pointed Heegaard diagrams we have been working with so far, only with one additional point w,

$$\mathcal{H} = (\Sigma, \boldsymbol{\alpha}, \boldsymbol{\beta}, z, w).$$

Here $(\Sigma, \boldsymbol{\alpha}, \boldsymbol{\beta})$ is a Heegaard diagram of Y. The diagram $\mathcal{H}$ encodes a knot K in the following way: Draw an oriented path from z to w on Σ, avoiding α-curves, and push it into the α-handlebody; then draw an oriented path from w to z on Σ, avoiding β-curves, and push it into the β-handlebody; K is the union of these two oriented paths. Formulated in an equivalent way, w and z each determine a gradient flow line from the index-3 critical point to the index-0 critical point, and K is the union of these two gradient flow lines.

Example 1.4.7 Figure 1.17 shows how one recovers a knot from a doubly pointed Heegaard diagram. The depicted Heegaard diagram encodes the left-handed trefoil.

Definition 1.4.8 (cf. Ozsváth and Szabó [OSz04b], Rasmussen [Ras03]) Suppose that Y is a $\mathbb{Z}$-homology sphere, which implies that there is only one Spin^c-structure on Y, which is torsion. Given a doubly pointed Heegaard diagram

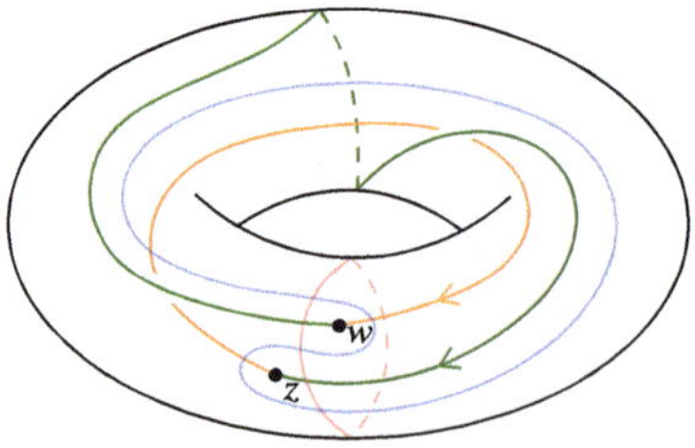

(a)

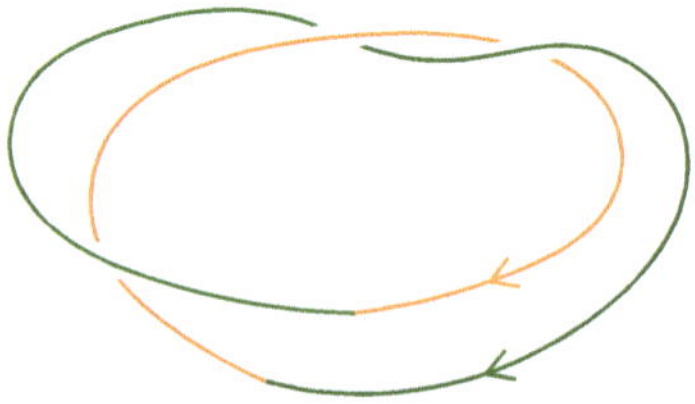

(b)

Fig. 1.17 Recovering a knot from a doubly pointed Heegaard diagram. (**a**) A doubly pointed Heegaard diagram. (**b**) The left-handed trefoil

$\mathcal{H} = (\Sigma, \boldsymbol{\alpha}, \boldsymbol{\beta}, z, w)$ of (Y, K), denote by $\mathcal{H}'$ the pointed Heegaard diagram $\mathcal{H} = (\Sigma, \boldsymbol{\alpha}, \boldsymbol{\beta}, z)$ of (Y, z). We define the *hat* and *minus knot Floer chain complexes* of $\mathcal{H}$,

$$\widehat{\mathrm{CFK}}(\mathcal{H}), \qquad \mathrm{CFK}^-(\mathcal{H}),$$

respectively as the chain complexes

$$\widehat{\mathrm{CF}}(\mathcal{H}'), \qquad \mathrm{CF}^-(\mathcal{H}'),$$

equipped with a relative $\mathbb{Z}$-filtration A, called the *Alexander filtration*, characterized by

$$A(\mathbf{x}) - A(\mathbf{y}) = n_z(B) - n_w(B)$$

for every $B \in \pi_2(\mathbf{x}, \mathbf{y})$, where n_z and n_w denote the multiplicities of z and w in B respectively. To ensure that $\mathrm{CFK}^-(\mathcal{H})$ is filtered, we set the filtration level of U to be -1.

One can also define the *infinity* and *plus knot Floer chain complexes* $\mathrm{CFK}^\infty(\mathcal{H})$ and $\mathrm{CFK}^+(\mathcal{H})$ similarly.

Note that $H_*(\mathrm{CFK}^\circ(Y, K))$ is again $\mathrm{HF}^\circ(Y)$, since we must ignore the filtration when taking the homology. Given a $\mathbb{Z}$-filtered chain complex $C = (C, \mathcal{F})$, recall that its *associated graded object* is the chain complex

$$\mathrm{g}C = \bigoplus_{i \in \mathbb{Z}} \mathcal{F}_i C / \mathcal{F}_{i-1} C,$$

with an internal relative $\mathbb{Z}$-grading given by i. (Colloquially, in this context, we "erase" any term from the differential that strictly lowers the filtration.) In our setting, the internal $\mathbb{Z}$-grading on $\mathrm{gCFK}^\circ(\mathcal{H})$ is called its *Alexander grading*. (It continues to have a homological relative *Maslov grading*.) We define the *knot Floer homology* of $\mathcal{H}$ to be

$$\mathrm{HFK}^\circ(\mathcal{H}) = H_*(\mathrm{gCFK}^\circ(\mathcal{H})).$$

When $Y = S^3$, we can fix an *absolute Maslov grading* M on $\mathcal{S}(\mathcal{H}')$ by requiring that

$$H_*(\widehat{\mathrm{CFK}}(\mathcal{H})) \cong \widehat{\mathrm{HF}}(S^3) \cong \mathbb{F}$$

be concentrated in grading 0. In this case, one can, with work, prove that the difference between the top and bottom Alexander gradings supported in $\widehat{\mathrm{HFK}}(\mathcal{H})$ is always $2g(K)$, where g denotes the Seifert genus [OSz04a]; there is an *absolute Alexander grading* A on $\mathcal{S}(\mathcal{H}')$ characterized by the condition that the top Alexander grading be $g(K)$ and the bottom $-g(K)$.

Of course, the point is that these are knot invariants.

Theorem 1.4.9 (Ozsváth and Szabó [OSz04b, Theorem 3.1], Rasmussen [Ras03, Theorem 1]) *Let $\mathcal{H}$ be a doubly pointed Heegaard diagram of (Y, K), where Y is a $\mathbb{Z}$-homology sphere and K is null-homotopic. Then, the filtered, graded chain homotopy type of* $\mathrm{CFK}^{\circ}(\mathcal{H})$ *is an invariant of (Y, K) for $\circ \in \{\frown, -, +, \infty\}$ (resp. for $\circ \in \{\frown, +\}$). Thus, we may write* $\mathrm{CFK}^{\circ}(Y, K)$ *for these chain complexes.*

Example 1.4.10 Using the doubly pointed Heegaard diagram in Fig. 1.17a, we can calculate the *hat* and *minus* knot Floer homologies of the left-handed trefoil to be

$$\widehat{\mathrm{HFK}}(S^3, \overline{T}_{2,3}) \cong \mathbb{F}^{\oplus 3}, \qquad \mathrm{HFK}^-(S^3, \overline{T}_{2,3}) \cong \mathbb{F}[U] \oplus {}^{\mathbb{F}[U]}\!/\!_{U}.$$

We conclude this discussion with some basic features of knot Floer homology:

- The *graded Euler characteristic* of $\widehat{\mathrm{HFK}}(S^3, K)$ is defined to be

$$\chi(\widehat{\mathrm{HFK}}(S^3, K)) = \sum_{M,A} (-1)^M t^A \cdot \mathrm{rk}_{\mathbb{F}} \widehat{\mathrm{HFK}}_M(S^3, K, A),$$

  where M and A denote the Maslov and Alexander gradings respectively. A key feature of knot Floer homology is that this turns out to be the symmetrized Alexander polynomial [OSz04b]:

$$\chi(\widehat{\mathrm{HFK}}(S^3, K)) = \Delta_K(t).$$

  (Recall that the Alexander polynomial is defined up to multiplication by a monomial; it can be symmetrized by requiring that $\Delta_K(t^{-1}) = \Delta_K(t)$.)
- From $\mathrm{CFK}^\infty(S^3, K)$, we can extract many other invariants of knots in S^3, e.g., $\tau(K)$ [OSz03b], $\varepsilon(K)$ [Hom14], $\Upsilon(K)$ [OSSz17a], the torsion order of $\widehat{\mathrm{HFK}}(S^3, K)$ [AE20b, AE20a], etc. Many of these are powerful invariants, and some of them turn out to be concordance invariants.
- Heegaard Floer homology can be used to study contact 3-manifolds (via open books), while knot Floer homology can be used for Legendrian and transverse knots. Namely, if a 3-manifold Y is equipped with a contact structure ξ, then there is an invariant $c(Y, \xi) \in \widehat{\mathrm{HF}}(-Y)$ [OSz05, HKM09]; likewise, given a Legendrian knot $\Lambda \subset (Y, \xi)$, there are also invariants $\mathcal{L}(Y, \xi, \Lambda) \in \mathrm{HFK}^-(-Y, \Lambda)$ and $\widehat{\mathcal{L}}(Y, \xi, \Lambda) \in \widehat{\mathrm{HFK}}(-Y, \Lambda)$ [LOSSz09, OSzT08].

Knot Floer homology has a vast amount of applications, and can itself be the subject of an entire lecture series; the above is but a condensed introduction that inevitably misses many other important features.

1.4.6 Multi-Pointed Heegaard and Knot Floer Homologies: Definition and Relationship with Grid Homology

Previously, Heegaard diagrams $\mathcal{H} = (\Sigma, \boldsymbol{\alpha}, \boldsymbol{\beta})$ have been assumed to have exactly $g(\Sigma)$ pairs of α- and β-circles, and $\{z\}$ and $\{w\}$ have been singleton sets of points. In fact, we can work with more general *multi-pointed Heegaard diagrams* of the form

$$\mathcal{H} = (\Sigma_g, \{\alpha_1, \dots, \alpha_{g+n-1}\}, \{\beta_1, \dots, \beta_{g+n-1}\}, \{z_1, \dots, z_n\}),$$

possibly also with $\{w_1, \dots, w_n\}$, for some $n \geq 1$. Such diagrams must satisfy the condition that each component of $\Sigma_g \setminus \boldsymbol{\alpha}$ contain exactly one z_i (and also exactly one w_j if w_j's exist), and likewise for $\Sigma_g \setminus \boldsymbol{\beta}$. For these multi-pointed Heegaard diagrams, we may define $\mathrm{CF}^\circ(\mathcal{H})$ or $\mathrm{CFK}^\circ(\mathcal{H})$ as before, except that we count holomorphic bigons in $\mathrm{Sym}^{g+n-1}(\Sigma)$ rather than in $\mathrm{Sym}^g(\Sigma)$.

The (filtered,) graded chain homotopy types of the resulting chain complexes turn out to coincide with those of the chain complexes defined with singly and doubly pointed Heegaard diagrams.[4] In the proof, one considers an additional Heegaard move, *quasi-(de)stabilization*, where one adds (or subtracts) a pair of α- and β-curves and a point z_i (and w_j if relevant).

For the story to come full circle, we note that grid diagrams are exactly multi-pointed Heegaard diagrams of knots $K \subset S^3$ that are nice in the sense of Sarkar and Wang [SW10].

Proposition 1.4.11 *Given a toroidal grid diagram $\mathbb{G} = (n, \mathbb{O}, \mathbb{X})$ of $K \subset S^3$, the quintuple*

$$\mathcal{H} = (T^2, \{\textit{horizontal circles}\}, \{\textit{vertical circles}\}, \mathbb{O}, \mathbb{X})$$

is a multi-pointed Heegaard diagram of (S^3, K).[5]

Thus, the knot Floer chain complex can be computed by combinatorially counting domains of the correct index, which are empty rectangles.

[4] However, for knot Floer homology, the issue of naturality becomes more subtle; see [Sar15, Juh16, Zem19b].

[5] There exist different conventions in the literature of this identification. Often, $\mathbb{O}$ and $\mathbf{w}$ are associated with the formal variable U while $\mathbb{X}$ and $\mathbf{z}$ are associated with the Alexander filtration. (The roles of $\mathbf{w}$ and $\mathbf{z}$ are swapped in our discussion of CFK^- above.) One way to resolve any discrepancy is to choose between $\mathcal{H}$ and $\mathcal{H}' = (-T^2, \{\text{vertical circles}\}, \{\text{horizontal circles}\}, \mathbb{X}, \mathbb{O})$, which is also a multi-pointed Heegaard diagram of (S^3, K) (rather than, say, of $(-S^3, K)$ or of $(S^3, -K)$).

Corollary 1.4.12 (Manolescu et al. [MOS09]) *Let $\mathbb{G} = (n, \mathbb{O}, \mathbb{X})$ be a grid diagram of $K \subset S^3$. Then there is a chain homotopy equivalence*

$$\mathcal{GC}^-(\mathcal{G}) \simeq \mathrm{CFK}^-(S^3, K)$$

of filtered, graded chain complexes over $\mathbb{F}[U]$.

References

[AE20a] Akram Alishahi and Eaman Eftekhary, *Knot Floer homology and the unknotting number*, Geom. Topol. **24** (2020), no. 5, 2435–2469. MR 4194296

[AE20b] ———, *Tangle Floer homology and cobordisms between tangles*, J. Topol. **13** (2020), no. 4, 1582–1657. MR 4186139

[Cro95] Peter R Cromwell, *Embedding knots and links in an open book i: Basic properties*, Topology and its Applications **64** (1995), no. 1, 37–58.

[Flo88] Andreas Floer, *The unregularized gradient flow of the symplectic action*, Comm. Pure Appl. Math. **41** (1988), no. 6, 775–813. MR 948771

[Gro85] M. Gromov, *Pseudo holomorphic curves in symplectic manifolds*, Invent. Math. **82** (1985), no. 2, 307–347. MR 809718

[HKM09] Ko Honda, William H. Kazez, and Gordana Matić, *On the contact class in Heegaard Floer homology*, J. Differential Geom. **83** (2009), no. 2, 289–311. MR 2577470

[HM17] Kristen Hendricks and Ciprian Manolescu, *Involutive Heegaard Floer homology*, Duke Math. J. **166** (2017), no. 7, 1211–1299. MR 3649355

[Hom14] Jennifer Hom, *The knot Floer complex and the smooth concordance group*, Comment. Math. Helv. **89** (2014), no. 3, 537–570. MR 3260841

[JTZ21] András Juhász, Dylan Thurston, and Ian Zemke, *Naturality and mapping class groups in Heegard Floer homology*, Mem. Amer. Math. Soc. **273** (2021), no. 1338, v+174. MR 4337438

[Juh16] András Juhász, *Cobordisms of sutured manifolds and the functoriality of link Floer homology*, Adv. Math. **299** (2016), 940–1038. MR 3519484

[Lip06] Robert Lipshitz, *A cylindrical reformulation of Heegaard Floer homology*, Geom. Topol. **10** (2006), 955–1096. MR 2240908

[LOSSz09] Paolo Lisca, Peter Ozsváth, András I. Stipsicz, and Zoltán Szabó, *Heegaard Floer invariants of Legendrian knots in contact three-manifolds*, J. Eur. Math. Soc. (JEMS) **11** (2009), no. 6, 1307–1363. MR 2557137

[LOT18] Robert Lipshitz, Peter S. Ozsváth, and Dylan P. Thurston, *Bordered Heegaard Floer homology*, Mem. Amer. Math. Soc. **254** (2018), no. 1216, viii+279. MR 3827056

[MO22] Ciprian Manolescu and Peter Ozsváth, *Heegaard Floer homology and integer surgeries on links*, preprint, version 5, 2022, `arXiv:1011.1317`.

[MOS09] Ciprian Manolescu, Peter Ozsváth, and Sucharit Sarkar, *A combinatorial description of knot Floer homology*, Ann. of Math. (2) **169** (2009), no. 2, 633–660. MR 2480614

[MOSzT07] Ciprian Manolescu, Peter Ozsváth, Zoltán Szabó, and Dylan Thurston, *On combinatorial link Floer homology*, Geom. Topol. **11** (2007), 2339–2412. MR 2372850

[Ném05] András Némethi, *On the Ozsváth-Szabó invariant of negative definite plumbed 3-manifolds*, Geom. Topol. **9** (2005), 991–1042. MR 2140997

[Ném08] ———, *Lattice cohomology of normal surface singularities*, Publ. Res. Inst. Math. Sci. **44** (2008), no. 2, 507–543. MR 2426357

[Oh93] Yong-Geun O h, *Floer cohomology of Lagrangian intersections and pseudo-holomorphic disks. I*, Comm. Pure Appl. Math. **46** (1993), no. 7, 949–993. MR 1223659

[OSSz14] Peter Ozsváth, András I. Stipsicz, and Zoltán Szabó, *A spectral sequence on lattice homology*, Quantum Topol. **5** (2014), no. 4, 487–521. MR 3317341

[OSSz15] Peter S. Ozsváth, András I. Stipsicz, and Zoltán Szabó, *Grid homology for knots and links*, Mathematical Surveys and Monographs, vol. 208, American Mathematical Society, Providence, RI, 2015. MR 3381987

[OSSz17a] ———, *Concordance homomorphisms from knot Floer homology*, Adv. Math. **315** (2017), 366–426. MR 3667589

[OSSz17b] ———, *Unoriented knot Floer homology and the unoriented four-ball genus*, Int. Math. Res. Not. IMRN (2017), no. 17, 5137–5181. MR 3694597

[OSz03a] Peter Ozsváth and Zoltán Szabó, *Absolutely graded Floer homologies and intersection forms for four-manifolds with boundary*, Adv. Math. **173** (2003), no. 2, 179–261. MR 1957829

[OSz03b] ———, *Knot Floer homology and the four-ball genus*, Geom. Topol. **7** (2003), 615–639. MR 2026543

[OSz03c] ———, *On the Floer homology of plumbed three-manifolds*, Geom. Topol. **7** (2003), 185–224. MR 1988284

[OSz04a] ———, *Holomorphic disks and genus bounds*, Geom. Topol. **8** (2004), 311–334. MR 2023281

[OSz04b] ———, *Holomorphic disks and knot invariants*, Adv. Math. **186** (2004), no. 1, 58–116. MR 2065507

[OSz04c] ———, *Holomorphic disks and three-manifold invariants: properties and applications*, Ann. of Math. (2) **159** (2004), no. 3, 1159–1245. MR 2113020

[OSz04d] ———, *Holomorphic disks and topological invariants for closed three-manifolds*, Ann. of Math. (2) **159** (2004), no. 3, 1027–1158. MR 2113019

[OSz05] ———, *Heegaard Floer homology and contact structures*, Duke Math. J. **129** (2005), no. 1, 39–61. MR 2153455

[OSz06] ———, *Holomorphic triangles and invariants for smooth four-manifolds*, Adv. Math. **202** (2006), no. 2, 326–400. MR 2222356

[OSz08] Peter S. Ozsváth and Zoltán Szabó, *Knot Floer homology and integer surgeries*, Algebr. Geom. Topol. **8** (2008), no. 1, 101–153. MR 2377279

[OSz11] ———, *Knot Floer homology and rational surgeries*, Algebr. Geom. Topol. **11** (2011), no. 1, 1–68. MR 2764036

[OSzT08] Peter Ozsváth, Zoltán Szabó, and Dylan Thurston, *Legendrian knots, transverse knots and combinatorial Floer homology*, Geom. Topol. **12** (2008), no. 2, 941–980. MR 2403802

[Ped18] Andrés Pedroza, *A quick view of Lagrangian Floer homology*, Geometrical themes inspired by the N-body problem, Lecture Notes in Math., vol. 2204, Springer, Cham, 2018, pp. 91–125. MR 3753117

[Per08a] Tim Perutz, *A remark on Kähler forms on symmetric products of Riemann surfaces*, preprint, version 2, 2008, `arXiv:math/0501547`.

[Per08b] Timothy Perutz, *Hamiltonian handleslides for Heegaard Floer homology*, Proceedings of Gökova Geometry-Topology Conference 2007, Gökova Geometry/Topology Conference (GGT), Gökova, 2008, pp. 15–35. MR 2509747

[Ras03] Jacob Andrew Rasmussen, *Floer homology and knot complements*, ProQuest LLC, Ann Arbor, MI, 2003, Thesis (Ph.D.)–Harvard University. MR 2704683

[Sar15] Sucharit Sarkar, *Moving basepoints and the induced automorphisms of link Floer homology*, Algebr. Geom. Topol. **15** (2015), no. 5, 2479–2515. MR 3426686

[SW10] Sucharit Sarkar and Jiajun Wang, *An algorithm for computing some Heegaard Floer homologies*, Ann. of Math. (2) **171** (2010), no. 2, 1213–1236. MR 2630063

[Tur97] Vladimir Turaev, *Torsion invariants of* Spin^c*-structures on* 3*-manifolds*, Math. Res. Lett. **4** (1997), no. 5, 679–695. MR 1484699

[Zem19a] Ian Zemke, *Link cobordisms and absolute gradings on link Floer homology*, Quantum Topol. **10** (2019), no. 2, 207–323. MR 3950650

[Zem19b] ———, *Link cobordisms and functoriality in link Floer homology*, J. Topol. **12** (2019), no. 1, 94–220. MR 3905679

[Zem23] ———, *The equivalence of lattice and Heegaard Floer homology*, preprint, version 3, 2023, arXiv:2111.14962.

Chapter 2
A Knot-Theoretic Tour of Dimension Four

Kyle Hayden and Márton Beke

Contents

2.1 Handles, Homotopy, and Homology 46
  2.1.1 Handles, Homotopy, and Homology 46
  2.1.2 The Intersection Form 50
2.2 Intersection Forms and Smooth 4-Manifolds 55
  2.2.1 Basic Properties for Intersection Forms 55
  2.2.2 Constraints on Intersection Forms 57
  2.2.3 Addendum: Rohlin's Invariant and Beyond 59
2.3 The Minimal Genus Problem 60
  2.3.1 The Minimal Genus Problem 61
  2.3.2 The Local Minimal Genus Problem 63
  2.3.3 An Exotic $\mathbb{R}^4$ 67
  2.3.4 Addendum and Additional Exercises 68
2.4 The Contact and Stein Setting 69
  2.4.1 Contact Geometry and the Slice-Bennequin Inequality 69
  2.4.2 Stein Surfaces 72
References 73

These notes provide a written account of the lecture series held as part of the *Singularities and low dimensional topology* winter school at the Rényi Institute in January 2023. They will provide more detail than the lectures themselves, as well as additional exercises and remarks related to more recent research for those who are interested. For a more complete treatment of the topics discussed, we refer the reader to the beautiful books by Gompf and Stipsticz [GS99] and Saveliev [Sav12].

We will begin by investigating the basic anatomy and algebraic topology of 4-manifolds. The first lecture will focus on using handle diagrams to determine

K. Hayden
Rutgers University Newark, Newark, NJ, USA
e-mail: kyle.hayden@rutgers.edu

M. Beke (✉)
Budapest University of Technology and Economics, Budapest, Hungary
e-mail: bekem@renyi.hu

© The Author(s), under exclusive license to Springer Nature Switzerland AG 2024
J. Fernández de Bobadilla et al. (eds.), *Singularities and Low Dimensional Topology*, Bolyai Society Mathematical Studies 30,
https://doi.org/10.1007/978-3-031-56611-0_2

algebro-topological data, and the second lecture will focus on the intersection form of a 4-manifold. As motivation for these first two lectures, our goal will be to assemble the ingredients of the following:

Theorem 2.0.1 (Freedman [Fre82]) *There exists a closed, simply connected, topological 4-manifold that does not admit a smooth structure.*

The third lecture will focus on studying surfaces in 4-manifolds and the minimal genus problem, which asks for the minimal genus of an embedded surface representing a fixed second homology class. This will culminate in a sketch of the following landmark theorem:

Theorem 2.0.2 (Donaldson [Don83], Freedman [Fre82]) *There exist smooth 4-manifolds that are homeomorphic but not diffeomorphic to* $\mathbb{R}^4$.

The fourth and final lecture will offer a brief look at contact structures on 3-manifolds and Stein structures on 4-manifolds. Constraints on smoothly embedded surfaces (in the form of an adjunction inequality) will provide a key tool for the minimal genus problem.

Conventions
Unless otherwise stated, homology is taken with $\mathbb{Z}$ coefficients. We will typically assume our manifolds are connected, except when considering disconnected n-manifolds as boundaries of connected $(n+1)$-manifolds (e.g., in the context of cobordism). The symbol $\cong$ will be used to denote whatever notion of equivalence is most natural in a given context (e.g., diffeomorphism for smooth manifolds, homeomorphism for topological manifolds, isomorphism for groups or lattices, etc.).

2.1 Handles, Homotopy, and Homology

The Goal
This lecture will focus on describing 4-manifolds and understanding their basic algebraic topology in terms of handle decompositions. As motivation for the first two lectures, our goal will be to assemble the ingredients of the following:

Theorem 2.1.1 (Freedman [Fre82]) *There exists a closed, simply connected, topological 4-manifold that does not admit a smooth structure.*

2.1.1 Handles, Homotopy, and Homology

Let X be a smooth 4-manifold equipped with a handle decomposition. Recall that an *n-dimensional k-handle* is a copy of $D^k \times D^{n-k}$, which may be attached to the boundary of an n-manifold X by fixing an embedding $\partial D^k \times D^{n-k} \hookrightarrow \partial X$. A *handle decomposition* of X is a decomposition of X into a union of n-manifolds $X_0 = \emptyset \subset X_1 \subset X_2 \subset \cdots$ such that X_i is obtained from X_{i-1} by a handle attachment.

Remark 2.1.2 (Anatomy of a Handle) The *core* of a k-handle $h = D^k \times D^{n-k}$ is the central $D^k \times 0$, its boundary is the *attaching sphere* $S^{k-1} \times 0$, and $S^{k-1} \times D^{n-k}$ is the *attaching region*. Dually, we call $0 \times D^{n-k}$ the *cocore*, its boundary $0 \times S^{n-k-1}$ the *belt sphere*, and the corresponding component $D^k \times S^{n-k-1}$ of ∂h is called the *belt region*.

For this lecture, let X be a smooth, compact 4-manifold equipped with a (finite) handle decomposition. As a matter of convenience, we will assume that X is connected, hence admits a handle decomposition that has a unique 0-handle. Moreover, we will assume that handles are attached in order of increasing index and that all handles of a given index are attached simultaneously. We will be focusing on 4-manifolds built without 3-handles, but we note that if a closed 4-manifold contains a single 4-handle, then the 3-handles are essentially uniquely determined by the handles of index ≤ 2 [LP72].

Our first goal is to use the handle decomposition to read off the algebraic topology of X. The key insight is that the handle decomposition "collapses" onto a cell decomposition for a topological space homotopy equivalent to the manifold X. Roughly speaking, each k-handle $D^k \times D^{4-k}$ collapses onto the k-cell $D^k \times 0$ (i.e., the handle's core disk), dragging all later cell attachments along by homotopy.

Example 2.1.3 If X consists only of one 0-handle and m handles of index $k = 1$ or $k = 2$, then X is homotopy equivalent to the m-fold wedge of k-spheres $\vee^m S^k$. (Collapse the 0-handle to a point, and all attaching regions $\partial D^k \times D^{4-k}$ collapse to that point, too.) Note that each S^k can be viewed as the union of the core k-disk of a k-handle and a (singular) k-disk formed as the cone of the attaching sphere inside $B^4 = \text{Cone}(S^3)$; this is depicted schematically for $m = 1$ and $k = 2$ in Fig. 2.1. Alternatively, it will often be useful to replace this singular k-disk with a smooth k-manifold in B^4 bounded by the attaching sphere in S^3.

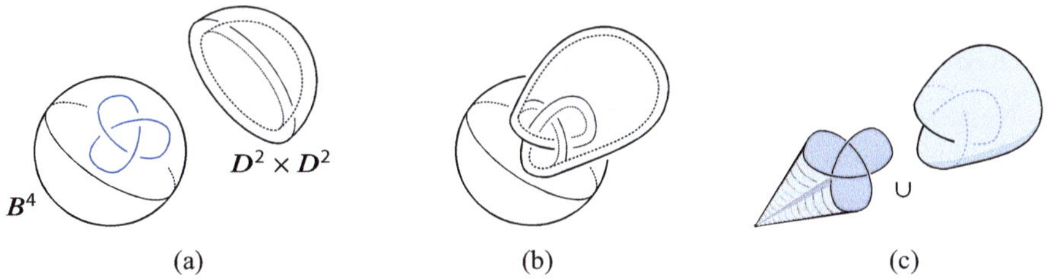

Fig. 2.1 Exhibiting a singular 2-sphere inside a 4-manifold built from one 0-handle and one 2-handle, known as a *knot trace* (cf. Example 2.1.3). (**a**) depicts the 0-handle B^4, the 2-handle $D^2 \times D^2$, and an attaching curve in ∂B^4; (**b**) depicts the resulting 4-manifold X; (**c**) depicts a singular 2-sphere in X

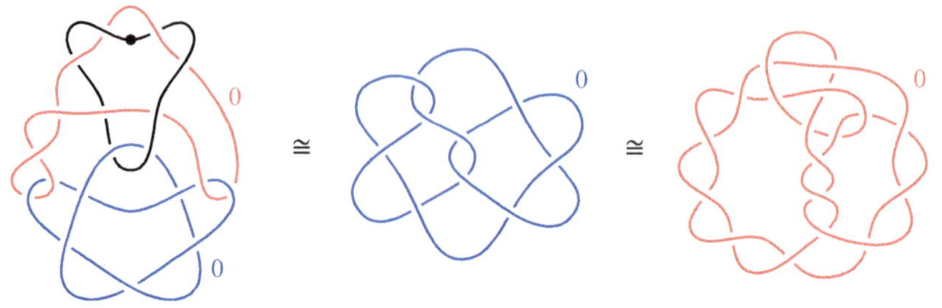

Fig. 2.2 Kirby diagrams for a 4-manifold that can be expressed as the 0-trace of two different knots

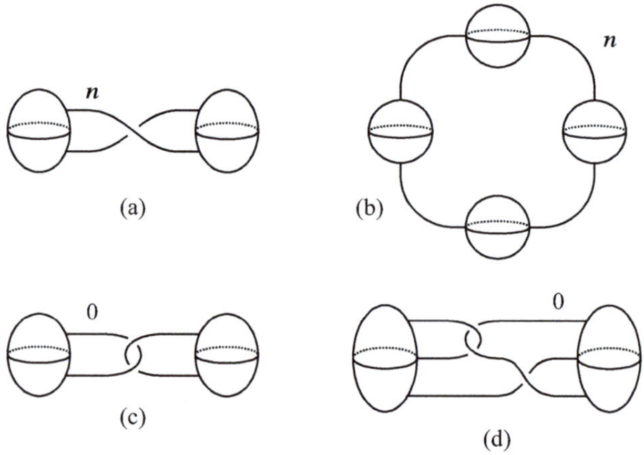

Fig. 2.3 Handle diagrams for the 4-manifolds in Examples 2.1.5, 2.1.6, and 2.1.8

Exercise 2.1.4 The 4-manifold obtained by attaching an n-framed 2-handle to B^4 along a knot K is commonly called the *n-trace of K* and denoted $X_n(K)$. Using handle slides and cancellations, show that the 4-manifold depicted on the left side of Fig. 2.2 (which uses dotted circle notation for 1-handles, as in [GS99, §5.4]) can be expressed as a 0-framed knot trace in two different ways. *(One of these knots is known as* 9_{42} *in the knot tables, while the other is identified as* `o9_34801` *in SnapPy [CDGW].)*

Next, we consider a collection of 4-manifolds that will serve as running examples for our discussion.

Example 2.1.5 Consider the 4-manifolds represented by the Kirby diagrams in Fig. 2.3: (a) the D^2-bundle over $\mathbb{R}P^2$ of Euler number n; (b) the D^2-bundle over T^2 of Euler number n; (c) a self-plumbing of $S^2 \times D^2$, which may be viewed as regular neighborhood of an immersed 2-sphere with square zero in a 4-manifold; (d) a contractible 4-manifold known as a *Mazur manifold*.

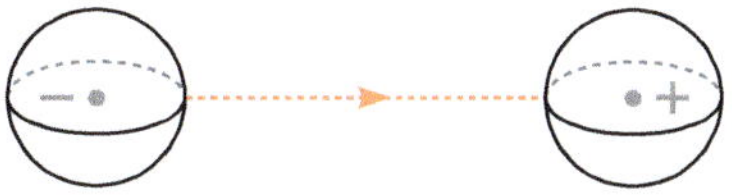

Fig. 2.4 The two spheres are identified, and the other part of the circle travels through the core of the handle which we don't picture

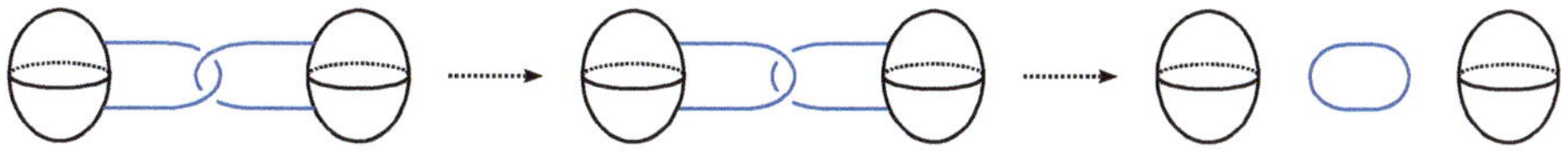

Fig. 2.5 Modifying an attaching curve by a homotopy that involves a crossing change

Now let X be any smooth, connected 4-manifold with a handle decomposition that has a unique 0-handle. Moreover, assume that handles are attached in order of increasing index and that all handles of a given index are attached simultaneously.

In general, $\pi_1(X)$ is determined by the union of handles of index ≤ 2. After choosing orientations for the handles' attaching spheres, we can extract a presentation of $\pi_1(X)$ as follows.

- *Generators* $\approx$ *1-handles:* Each generator is represented by an oriented *core* curve that passes once over the given 1-handle (Fig. 2.4).
- *Relations* $\approx$ *2-handles:* Each relation is determined by a 2-handle by expressing its oriented *attaching* curve (up to homotopy) as a word in the 1-handles' core curves (i.e., the order and sign with which 1-handles are encountered as one traverses the attaching curve).

We note that changing the attaching sphere's orientations will affect the presentation of the group but not its isomorphism type. Also note that the relations are independent of the framings of the 2-handles.

Example 2.1.6 For the 4-manifolds in Fig. 2.3: (a) $\pi_1 \cong \langle x \mid x^2 \rangle \cong \mathbb{Z}/2\mathbb{Z}$ (b) $\pi_1 \cong \langle x, y \mid xyx^{-1}y^{-1} \rangle \cong \mathbb{Z} \oplus \mathbb{Z}$ (c) $\pi_1 \cong \langle x \mid xx^{-1} \rangle \cong \mathbb{Z}$ (d) $\pi_1 \cong \langle x \mid xx^{-1}x \rangle \cong 1$

When calculating $\pi_1(X)$ from a handle diagram, it can be convenient to modify the 2-handles' attaching curves up to homotopy. For example, Fig. 2.5 uses a crossing change to show that the attaching curve for the 2-handle in Fig. 2.3c is nullhomotopic, hence contributes a trivial relation.

Exercise 2.1.7

(a) Use handle diagrams to build a 4-manifold X such that $\pi_1(X)$ is nonabelian.
(b) Prove that every finitely presented group arises as $\pi_1(X)$ for a smooth, compact, orientable 4-manifold X.
(c) Upgrade (b) to achieve a *closed* 4-manifold. (Hint: Don't start over.)

To calculate $H_1(X)$, we simply abelianize $\pi_1(X)$. The homology groups $H_k(X)$ with $k \geq 2$ can be calculated using cellular homology: the kth chain group $C_k(X)$ is generated by k-handles, and the boundary formula for a k-handle h is given by

$$\partial_k h = \sum (B_i \cdot A)\, h_i$$

where the sum ranges over the $(k-1)$-handles h_i and A denotes the attaching sphere of the k-handle h, B_i denotes the belt sphere of h_i, and $B_i \cdot A$ denotes their signed intersection.

Example 2.1.8 For the 4-manifolds in parts (a) and (d) of Fig. 2.3, the groups $H_2(X)$ are trivial. In parts (b) and (c), we have $H_2(X) \cong \mathbb{Z}$. Note that, because these examples have no 3- or 4-handles, the chain groups $C_3(X)$ and $C_4(X)$ are trivial; as a consequence, we have $H_2(X) \cong \ker \partial_2 \subset C_2(X)$.

Remark 2.1.9 We have now shown that the manifold X in Fig. 2.3d has $\pi_1(X) = 1$ and $H_k(X) = 1$ for all $k \geq 1$, hence applying the Hurewicz and Whitehead theorems implies that X is contractible.

2.1.2 *The Intersection Form*

The most important algebro-topological invariant of a 4-manifold X is its cohomology ring, where the multiplicative structure is known as the *intersection form* of X.

Definition 2.1.10 Given a compact, orientable 4-manifold X, fix an orientation on X by choosing a fundamental class $[X]$ generating $H_4(X, \partial X) \cong \mathbb{Z}$. The *intersection form* of X is the symmetric bilinear form

$$Q_X : H^2(X, \partial X) \otimes H^2(X, \partial X) \to \mathbb{Z} \qquad (a, b) \mapsto \langle a \smile b, [X] \rangle.$$

Remark 2.1.11 Note that the cup product is symmetric on $H^2(X, \partial X; \mathbb{Z})$. Analogous forms can be defined on H^n for any $2n$-dimensional manifold; the form will be symmetric if n is even and skew-symmetric if n is odd.

The next result shows that intersection forms completely determine the homeomorphism type of closed, simply connected 4-manifolds.

Theorem 2.1.12 (Freedman [Fre82]) *A pair of smooth, closed, simply connected 4-manifolds X and X' are homeomorphic $\iff$ Q_{X_1} and Q_{X_2} are isomorphic.*

How do we calculate intersection forms? Note that, by Poincaré-Lefschetz duality, we have $H^2(X, \partial X) \cong H_2(X)$, so we can view this as a bilinear form on homology

$$Q_X : H_2(X) \otimes H_2(X) \to \mathbb{Z} \qquad (\alpha, \beta) \mapsto \langle PD(\alpha) \smile PD(\beta), [X] \rangle.$$

When X is a smooth manifold, we can represent classes in $H_2(X)$ by smoothly embedded, oriented surfaces (Exercise 2.1.20) and recast Q_X as counting signed intersections between them (cf. Proposition 2.1.19). We will implement this using a diagrammatic approach below.

2.1.2.1 Linking Numbers

Let K_1 and K_2 be disjoint, oriented knots in S^3. A basic exercise shows that any knot complement in S^3 has $H_1 \cong \mathbb{Z}$ and is generated by the meridian of K_1, leading to the following definition.

Definition 2.1.13 Given oriented knots $K_1, K_2 \subset S^3$, their *linking number* $\mathrm{lk}(K_1, K_2)$ is the integer corresponding to the class $[K_2] \in H_1(S^3 \setminus K_1) \cong \mathbb{Z}$, where $1 \in \mathbb{Z}$ corresponds to a positively oriented meridian of K_1.

Remark 2.1.14 The linking number can be expressed in other ways:

(a) It can be computed diagrammatically via the formula

$$\mathrm{lk}(K_1, K_2) = \frac{1}{2} \sum \{\text{signed crossings between } K_1 \text{ and } K_2\}$$

where the signs are determined via the right-hand rule as below. This makes it clear that the linking number is symmetric: $\mathrm{lk}(K_1, K_2) = \mathrm{lk}(K_2, K_1)$.

(b) The class $[K_2] \in H_1(S^3 \setminus K_1)$ can be calculated by choosing an oriented Seifert surface Σ_1 for the oriented knot K_1, perturbing K_2 to be transverse to Σ_1, then counting the signed intersection number $K_2 \cdot \Sigma_1$.

Example 2.1.15 Consider the oriented link $L = K_1 \cup K_2 \cup K_3$ shown in Fig. 2.6. A direct calculation using either of the above methods shows

$$\mathrm{lk}(K_1, K_2) = 2 \qquad \mathrm{lk}(K_1, K_3) = 0 \qquad \mathrm{lk}(K_2, K_3) = 0.$$

In particular, note that all four crossings between K_1, K_2 are positive, so $\mathrm{lk}(K_1, K_2) = 2$. The component K_3 is unknotted and bounds an obvious disk in the page; it has two oppositely-signed intersection points with K_1, so $\mathrm{lk}(K_1, K_3) = 0$. Finally, we also have $\mathrm{lk}(K_2, K_3) = 0$ because K_2 and K_3 have no crossings.

Framings on link components can be interpreted as self-linking numbers, enabling us to organize the linking data of a framed, oriented link into a matrix as follows.

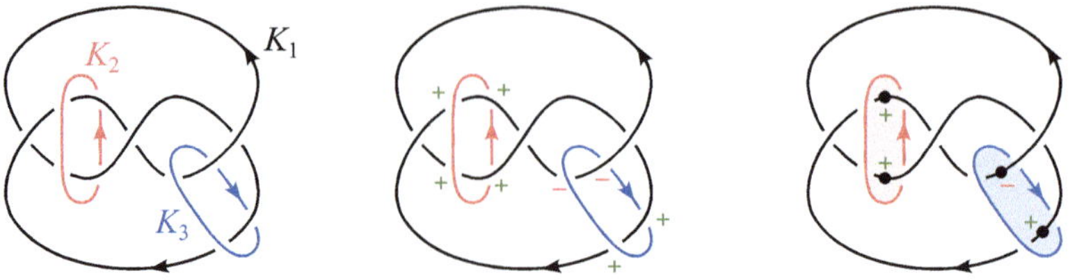

Fig. 2.6 Calculating linking numbers for Example 2.1.15

Fig. 2.7 A framed, oriented link

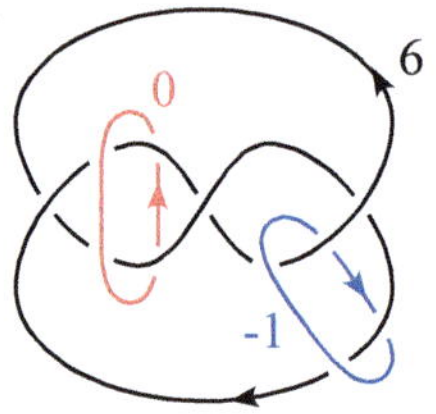

Definition 2.1.16 Given a framed, oriented n-component link $L \subset S^3$ (e.g., attaching data for 2-handles) with components K_i, its *linking matrix* is the symmetric $n \times n$ matrix A whose ijth entry is

$$a_{ij} = \begin{cases} \mathrm{lk}(K_i, K_j) & i \neq j \\ \mathrm{framing}(K_i) & i = j. \end{cases}$$

Remark 2.1.17 A more uniform definition could be obtained by using the framings to define parallel pushoffs K_i' for each K_i, then setting $a_{ij} = \mathrm{lk}(K_i, K_j')$ for all i and j.

Example 2.1.18 The diagram in Fig. 2.7 shows a copy of the link from Example 2.1.15 equipped with framings on the link components. With the orientations and notation from Fig. 2.6, the linking matrix is

$$\begin{bmatrix} 6 & 2 & 0 \\ 2 & 0 & 0 \\ 0 & 0 & -1 \end{bmatrix}.$$

2.1.2.2 Calculating Intersection Forms from Handle Diagrams

The next proposition explains our interest in the linking matrix:

Proposition 2.1.19 (Linking Matrices and Intersection Forms) *If X is obtained from B^4 by attaching 2-handles along a framed, oriented n-component link L, then $H_2(X) \cong \mathbb{Z}^n$ and Q_X is presented by the linking matrix of L.*

Proof Sketch As mentioned in Example 2.1.3, X is homotopy equivalent to $\vee^n S^2$, implying $H_2(X) \cong \mathbb{Z}^n$. The intersection form Q_X is determined by its behavior on any chosen basis for $H_2(X)$, and we can choose a natural one as follows. For each component K_i in L, fix an oriented Seifert surface $\Sigma_i \subset S^3$ for K_i. (For later convenience, we will assume Σ_i is in general position with respect to $L \setminus K_i$ so that it meets each other link component transversely.) Let $\hat{\Sigma}_i$ be the union of Σ_i with the core disk $D_i = D^2 \times 0$ inside the 2-handle attached along K_i. The homology classes $[\hat{\Sigma}_i]$ of these closed surfaces $\hat{\Sigma}_i$ form a basis for $H_2(X)$. (Indeed, the homotopy equivalence collapsing X to $\vee^n S^2$ carries $\hat{\Sigma}_i$ to the ith wedge summand S^2 via a degree-one map.)

For $i \neq j$, any intersection between $\hat{\Sigma}_i$ and $\hat{\Sigma}_j$ must occur in S^3. Let Σ'_j be a surface obtained from Σ_j by pushing the interior $\mathring{\Sigma}_j$ slightly into B^4 so that $\Sigma'_j \cap \partial B^4 = K_j$. This modified closed surface $\hat{\Sigma}'_j = \Sigma'_j \cup_{K_j} D_j$ satisfies $[\hat{\Sigma}'_j] = [\hat{\Sigma}_j]$ and is transverse to $\hat{\Sigma}_i$, meeting it only where K_j intersects Σ_i. Thus we have

$$[\hat{\Sigma}_i] \cdot [\hat{\Sigma}_j] = [\hat{\Sigma}_i] \cdot [\hat{\Sigma}'_j] = \hat{\Sigma}_i \cdot \hat{\Sigma}'_j = \Sigma_i \cdot K_j = \text{lk}(K_i, K_j).$$

To find $[\hat{\Sigma}_i] \cdot [\hat{\Sigma}_i]$, we need a perturbed copy of $\hat{\Sigma}_i$ that is transverse to $\hat{\Sigma}_i$. In the 2-handle $D^2 \times D^2$ attached along K_i, choose a perturbed copy of the core disk, expressed as $D'_i = D^2 \times \{\varepsilon\}$ for a point $\varepsilon \in D^2 \setminus \{0\}$. The intersection $D'_i \cap S^3$ is an n-framed pushoff of K_i, which we will denote by K'_i. Let Σ'_i denote a copy of Σ_i dragged along during the isotopy from K_i to K'_i, with its interior pushed into B^4 as above. Then $\hat{\Sigma}'_i = \Sigma'_i \cup D'_i$ satisfies $[\hat{\Sigma}'_i] = [\hat{\Sigma}_i]$ and is transverse to $\hat{\Sigma}_i$, allowing us to compute

$$[\hat{\Sigma}_i] \cdot [\hat{\Sigma}_i] = [\hat{\Sigma}_i] \cdot [\hat{\Sigma}'_i] = \hat{\Sigma}_i \cdot \hat{\Sigma}'_j = \Sigma_i \cdot K'_i = \text{lk}(K_i, K'_i) = \text{framing}(K_i) \in \mathbb{Z},$$

as desired. □

While it may seem restrictive, many interesting examples of 4-manifolds can be built by attaching 2-handles to B^4. If a 4-manifold X has a unique 0-handle and some number of 4-handles, we may still apply Proposition 2.1.19 because removing (the interiors of) the 4-handles does not affect $H^2(X)$; see Exercise 2.1.23 below. Similar techniques apply in the presence of 1- and 3-handles. In this case, the 2-handles' attaching curves still provide a basis for the chain group $C_2(X)$; we must extract a basis for $H_2(X)$ as in Sect. 2.1.1, with each basis element expressed as a linear combination of the 2-handles' attaching curves. Restricting the linking form to the subspace spanned by these elements yields a presentation of the intersection form on $H_2(X)$.

Exercise 2.1.20 Let X be any smooth, compact 4-manifold (possibly with boundary). Prove that each class in $H_2(X)$ is represented by a smoothly embedded surface in X. *Hint: After choosing a handle structure on X, each class in $H_2(X)$ corresponds to some linear combination of 2-handles. To build a representative*

Fig. 2.8 A framed Hopf link

surface, begin with an appropriate collection of disjoint copies of the cores of these 2-handles, then build the rest of your surface (while ensuring it is embedded).

Remark 2.1.21 We consider intersection forms up to congruence: If we change our basis for $\mathbb{Z}^n$ by a matrix A, then any bilinear form Q on $\mathbb{Z}^n$ transforms as $Q \rightsquigarrow A^T Q A$.

Exercise 2.1.22 Let M_n denote the 4-manifold built from one 0-handle and two 2-handles as given in Fig. 2.8.

(a) Show that $Q_{M_n} \cong \begin{bmatrix} 0 & 1 \\ 1 & n \end{bmatrix}$ (for a natural choice of basis).
(b) Show that $\partial M_n \cong S^3$.
(c) In either order, do the following:

(*i*) *Linear algebra:* Find a basis for $H_2(M_n; \mathbb{Z})$ in which Q_{M_n} is presented by

$$\begin{bmatrix} 0 & 1 \\ 1 & 0 \end{bmatrix} \text{ if } n \text{ is even} \quad \text{or} \quad \begin{bmatrix} 0 & 1 \\ 1 & 1 \end{bmatrix} \text{ if } n \text{ is odd.}$$

(*ii*) *Kirby calculus:* Use handle slides to show $M_n \cong M_m \Leftrightarrow m \equiv n \pmod 2$.

The first part of the next exercise implies that we can also use Proposition 2.1.19 for 4-manifolds built from one 0-handle, some number of 2-handles, and a single 4-handle.

Exercise 2.1.23

(a) Given a closed 4-manifold X, show that $H_2(X \setminus \mathring{B}^4) \cong H_2(X)$ and $Q_{X \setminus B^4} \cong Q_X$.
(b) Given 4-manifolds X and X', show that $H_2(X \# X') \cong H_2(X) \oplus H_2(X')$ and $Q_{X \# X'} \cong Q_X \oplus Q_{X'}$.
(c) Generalize this to the case where $\partial X = -\partial X'$ is a 3-manifold Y with $H_1(Y) = 0$ and we glue $X \cup_Y -X'$.

Example 2.1.24

(a) $\pm\mathbb{C}P^2$ (where $-\mathbb{C}P^2 = \overline{\mathbb{C}P}^2$) is obtained from B^4 by attaching a ± 1-framed 2-handle along an unknot, then attaching a 4-handle, so $Q_{\mathbb{C}P^2} = \langle \pm 1 \rangle$.
(b) $S^2 \times S^2$ corresponds to a 0-framed Hopf link (i.e., $n = 0$ in Exercise 2.1.22) plus a 4-handle, so $Q_{S^2 \times S^2}$ is given by the *standard hyperbolic form*

$$H = \begin{bmatrix} 0 & 1 \\ 1 & 0 \end{bmatrix}.$$

Exercise 2.1.25 Prove (or at least convince yourself) that every symmetric matrix Q over $\mathbb{Z}$ arises as the linking matrix of a framed link in S^3, and hence that Q arises as the intersection form of a compact 4-manifold with boundary.

2.2 Intersection Forms and Smooth 4-Manifolds

I once had the pleasure of taking a topology class taught by Paul Melvin. I was an undergrad, and I confessed to him that I couldn't really picture 3-manifolds other than $\mathbb{R}^3$—and maybe S^3, too. He nodded consolingly and paused a while, then said, "Well, I like to picture them as the boundaries of 4-manifolds."

This lecture will take a closer look at the intersection forms of 4-manifolds, including 4-manifolds with boundary. And Paul was right: we will encounter an important interplay between 3- and 4-manifold topology, beginning with Rohlin's theorem [Roh52] and the Poincaré homology sphere, and connecting all the way through to Manolescu's relatively recent disproof of the Triangulation Conjecture [Man16].

The Goal
Our goal is to outline the existence of closed, simply connected topological 4-manifolds that do not admit smooth structures. We will be able to sketch this after laying out some (elementary) properties of symmetric bilinear forms and citing some (non-elementary!) results about the ones that arise as intersection forms of smooth, closed, simply-connected 4-manifolds.

2.2.1 Basic Properties for Intersection Forms

We begin with observations and definitions that apply to any symmetric bilinear form Q on $\mathbb{Z}^n$.

- The *rank* of Q is simply n, the dimension of the underlying lattice.
- Fix a basis for $\mathbb{Z}^n$ to view Q as a symmetric matrix. The eigenvalues of a symmetric matrix are real, so we can let $b^+(Q)$ and $b^-(Q)$ denote the numbers of (strictly) positive and (strictly) negative eigenvalues, respectively, counted with multiplicity. Note that $b^\pm$ are independent of the basis for $\mathbb{Z}^n$, hence are invariants of Q itself.
- The *signature* of Q is $\sigma(Q) = b^+(Q) - b^-(Q)$.
- Q is $(\pm)$ *definite* if $\sigma(Q) = \pm\,\mathrm{rank}(Q)$ and is *indefinite* otherwise.
- Q is *unimodular* if $\det Q = \pm 1$, which ensures Q is invertible over $\mathbb{Z}$. More generally, we say Q is *nondegenerate* if $\det Q \neq 0$, which only ensures Q is invertible over $\mathbb{Q}$.
- The *type* of Q is *even* if $Q(x, x) \equiv 0 \mod 2$ for all $x \in \mathbb{Z}^n$ and is *odd* otherwise.

Remark 2.2.1 The determinant of a bilinear form over $\mathbb{Z}$ is well-defined. This is because any change-of-basis matrix A over $\mathbb{Z}$ is unimodular, hence $\det(Q) = \det(A^T QA)$. When working with bilinear forms over $\mathbb{Q}$ or $\mathbb{R}$, this fails, but the *sign* of the determinant is still well-defined.

Example 2.2.2 For the 4-manifold X from Example 2.1.15, we had

$$Q_X = \begin{bmatrix} 6 & 2 & 0 \\ 2 & 0 & 0 \\ 0 & 0 & -1 \end{bmatrix}.$$

Observe that $\det(Q_X) = 4$, so Q_X is nondegenerate but not unimodular. We can easily see that -1 is an eigenvalue, so the positive determinant implies that we must have $b_- = 2$ and $b_+ = 1$. Therefore $\sigma(Q_X) = -1$, and Q_X is indefinite. Since $e_3 \cdot e_3 = -1 \not\equiv 0 \bmod 2$, Q_X is odd.

Exercise 2.2.3

(a) Repeat the above for the intersection forms on $\pm\mathbb{C}P^2$ and $S^2 \times S^2$.
(b) How do these invariants behave under connected sums of 4-manifolds?

Exercise 2.2.4 A form Q on $\mathbb{Z}^n$ is even $\iff$ every matrix for Q has even diagonal $\iff$ at least one matrix for Q has even diagonal.

Example 2.2.5 Take the 4-manifold X in Example 2.1.18, whose intersection form was not even, and blow down the (-1)-framed unknotted component. We obtain the handle diagram shown in Fig. 2.9, representing a new 4-manifold X' with $\partial X' = \partial X$. Using Remark 2.2.1 and the fact that all framing coefficients in the diagram are even, we conclude that X' is even.

We will not prove the following result, but we will briefly make use of it later.

Theorem 2.2.6 *Every closed, connected, orientable 3-manifold bounds a smooth, simply connected 4-manifold X built from one 0-handle and a number of 2-handles. Moreover, we may choose X so that Q_X is even.*

The key idea behind realizing 3-manifolds as boundaries of 4-manifolds is to show that any closed, orientable 3-manifold Y can be realized as Dehn surgery on an integrally-framed link L in S^3. Then view L as instructions for attaching 2-handles

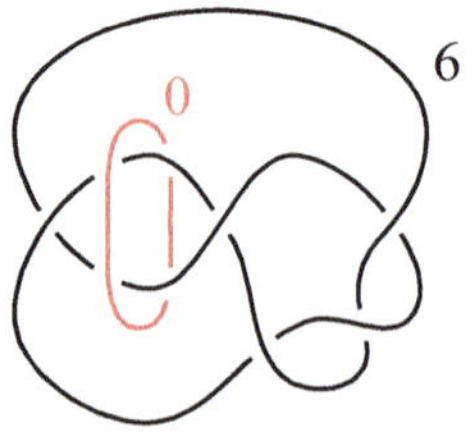

Fig. 2.9 A 4-manifold with even intersection form

to B^4, yielding a 4-manifold X with $\partial X = Y$. If X has Q_X odd, it turns out that there must be a particular type of non-empty sublink of L (called a "characteristic sublink") that is responsible for Q_X being odd, like the (-1)-framed unknot in Example 2.2.5. As in that example, there is always a sequence of blow-ups and blow-downs that eventually eliminate all nonempty characteristic sublinks, resulting in a new 4-manifold with even intersection form.

The simplest symmetric bilinear form that is unimodular, positive definite, and even turns out to be an 8-dimensional form known as E_8, presented by the matrix

$$E_8 = \begin{bmatrix} 2&1&0&0&0&0&0&0 \\ 1&2&1&0&0&0&0&0 \\ 0&1&2&1&0&0&0&0 \\ 0&0&1&2&1&0&0&0 \\ 0&0&0&1&2&1&0&1 \\ 0&0&0&0&1&2&1&0 \\ 0&0&0&0&0&1&2&0 \\ 0&0&0&0&1&0&0&2 \end{bmatrix}.$$

Example 2.2.7 We can realize E_8 as the intersection form of the compact 4-manifold with boundary shown in the top left corner of Fig. 2.10, known as the E_8-plumbing manifold. The boundary of the E_8-plumbing manifold is known as the *Poincaré homology sphere*, denoted P. In Fig. 2.10, we see that P can also be obtained from $(+1)$-framed Dehn surgery on the right-handed trefoil.

Exercise 2.2.8 Carefully check the handle calculus in Fig. 2.10, paying attention to the framings (especially near the end).

Definition 2.2.9 Let Y be a closed 3-manifold. We say that Y is an *integer homology 3-sphere* (abbreviated as $\mathbb{Z}HS^3$) if $H_*(Y;\mathbb{Z}) \cong H_*(S^3;\mathbb{Z})$. More generally, given a coefficient ring R, we say Y is an *R-homology 3-sphere* if $H_*(Y;R) \cong H_*(S^3;R)$.

Exercise 2.2.10

(a) Q_X is unimodular ($\det(Q_X) = \pm 1$) if and only if ∂X is a $\mathbb{Z}HS^3$.
(b) Q_X is nondegenerate ($\det(Q_X) \neq 0$) if and only if ∂X is a $\mathbb{Q}HS^3$.

2.2.2 *Constraints on Intersection Forms*

In what follows, Q denotes a unimodular, symmetric bilinear form over $\mathbb{Z}$. First, a constraint that's purely algebraic:

Theorem 2.2.11 *If Q is even, then $\sigma(Q)$ is divisible by 8.*

Note that the E_8 form satisfies this, of course, because $\sigma(E_8) = 8$. Rohlin proved a stronger constraint for intersection forms of smooth 4-manifolds.

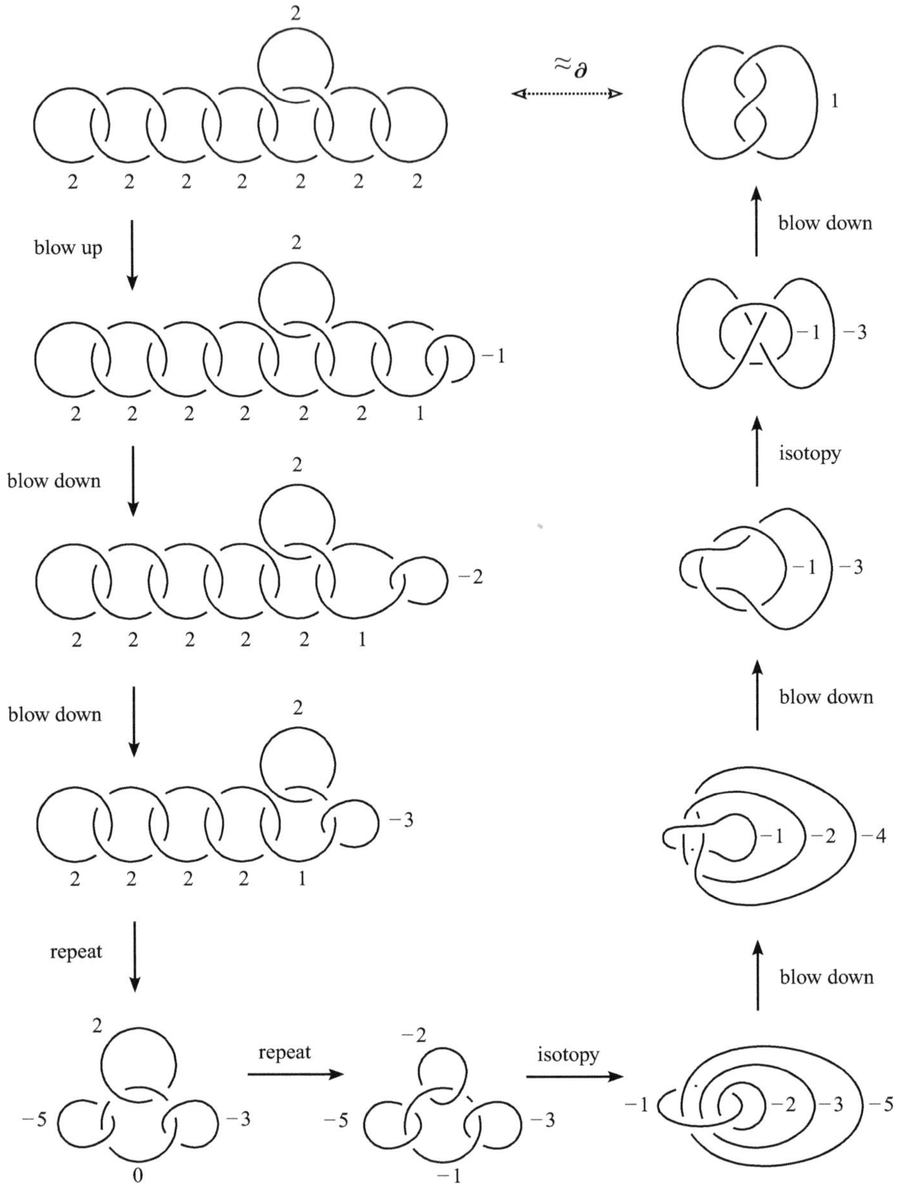

Fig. 2.10 Relating the boundaries of the E_8-plumbing manifold and surgery on the trefoil knot

Theorem 2.2.12 (Rohlin [Roh52]) *If X is a smooth, closed, simply connected 4-manifold with even intersection form Q_X, then $\sigma(Q_X)$ is divisible by 16.*

Remark 2.2.13 Rohlin's theorem holds for any closed *spin* 4-manifold. For a simply connected 4-manifold X (possibly with nonempty boundary), the spin condition is equivalent to Q_X being even.

Note that Rohlin's theorem does not hold for 4-manifolds with nonempty boundary; indeed, the E_8-plumbing manifold from Example 2.2.7 provides a counterexample. Moreover, it follows that the Poincaré homology sphere cannot bound a smooth integer homology ball, otherwise we could glue it to the E_8-plumbing manifold to produce a smooth, closed 4-manifold that violates Rohlin's theorem (see Exercise 2.1.23(c)). Remarkably, Freedman showed that this "closing off" strategy can always be applied in the topological setting.

Theorem 2.2.14 (Freedman [Fre82]) *Every integer homology sphere bounds a compact, contractible topological 4-manifold.*

Given an integer homology 3-sphere Y, the proof of this theorem in [FQ90] goes roughly as follows: Beginning with $Y \times [-1, 1]$, one can perform a sequence of surgeries on loops and topologically embedded spheres in the interior of $Y \times [-1, 1]$ (i.e., replacing copies of $S^1 \times D^3$ with $D^2 \times S^2$, or vice versa) to produce a new 4-manifold W with $\partial W = Y \sqcup -Y$, $\pi_1(W) = 0$, and $H_2(W) = 0$. Next, glue infinitely many of these end-to-end to produce an infinite stack $W \cup W \cup \cdots$ and take its 1-point compactification. The resulting space has boundary Y, is contractible, and can be shown to be a topological manifold (i.e., even at the compactification point).

Proof Sketch for Theorem 2.0.1 Let X denote the E_8-plumbing manifold. Since E_8 is unimodular, ∂X is a $\mathbb{Z}HS^3$ and thus bounds some contractible topological 4-manifold C (Theorem 2.2.14). Then $X \cup -C$ is a closed, simply connected topological 4-manifold that still has intersection form E_8 (because C is contractible). By Rohlin's theorem (Theorem 2.2.12), this 4-manifold $X \cup -C$ cannot admit a smooth structure. □

Exercise 2.2.15 Find another closed, simply connected topological 4-manifold (other than E_8) that does not admit a smooth structure.

2.2.3 Addendum: Rohlin's Invariant and Beyond

As mentioned in Theorem 2.2.6, every closed, orientable, connected 3-manifold Y bounds a smooth, simply connected, compact 4-manifold with Q_X even. Provisionally, we can then define a quantity

$$\mu(Y) = \frac{1}{8}\sigma(X) \mod 2.$$

When Y is a $\mathbb{Z}HS^3$, this quantity is an invariant of Y: If Y bounds X and X' with the desired properties, then applying Rohlin's theorem to $X \cup -X'$ gives

$$\sigma(X) - \sigma(X') = \sigma(X \cup -X') \equiv 0 \mod 16 \implies \frac{1}{8}\sigma(X) \equiv \frac{1}{8}\sigma(X') \mod 2.$$

Remark 2.2.16 We can obtain an invariant in the general setting where $H_1(Y) \neq 0$ by considering *spin* structures. The even type of X ensures that X admits a spin structure, and it is unique if $H^1(X; \mathbb{Z}/2\mathbb{Z}) = 0$. The spin structure on X restricts to a spin structure s on Y. The quantity $\mu(Y, s)$ is defined as above, but only using 4-manifolds X over which s extends.

Example 2.2.17 For the Poincaré homology sphere $P = \Sigma(2, 3, 5)$ we can use the E_8-plumbing as X, so

$$\mu(P) = \frac{1}{8}\sigma(E_8) = \frac{1}{8}(8) = 1 \pmod 2)$$

Exercise 2.2.18 If Y is a $\mathbb{Z}HS^3$ with $\mu(Y) \equiv 1$, then Y cannot bound a smooth integral homology 4-ball ($\mathbb{Z}HB^4$).

A priori, the obstruction from Rohlin's invariant μ does not obstruct $Y\#Y$ from bounding a $\mathbb{Z}HB^4$. However, in 2013, Manolescu used Pin(2)-equivariant Seiberg-Witten theory to prove:

Theorem 2.2.19 (Manolescu [Man16]) *If Y is a $\mathbb{Z}HS^3$ with $\mu(Y) \equiv 1 \mod 2$, then $Y\#Y$ does not bound a $\mathbb{Z}HB^4$.*

Combining this with work of Galewski-Stern and Matumoto, this enabled Manolescu to resolve the higher-dimensional Triangulation Conjecture:

Theorem 2.2.20 ([Man16]) *In all dimensions ≥ 5, there exists a closed, connected, topological manifold that is not triangulable (i.e., not homeomorphic to any simplicial complex).*

2.3 The Minimal Genus Problem

Previous lectures covered the core algebraic topology of smooth 4-manifolds and surveyed some important theorems that (to a significant degree) reduce the study of closed, simply connected, topological 4-manifolds to algebraic topology. In this lecture, we will see how embedded surfaces can shed light on the topology of a 4-manifold X—seeing beyond the algebraic topology and continuous topology, peering directly into the smooth structure on X. The central question in this lecture is motivated by the fact that any second homology class in a smooth 4-manifold can be represented by a smoothly embedded surface (cf. Exercise 2.1.20).

Problem 2.3.1 (The Minimal Genus Problem) Given a smooth 4-manifold X and a class $\alpha \in H_2(X; \mathbb{Z})$, what is the minimal genus realized by an closed, orientable, smoothly embedded surface $\Sigma \subset X$ with $[\Sigma] = \alpha \in H_2(X; \mathbb{Z})$?

The Goal
We will practice describing embedded surfaces (and their tubular neighborhoods) in 4-manifolds. We will review various constraints on their genera, culminating in a sketch of the following landmark theorem.

Theorem 2.3.2 (Donaldson [Don83], Freedman [Fre82]) *There exist smooth 4-manifolds that are homeomorphic but not diffeomorphic to* $\mathbb{R}^4$.

In contrast, in all other dimensions $n \neq 4$, there is a unique smooth structure on $\mathbb{R}^n$.

2.3.1 The Minimal Genus Problem

We begin with a few examples.

Example 2.3.3 For $X = S^2 \times S^2$, the natural generators $\alpha = [S^2 \times \{pt\}]$ and $\beta = [\{pt\} \times S^2]$ of $H_2(X) \cong \mathbb{Z} \oplus \mathbb{Z}$ are represented by smoothly embedded 2-spheres.

Exercise 2.3.4 Consider $S^2 \times S^2$ and the generators $\alpha, \beta \in H_2(S^2 \times S^2)$ as above.

(a) What is the homology class of the diagonal embedding $S^2 \hookrightarrow S^2 \times S^2$ given by $x \mapsto (x, x)$? What about $x \mapsto (x, -x)$ and $x \mapsto (-x, x)$?
(b) For $\alpha, \beta \in H_2(S^2 \times S^2)$ as above, try to find or estimate the minimal genus of $\alpha + n\beta$ for $n \in \mathbb{Z}$.

Let us consider another important 4-manifold: $\mathbb{C}P^2$. The generator of $H_2(\mathbb{C}P^2) \cong \mathbb{Z}$ is represented by the complex projective line $\mathbb{C}P^1 \subset \mathbb{C}P^2$, which is a smooth 2-sphere. What about the class $d[\mathbb{C}P^1] \in H_2(\mathbb{C}P^2)$ for $d \geq 2$? The next result asserts that a minimal representative is always realized by a degree-d complex curve Σ_d given by

$$\{[x : y : z] \in \mathbb{C}P^2 : x^d + y^d + z^d = 0\}.$$

Theorem 2.3.5 (Thom Conjecture, Kronheimer-Mrowka [KM94]) *If* Σ *is a smooth oriented surface embedded in* $\mathbb{C}P^2$ *representing* $d[\mathbb{C}P^1] \in H_2(\mathbb{C}P^2)$ *with* $d \neq 0$*, then*

$$g(\Sigma) \geq (d-1)(d-2)/2.$$

For example, if $d = 2$, we get $g = 0$, so $2[\mathbb{C}P^1]$ is also represented by a smooth 2-sphere. This is illustrated in Fig. 2.11 using Kirby diagrams: Beginning with the standard diagram of $\mathbb{C}P^2$, take two disjoint copies of the core disk of the

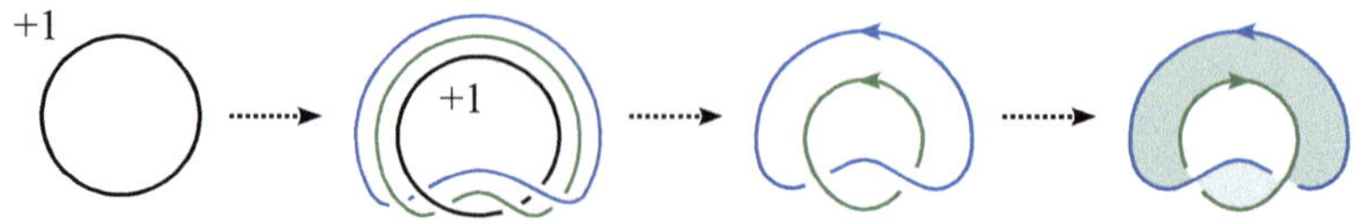

Fig. 2.11 Constructing a smooth 2-sphere in $\mathbb{C}P^2$

2-handle (equipped with the same orientation). The boundaries of these disks form an oriented Hopf link L in the boundary of the 0-handle, and we observe that L bounds an annulus. Gluing this annulus to the two disks yields a smooth 2-sphere in the class $2[\mathbb{C}P^1]$.

While Theorem 2.3.5 was proven using gauge theory, the special case of $d = 3$ (i.e., $g = 1$) can be obtained using Rohlin's theorem (Theorem 2.2.12):

Proposition 2.3.6 *The class* $3[\mathbb{C}P^1] \in H_2(\mathbb{C}P^2)$ *cannot be represented by a smoothly embedded 2-sphere.*

Proof Towards a contradiction, suppose $\Sigma \subset \mathbb{C}P^2$ is a smooth 2-sphere with $[\Sigma] = 3[\mathbb{C}P^1]$. Note that

$$[\Sigma] \cdot [\Sigma] = (3[\mathbb{C}P^1]) \cdot (3[\mathbb{C}P^1]) = 9([\mathbb{C}P^1] \cdot [\mathbb{C}P^1]) = 9,$$

so Σ has a tubular neighborhood diffeomorphic to the D^2-bundle over S^2 with Euler number 9. This has the Kirby diagram shown on the left side of the figure below. Blowing up $\mathbb{C}P^2$ eight times yields $\mathbb{C}P^2 \# 8\overline{\mathbb{C}P}^2$. In this new 4-manifold, we can consider the "proper transform" of Σ, a smooth 2-sphere Σ' of square $+1$ in the homology class $3[\mathbb{C}P^1] + \sum_{i=1}^{8} e_i$, where e_i is the homology class of $\overline{\mathbb{C}P}^1$ in the ith copy of $\overline{\mathbb{C}P}^2$; this sphere Σ' is depicted on the right side of the figure below as the union of a gray disk and the core of the $(+1)$-framed 2-handle.

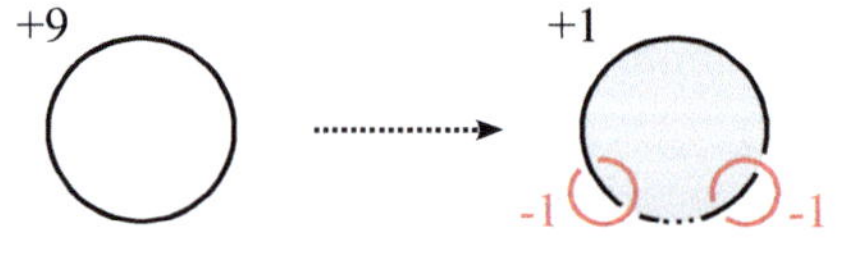

Let X denote the smooth, closed, simply connected 4-manifold obtained by blowing down Σ' in $\mathbb{C}P^2 \# 8\,\overline{\mathbb{C}P}^2$. It can be shown that the intersection form of Q_X is the E_8 lattice (Exercise 2.3.7), hence X has $\sigma = -8$ with an even intersection form, contradicting Rohlin's theorem. □

Exercise 2.3.7 Let us calculate the intersection form of X. Since Σ' is a sphere of square $+1$, it has a tubular neighborhood bounded by S^3, hence $H_2(\mathbb{C}P^2 \# 8\overline{\mathbb{C}P}^2)$ splits into a subspace V spanned by $v_0 := [\Sigma']$ and an orthogonal subspace $V^\perp$. After blowing down, we have $H_2(X) \cong \mathbb{Z}^8$ and Q_X is the restriction of $Q_{\mathbb{C}P^2 \# 8\,\overline{\mathbb{C}P}^2}$ to $V^\perp$.

(a) Show that the classes e_i do not lie in $V^\perp$, but the classes $v_i := e_i - e_{i+1}$ do indeed lie in $V^\perp$ for $i = 1, \dots, 7$. Also show that $v_i \cdot v_i = -2$.
(b) Show that the class $[\mathbb{C}P^1]$ does not lie in $V^\perp$, then find a linear combination $v_8 := [\mathbb{C}P^1] + \sum_{i=1}^{8} c_i e_i$ for some $c_i \in \mathbb{Z}$ that lies in $V^\perp$. Can you find a solution such that $v_8 \cdot v_8 = -2$?
(c) Verify that, with your choice of v_8 from (b), the collection $\{v_0, v_1, \dots, v_8\}$ gives a basis for $H_2(\mathbb{C}P^2 \# 8\overline{\mathbb{C}P}^2)$ (e.g., by showing that the associated change-of-basis matrix B has determinant ± 1).
(d) In this new basis (up to reordering), show that $Q_{\mathbb{C}P^2 \# 8\overline{\mathbb{C}P}^2}$ is given by $\begin{bmatrix} 1 & 0 \\ 0 & E_8 \end{bmatrix}$.
(Hint: Don't forget about Remark 2.1.21.)
(e) Conclude that $Q_X \cong E_8$.

This is an example of a more general approach to minimal genus problems: If we wish to show that a second homology class α in a 4-manifold X cannot be represented by a smooth surface of genus g, suppose that such a surface exists and use it to construct an auxiliary smooth 4-manifold that violates Rohlin's theorem or another smooth obstruction. An example of this concerns surfaces $\Sigma \subset X$ whose homology classes are *characteristic*, i.e., $[\Sigma] \cdot [\Sigma'] \equiv [\Sigma'] \cdot [\Sigma'] \bmod 2$ for all surfaces $\Sigma' \subset X$.

Theorem 2.3.8 (Kervaire-Milnor [KM61]) *Let X be a smooth, closed, simply connected 4-manifold. If $\Sigma \subset X$ is a smoothly embedded 2-sphere that is characteristic, then*

$$[\Sigma] \cdot [\Sigma] \equiv 0 \mod 16.$$

For completeness, we note that a vast generalization of the Thom Conjecture is true. Indeed, the moral of the Thom Conjecture is that nonsingular complex curves in $\mathbb{C}P^2$ are genus-minimizing in their homology classes (and in this case the genus happens to have a simple expression in terms of the degree of the homology class). More generally, we have the following:

Theorem 2.3.9 (Symplectic Thom Conjecture, Ozsváth-Szabó [OS00]) *An embedded symplectic surface in a closed, symplectic four-manifold is genus-minimizing in its homology class.*

2.3.2 *The Local Minimal Genus Problem*

Every knot K in S^3 bounds embedded surfaces inside S^3, but we can often find simpler surfaces bounded by K if we allow the interior of the surface to be embedded in B^4. This motivates the following definition:

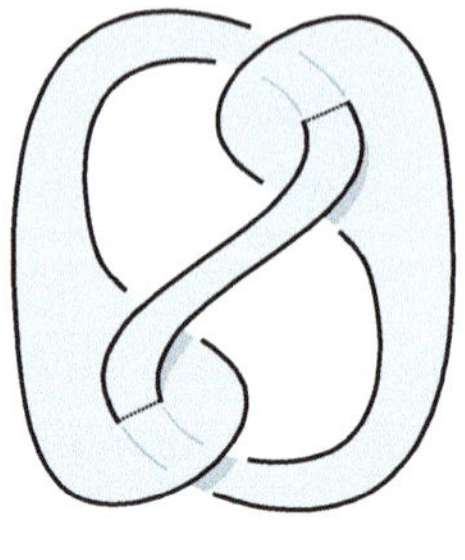

Fig. 2.12 A ribbon-immersed disk bounded by the stevedore knot

Definition 2.3.10 Given a knot K in S^3, a *slice surface* for K is a compact, connected, orientable surface $(\Sigma, \partial\Sigma) \hookrightarrow (B^4, S^3)$ with $\partial\Sigma = K$. The *slice genus* of K is

$$g_4(K) = \min\{g(\Sigma) \mid \Sigma \text{ is a slice surface for } K\}.$$

We say a knot K is *slice* if it has slice genus zero. Historically, this is motivated by the observation that slice knots are precisely those knots that arise as the transverse intersection (or slice) of an embedded 2-sphere in S^4 along the equatorial $S^3 \subset S^4$.

Example 2.3.11 A slice disk bounded by the stevedore knot is depicted in Fig. 2.12. Here the disk is represented diagrammatically as a self-transverse immersed disk in S^3; the double-points occur as "ribbon" intersections, and we may obtain an embedded surface in B^4 by pushing the interior of the disk into B^4 such that the intersecting sheets occur at different heights.

Example 2.3.12 The trefoil knot is *not* slice, i.e., does not have slice genus zero. (Its slice genus is one.) To see this, consider a cuspidal degree-3 curve $\Sigma \subset \mathbb{C}P^2$; the trefoil arises as the link of the unique singularity of Σ. If the trefoil were smoothly slice, one could delete the conic singularity from Σ and glue in the smooth slice disk to produce a smooth 2-sphere in the same homology class as Σ, namely $3[\mathbb{C}P^1] \in H_2(\mathbb{C}P^2)$. However, this would contradict Proposition 2.3.6.

Remark 2.3.13 As discussed in [ACM⁺23], a related strategy can be used to show that the figure-eight knot is not smoothly slice; if it were, one could construct a smooth 2-sphere representing the class $(1, 3)$ in $H_2(\mathbb{C}P^2\#\mathbb{C}P^2; \mathbb{Z}) \cong \mathbb{Z} \oplus \mathbb{Z}$, the existence of which can be ruled out using Rohlin's theorem. Remarkably, Aceto-Castro-Miller-Park-Stipsicz [ACM⁺23] show that a refinement of this approach can be used to show that the $(2, 1)$-cable of the figure-eight knot is not smoothly slice, providing an alternate solution to a longstanding question (originally resolved in [DKM⁺22]). In place of Rohlin's theorem, this refinement uses obstructions from Seiberg-Witten theory on smooth spin 4-manifolds equipped with a $\mathbb{Z}/2\mathbb{Z}$-symmetry.

We can also consider embedded surfaces in the continuous setting. To avoid pathological behavior, it is common to work with a somewhat restricted class of topologically embedded, locally flat surfaces:

Definition 2.3.14 (Topological Embeddings) Let Σ be a surface and X a topological 4-manifold, each possibly with boundary. A map $\iota : \Sigma \hookrightarrow X$ is a *topological embedding* if ι is a homeomorphism onto its image. The embedding is *locally flat* if every point $p \in \iota(\Sigma) \subset X$ has a neighborhood in which the pair $(X, \iota(\Sigma))$ is homeomorphic to $(\mathbb{R}^4, \mathbb{R}^2 \times 0)$. For brevity, we say a knot $K \subset S^3$ is *topologically slice* if it bounds a topologically embedded, locally flat disk in B^4.

In many cases, there are more direct obstructions to sliceness (in both the smooth and topological settings). A key obstruction dates back to work of Fox and Milnor, who studied the question of which knots and links arise as links of singularities of piecewise-linear 2-spheres in S^4 [FM66]. If knots $K_1, \dots, K_n$ form the links of singularities of such a sphere $\Sigma \subset S^4$, then the knot $K_1\#\cdots\#K_n$ bounds a nonsingular disk in B^4. (To see this, take a path γ that joins the singularities, then remove a neighborhood $U \cong \mathring{B}^4$ of γ from Σ to obtain a nonsingular disk $D \subset \Sigma$ that lies in $S^4 \setminus U \cong B^4$ and is bounded by $K_1\#\cdots\#K_n$.) Fox and Milnor showed that the Alexander polynomial provides a slicing obstruction:

Theorem 2.3.15 (Fox-Milnor [FM66]) *If K is (smoothly or topologically) slice, then its Alexander polynomial factors as $\Delta_K(t) = f(t)f(1/t)$ for some polynomial $f \in \mathbb{Z}[t]$.*

Corollary 2.3.16 *If K is (smoothly or topologically) slice, then the value $|\Delta_K(-1)|$, known as the determinant of K, is a square.*

Example 2.3.17 The trefoil knot has Alexander polynomial $\Delta_K(t) = t - 1 + t^{-1}$, hence has $|\Delta_K(-1)| = 3$ and is not smoothly or topologically slice.

Exercise 2.3.18 Find the Alexander polynomial of the knot 9_{42} shown in the middle of Fig. 2.2 and use it to prove that 9_{42} is not (smoothly or topologically) slice.

Remarkably, Freedman showed that the Alexander polynomial can provide a sufficient condition for topological slicing:

Theorem 2.3.19 (Freedman [Fre82]) *If $\Delta_K(t) \equiv 1$, then K is topologically slice.*

Example 2.3.20 The pretzel knot $P(3, -5, -7)$ depicted in Fig. 2.13 has trivial Alexander polynomial $\Delta(t) \equiv 1$, hence is topologically slice.

This contrasts with the smooth setting:

Proposition 2.3.21 ([Rud93]) *The pretzel knot $P(3, -5, -7)$ is not smoothly slice.*

Proof The strategy will build on the approach from previous examples: We will first embed a genus-one Seifert surface bounded by $P(3, -5, -7)$ from Fig. 2.13 into a smooth, closed surface Σ of degree $d = 6$ in $\mathbb{C}P^2$ that minimizes the genus in its homology class. Then we will show that, if $P(3, -5, -7)$ were smoothly slice, we

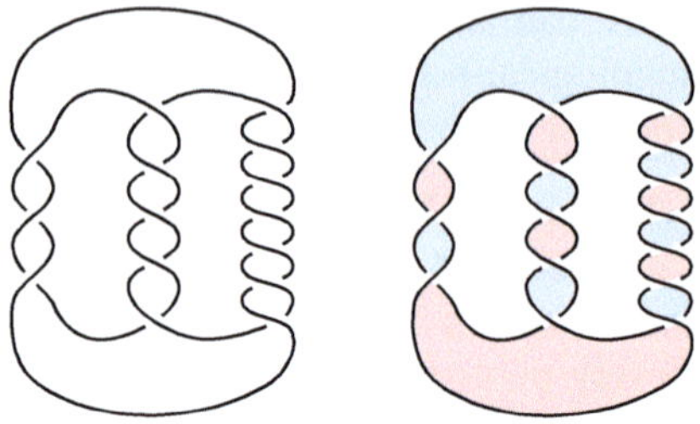

Fig. 2.13 The pretzel knot $P(3, -5, -7)$ and a genus-one Seifert surface

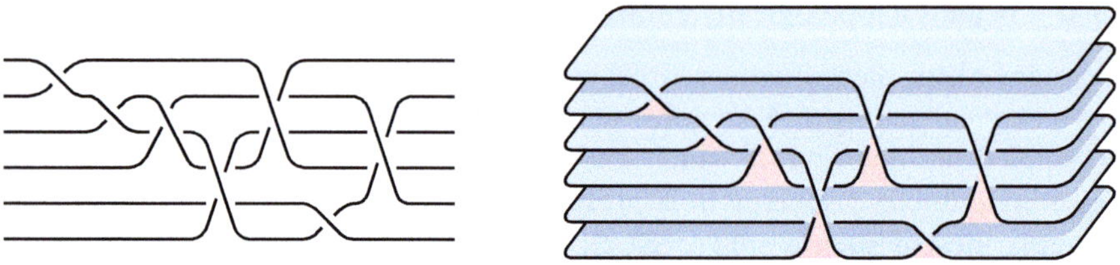

Fig. 2.14 Expressing the pretzel knot $P(3, -5, -7)$ as a braid, together with a braided Seifert surface

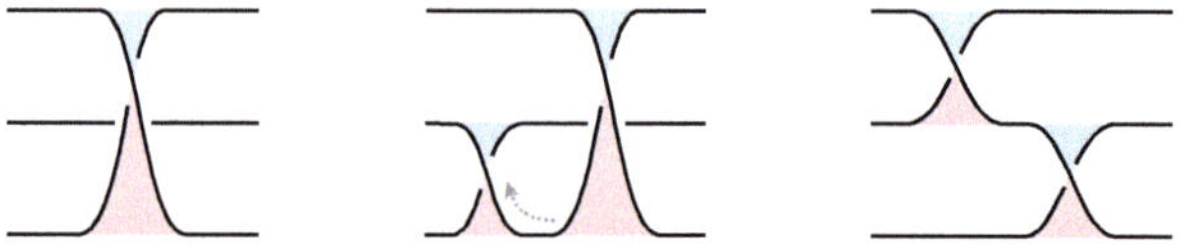

Fig. 2.15 Adding a band and performing a band slide

could modify Σ to produce a smooth surface of smaller genus in the same homology class, violating Theorem 2.3.5.

To that end, we observe that the pretzel knot $P(3, -5, -7)$ can be expressed as the closure of a braid (in particular, a *strongly quasipositive braid* [Rud83, Rud90]) as illustrated in Fig. 2.14. The right side of Fig. 2.14 depicts a natural genus-one Seifert surface F associated to this braid.

By attaching bands to F and, where necessary, sliding the original bands over the new bands as in Fig. 2.15, we can embed F into the standard Seifert surface F' for the torus link $T(6, 6)$ depicted in Fig. 2.16. Note that $g(F') = 10$. This torus link arises as the link of the unique singularity for a cuspidal degree-6 curve in $\mathbb{C}P^2$. We may replace a neighborhood of the singularity with F' (and smooth corners) to produce a smooth surface Σ of genus $g(\Sigma) = g(F') = 10$ representing $6[\mathbb{C}P^1] \in H_2(\mathbb{C}P^2)$.

If $P(3, -5, -7)$ were smoothly slice, then we could replace the genus-one subsurface $F \subset \Sigma$ with a slice disk and produce a smooth, closed surface of genus 9 representing $6[\mathbb{C}P^1] \in H_2(\mathbb{C}P^2)$, violating Theorem 2.3.5. □

Remark 2.3.22 Combined with Example 2.3.20, the preceding argument shows that the Thom Conjecture fails in the topological category. In particular, since $P(3, -5, -7)$ is topologically slice, the construction above can be used to produce

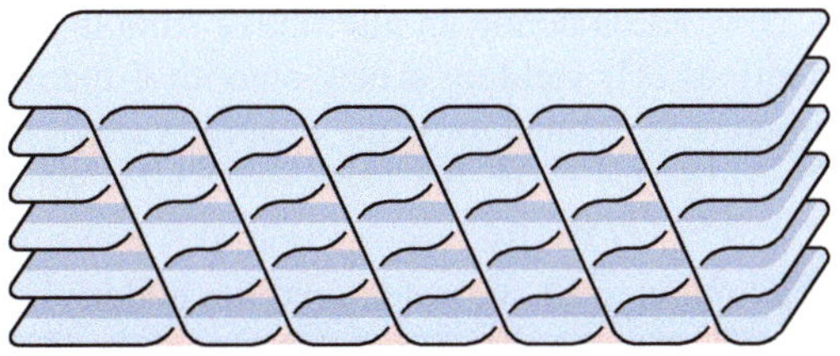

Fig. 2.16 The standard Seifert surface for the torus link $T(6, 6)$

a topologically embedded, locally flat surface of genus 9 representing $6[\mathbb{C}P^1] \in H_2(\mathbb{C}P^2)$.

Remark 2.3.23 For completeness, here is the full statement from [Fre82, Theorem 1.13]: *Consider a smooth embedding $f : S^1 \hookrightarrow S^3$. The knot $K = f(S^1)$ has Alexander polynomial $\Delta_K(t) = 1$ if and only if f extends to some topological embedding $F : D^2 \hookrightarrow B^4$ such that*

1. *F is smooth except at a single point $p \in D^2$ which is nevertheless local-homotopically unknotted (or, equivalently, the end of $B^4 \setminus F(D^2)$ is proper homotopy equivalent to the standard end $S^1 \times D^2 \times [0, \infty)$);*
2. *$\pi_1\big(B^4 \setminus F(D^2)\big) \cong \mathbb{Z}$.*

2.3.3 An Exotic $\mathbb{R}^4$

To sketch the existence of an exotic $\mathbb{R}^4$, we will need a result that ensures the 4-manifold we construct admits a smooth structure at all.

Theorem 2.3.24 (Quinn [Qui82]) *Any connected noncompact topological 4-manifold admits a smooth structure (which extends the unique smooth structure up to diffeomorphism on any boundary components).*

This result in hand, we can sketch the existence of an exotic $\mathbb{R}^4$.

Proof Sketch for Theorem 2.3.2 Let K be any knot in S^3 that is topologically but not smoothly slice in B^4. Viewing S^4 as a union of two 4-balls B^4_- and B^4_+, we write

$$\mathbb{R}^4 = S^4 \setminus \infty = B^4_- \cup (B^4_+ \setminus 0)$$

Let K be embedded in the equatorial $S^3 = \partial B^4_\pm$, and let $\Delta \subset (B^4_+ \setminus 0)$ be a topological slice disk bounded by K. Since Δ is a topologically embedded locally flat disk, it has a tubular neighborhood $N(\Delta) \cong \Delta \times D^2$. We may view $N(\Delta)$ as a 2-handle and attach it to B^4_- to produce a topological embedding of a 4-manifold $X = B^4 \cup_{N(K)} D^2 \times D^2$ into $\mathbb{R}^4$. (This is an example of a *knot trace*; see Exercise 2.1.4.) Let $f : X \hookrightarrow \mathbb{R}^4$ denote this embedding and set $E = \mathbb{R}^4 \setminus f(\mathring{X})$.

Now let us define an alternative smooth structure on the underlying topological manifold $\mathbb{R}^4$, yielding a new smooth 4-manifold $\mathcal{R}$: On $f(X) \subset \mathbb{R}^4$, the smooth structure is defined to be the pushforward of the standard smooth structure on $X = B^4 \cup_{N(K)} D^2 \times D^2$. By Theorem 2.3.24, we can extend this to a smooth structure $E = \mathbb{R}^4 \setminus f(\mathring{X})$. By construction, this new smooth 4-manifold $\mathcal{R}$ is homeomorphic to $\mathbb{R}^4$ (since we have not altered the underlying topological manifold).

Now observe that $\mathcal{R}$ contains an embedded 2-sphere formed from the cone on K inside B^4_- and a copy of the core disk $D^2 \times 0$ inside the 2-handle in $D^2 \times D^2$. This 2-sphere is smooth away from its unique conical singularity over K. If $\mathcal{R}$ were diffeomorphic to $\mathbb{R}^4$, then the Fox-Milnor observation discussed above would imply that K is smoothly slice in B^4, a contradiction. We conclude that $\mathcal{R}$ is an exotic $\mathbb{R}^4$. □

As a historical remark, we note that the existence of an exotic $\mathbb{R}^4$ was originally proven using a different construction and obstruction. In particular, an exotic $\mathbb{R}^4$ is found as a certain subspace of the *Kummer surface*, commonly known as $K3$, which is given by the equation

$$K3 = \{[z_0 : z_1 : z_2 : z_3] \in \mathbb{C}P^3 \mid z_0^4 + z_1^4 + z_2^4 + z_3^4 = 0\}.$$

The relevant obstruction comes from Donaldon's Diagonalization Theorem:

Theorem 2.3.25 (Donaldson [Don83]) *Let X be a smooth, closed, oriented 4-manifold. If the intersection form Q_X is definite, then Q_X is diagonalizable (hence equivalent to $\pm I_n$).*

2.3.4 Addendum and Additional Exercises

We conclude this lecture with some additional material and exercises that focus on the relationship between smooth and singular surfaces.

Exercise 2.3.26 (Neighborhoods of PL Spheres) Suppose that Σ is a 2-sphere in a smooth 4-manifold X that is smoothly embedded away from a finite number of singularities, each modeled on the cone of a knot in B^4. Show that Σ has a neighborhood $N(\Sigma)$ that is diffeomorphic to the n-framed trace of some knot $K \subset S^3$, where n is the self-intersection number $[\Sigma] \cdot [\Sigma]$.

The next exercise presents a useful lemma that is closely related to Fox and Milnor's observation about piecewise-linear 2-spheres in $\mathbb{R}^4$ [FM66].

Exercise 2.3.27 (Trace Embedding Lemma) Let $X(K)$ denote the 0-framed trace of a knot $K \subset S^3$. Show that $X(K)$ embeds smoothly into $\mathbb{R}^4$ if and only if K is slice.

Remark 2.3.28 In [Pic20], Piccirillo used the Trace Embedding Lemma to help resolve a longstanding question by showing that the Conway knot (denoted $11n34$)

is not slice. All previously-known slicing obstructions fail to obstruct the Conway knot from being slice directly, including the s-invariant (which vanishes for all slice knots) defined by Rasmussen [Ras10] via Khovanov homology [Kho00]. Piccirillo's elegant proof circumvents this by identifying a second knot K whose 0-trace is diffeomorphic to that of the Conway knot, and then she shows that $s(K) \neq 0$; this implies K is not slice, hence $X_0(K)$ cannot embed smoothly in $\mathbb{R}^4$. Since $X_0(K) \cong X_0(11n34)$, this in turn implies that the Conway knot is not slice either.

Exercise 2.3.29 As a toy example of Piccirillo's strategy, work out the following alternative (and overkill!) proof that the knot 9_{42} is not slice (cf. Exercise 2.3.18): Use software to calculate the s-invariants of the knots in Fig. 2.2.[1] You should find that the first knot $K = 9_{42}$ has $s(K) = 0$, whereas the second knot K' has $s(K') \neq 0$. Apply the Trace Embedding Lemma to conclude that neither knot is slice.

2.4 The Contact and Stein Setting

Next we will consider the geometric setting of Stein 4-manifolds and contact 3-manifolds arising as their boundaries.

2.4.1 *Contact Geometry and the Slice-Bennequin Inequality*

Recall that a *contact structure* on a 3-manifold Y is a 2-plane distribution $\xi \subset TY$ that can be locally expressed as the kernel of a 1-form α such that $\alpha \wedge d\alpha \neq 0$; we call α a *contact form.*

Example 2.4.1 The standard contact structure on $\mathbb{R}^3$ is given by $\xi = \ker(dz + x\,dy)$. This has a natural extension to a standard contact structure on S^3; the latter can also be realized as the set of affine complex lines that are tangent to the unit $S^3 \subset \mathbb{C}^2$.

A knot K in a contact 3-manifold (Y, ξ) is called *Legendrian*, if it is tangent to ξ. Diagrammatically, Legendrian knots in the standard contact $\mathbb{R}^3$ are typically studied via the "front projection" $(x, y, z) \mapsto (y, z)$.

Example 2.4.2 The front projection for a Legendrian representative of the (right-handed) trefoil is shown in Fig. 2.17.

The differential equation $x = -dz/dy$ imposes two basic features on front projection: cusps instead of vertical tangencies and overcrossing conditions. To state

[1] For example, use SnapPy [CDGW] to find a planar diagram or Dowker-Thistlethwaite code for each knot, then calculate their s-invariants using the program KnotJob [Sch21] or the KnotTheory' package [Kno] in Mathematica [Wol].

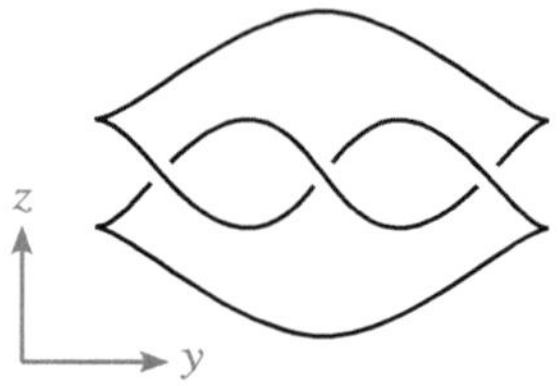

Fig. 2.17 A Legendrian representative of the right-handed trefoil

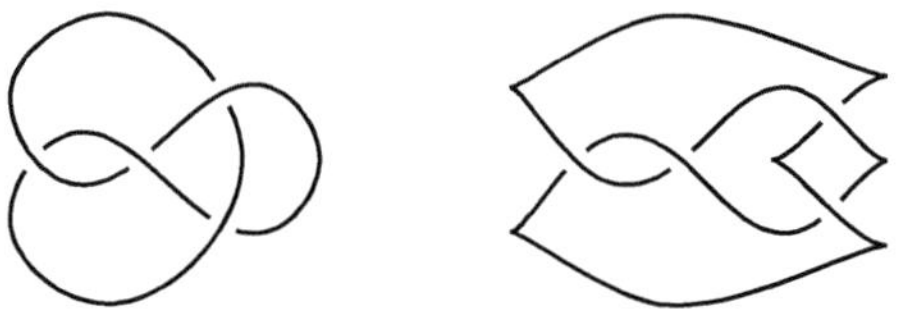

Fig. 2.18 Bringing the figure-eight knot into Legendrian position

them more precisely, suppose a Legendrian knot is parametrized via $t \mapsto \sigma(t) = (x(t), y(t), z(t))$ where $\sigma'(t)$ is nowhere-vanishing:

1. Front projections have cusps instead of vertical tangencies. Such points occur when $y'(t) = 0$, which forces $z'(t) = 0$, in turn implying that $\sigma'(t)$ points in the $\partial/\partial x$ direction whenever $y'(t) = 0$.
2. At crossings (which are required to project to a pair of transverse arcs), the overstrand has more negative slope. This is because the slope dz/dy corresponds to the negative of the x-coordinate.

It turns out that essentially any planar diagram satisfying these conditions is realized as the projection of a Legendrian knot or link (unique up to Legendrian isotopy). In a related vein, one can prove:

Proposition 2.4.3 *Every knot can be made Legendrian by a small isotopy.*

Example 2.4.4 We illustrate this for the figure-eight knot, also denoted 4_1. Vertical tangencies are replaced with cusps. Two of the original crossings satisfy the overstrand condition, and we add zig-zags to fix the remaining two crossings (Fig. 2.18).

Legendrian knots have two "classical" invariants.

Definition 2.4.5 Given an oriented Legendrian knot diagram D, define

- the *Thurston-Bennequin number* $\mathrm{tb}(D) := \mathrm{writhe}(D) - \frac{1}{2}\#\{\text{cusps}\}$, where the writhe is the difference between the numbers of positive and negative crossings;
- the *rotation number* $\mathrm{rot}(D) := \frac{1}{2}\left(\#\{\text{down cusps}\} - \#\{\text{up cusps}\}\right)$.

These two quantities have intrinsic geometric definitions as well, which is convenient for establishing their invariance under Legendrian isotopy (without appealing to a Legendrian version of Reidemeister's theorem). Roughly speaking, the rotation number measures the rotation of the tangent vector to the knot inside the contact planes relative to $\partial/\partial x$, and the Thurston-Bennequin number measures

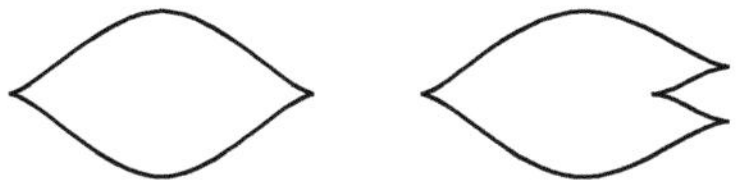

Fig. 2.19 Two diagrams for the unknot with different Thurston-Bennequin and rotation numbers

the linking between the knot and a pushoff along a vector field which is transverse to the contact planes.

Exercise 2.4.6 Fix an orientation on the Legendrian trefoil K in Fig. 2.17 and verify that it has $\mathrm{tb} = 1$ and $\mathrm{rot} = 0$.

Exercise 2.4.7

(a) For the various possible orientations, find the Thurston-Bennequin and rotation numbers of the Legendrian unknots in Fig. 2.19.
(b) In general, how does a change of orientation affect these invariants?
(c) How does the introduction of zig-zags (called *stabilization*) affect these invariants?

These classical invariants can be used to articulate a constraint on the (smooth) slice genus of a knot; the strongest possible constraint is then obtained by minimizing over all Legendrian representatives of a given knot.

Theorem 2.4.8 (Slice-Bennequin Inequality) *Given a Legendrian representative of K, we have*

$$2g_4(K) - 1 \geq \mathrm{tb}(K) + |\,\mathrm{rot}(K)|,$$

where g_4 denotes the smooth slice genus.

The slice-Bennequin inequality was originally established by Rudolph [Rud93] using the Thom conjecture (Theorem 2.3.5) [KM94]. We will hint at an alternative proof further below using the "adjunction inequality" for surfaces in Stein domains.

Example 2.4.9 For each orientation, the Legendrian figure-eight knot in Fig. 2.18 has

$$\mathrm{tb} = \mathrm{writhe} - \tfrac{1}{2}\#\{\mathrm{cusps}\} = (2-2) - \tfrac{1}{2}(6) = -3$$

$$\mathrm{rot} = \frac{1}{2}\left(\#\{\text{down cusps}\} - \#\{\text{up cusps}\}\right) = \frac{1}{2}(3-3) = 0.$$

The slice-Bennequin inequality then implies $2g_4 - 1 \geq -3 + |0|$, which reduces to $g_4 \geq -1$. In this case, the slice-Bennequin inequality fails to provide a useful obstruction. Fortunately, there are many examples in which the bound is effective and superior to classical tools, such as the slicing obstructions from the Alexander polynomial.

Fig. 2.20 A Legendrian representative of the pretzel knot $P(3, -5, -7)$

Exercise 2.4.10 Find a Legendrian representative of the pretzel knot $P(3, -5, -7)$ that has tb $= 1$ and rot $= 0$. *(One solution is given in Fig. 2.20.)* Then prove $P(3, -5, -7)$ is not (smoothly) slice.

2.4.2 Stein Surfaces

Recall Whitney's embedding theorem, which states that any smooth n-manifold can be smoothly embedded into $\mathbb{R}^{2n}$. In contrast, many complex manifolds do not admit holomorphic embeddings into $\mathbb{C}^n$ for any n. (For example, embeddings of closed complex manifolds Z are obstructed by the maximum principle for holomorphic maps, since the coordinate projections $Z \hookrightarrow \mathbb{C}^n \to \mathbb{C}$ must be constant.) The class of complex manifolds that do admit embeddings into some $\mathbb{C}^n$ are called *Stein manifolds*.

Definition 2.4.11 A *Stein surface* Z is a complex surface ($\dim_{\mathbb{C}} Z = 2$) admitting a proper holomorphic embedding into $\mathbb{C}^n$ for some n.

Remark 2.4.12 This is not the original definition (which involves the notions of holomorphic convexity and separability), but it is an elegant alternative characterization.

A natural compact analogue is a *Stein domain*. For our purposes, we will view these as the compact 4-manifolds X (with boundary) arising as subsets $Z \cap B^{2n}(r)$ where $Z \subset \mathbb{C}^n$ is a Stein surface and $B^{2n}(r)$ is a ball of radius $r > 0$. The boundary inherits a natural contact structure by considering the complex tangencies. (In particular, at each point $x \in \partial X$, the contact plane ξ_x is the unique plane in the 3-dimensional tangent space $T_x \partial X$ that is fixed under complex multiplication in the complex vector space $T_x\mathbb{C}^n$.) We also note that manifolds of this form have a natural symplectic structure.

The following important property of Stein domains brings us back to the minimal genus problem considered in the previous section.

Theorem 2.4.13 (Adjunction Inequality [LM98]) *If X is a Stein domain and $\Sigma \subset X$ is a smoothly embedded surface representing a nonzero class in $H_2(X)$, then*

$$[\Sigma] \cdot [\Sigma] + \left|\langle c_1(X, [\Sigma])\rangle\right| \leq 2g(\Sigma) - 2,$$

where $c_1(X)$ denotes the first Chern class of X as a complex manifold.

How do we calculate this in practice? Work of Eliashberg [Eli90, Eli93] and Gompf [Gom98] establishes a handle theory for Stein domains that admits a convenient diagrammatic framework. For our purposes, we focus on the case of 4-manifolds built with a single 0-handle and some number of 2-handles.

Theorem 2.4.14 (Eliashberg [Eli90, Eli93], cf. Weinstein [Wei91]) *Let X be a 4-manifold obtained by attaching 2-handles to B^4 along a framed, oriented Legendrian link $L = \sqcup_1^n K_i$ in the standard contact S^3. If the framing on each component K_i equals $\mathrm{tb}(K_i) - 1$, then X admits a Stein structure.*

Moreover, the first Chern class $c_1(X) \in H^2(X)$ is determined by the rotation numbers of L via the formula

$$\langle c_1(X), h_i\rangle = \mathrm{rot}(K_i),$$

where h_i is the generator of $H_2(X)$ corresponding to the 2-handle attached along K_i (with orientation).

Note that reversing the orientation of K_i has the effect of replacing h_i with $-h_i$.

Example 2.4.15 Since the Legendrian trefoil K in Fig. 2.17 has $\mathrm{tb}(K) = 1$, its 0-framed knot trace $X = X_0(K)$ admits a Stein structure. Moreover, since $\mathrm{rot}(K) = 0$, we see that the class $c_1(X) \in H^2(X) \cong \mathbb{Z}$ vanishes. Applying the adjunction inequality, we see that any smoothly embedded surface $\Sigma \subset X$ representing a nontrivial homology class satisfies $g(\Sigma) \geq 1$ (using the fact that $[\Sigma] \cdot [\Sigma] = 0$).

Exercise 2.4.16 Given a Legendrian knot K in the standard contact S^3, use Theorem 2.4.14 and the adjunction inequality (Theorem 2.4.13) to establish the slice-Bennequin inequality for K (Theorem 2.4.8).

By similar methods we can get constraints on the genera of surfaces in many other 4-manifolds as well (cf. [AM97]).

References

[ACM⁺23] Paolo Aceto, Nickolas A Castro, Maggie Miller, JungHwan Park, and András Stipsicz. Slice obstructions from genus bounds in definite 4-manifolds. *arXiv preprint arXiv:2303.10587*, 2023.

[AM97] S. Akbulut and R. Matveyev. Exotic structures and adjunction inequality. *Turkish J. Math.*, 21(1):47–53, 1997.

[CDGW] Marc Culler, Nathan M. Dunfield, Matthias Goerner, and Jeffrey R. Weeks. SnapPy, a computer program for studying the geometry and topology of 3-manifolds. http://snappy.computop.org.

[DKM+22] Irving Dai, Sungkyung Kang, Abhishek Mallick, JungHwan Park, and Matthew Stoffregen. The (2, 1)-cable of the figure-eight knot is not smoothly slice. *arXiv preprint arXiv:2207.14187*, 2022.

[Don83] S. K. Donaldson. An application of gauge theory to four-dimensional topology. *J. Differential Geom.*, 18(2):279–315, 1983.

[Eli90] Ya. Eliashberg. Topological characterization of Stein manifolds of dimension > 2. *Internat. J. Math.*, 1(1):29–46, 1990.

[Eli93] Ya. Eliashberg. Legendrian and transversal knots in tight contact 3-manifolds. In *Topological methods in modern mathematics (Stony Brook, NY, 1991)*, pages 171–193. Publish or Perish, Houston, TX, 1993.

[FM66] Ralph H. Fox and John W. Milnor. Singularities of 2-spheres in 4-space and cobordism of knots. *Osaka Math. J.*, 3:257–267, 1966.

[FQ90] Michael H. Freedman and Frank Quinn. *Topology of 4-manifolds*, volume 39 of *Princeton Mathematical Series*. Princeton University Press, Princeton, NJ, 1990.

[Fre82] Michael Hartley Freedman. The topology of four-dimensional manifolds. *J. Differential Geom.*, 17(3):357–453, 1982.

[Gom98] R. Gompf. Handlebody construction of Stein surfaces. *Ann. of Math. (2)*, 148(2):619–693, 1998.

[GS99] Robert E. Gompf and András I. Stipsicz. *4-manifolds and Kirby calculus*, volume 20 of *Graduate Studies in Mathematics*. Amer. Math. Society, Providence, RI, 1999.

[Kho00] Mikhail Khovanov. A categorification of the Jones polynomial. *Duke Math. J.*, 101(3):359–426, 2000.

[KM61] Michel A. Kervaire and John W. Milnor. On 2-spheres in 4-manifolds. *Proceedings of the National Academy of Sciences of the United States of America*, 47(10):1651–1657, 1961.

[KM94] P. B. Kronheimer and T. S. Mrowka. The genus of embedded surfaces in the projective plane. *Mathematical Research Letters*, Volume 1(6):797–808, 1994.

[Kno] `KnotTheory'`. Mathematica Package available at `katlas.math.toronto.edu`, accessed June 2013.

[LM98] P. Lisca and G. Matić. Stein 4-manifolds with boundary and contact structures. *Topology Appl.*, 88:55–66, 1998.

[LP72] François Laudenbach and Valentin Poénaru. A note on 4-dimensional handlebodies. *Bull. Soc. Math. France*, 100:337–344, 1972.

[Man16] Ciprian Manolescu. Pin (2)-equivariant Seiberg-Witten Floer homology and the triangulation conjecture. *Journal of the American Mathematical Society*, 29(1):147–176, 2016.

[OS00] Peter Ozsvath and Zoltan Szabo. The symplectic Thom conjecture. *Annals of Mathematics*, 151(1):93–124, 2000.

[Pic20] Lisa Piccirillo. The Conway knot is not slice. *Ann. of Math. (2)*, 191(2):581–591, 2020.

[Qui82] Frank Quinn. Ends of maps. III. Dimensions 4 and 5. *Journal of Differential Geometry*, 17(3):503–521, 1982.

[Ras10] Jacob Rasmussen. Khovanov homology and the slice genus. *Invent. Math.*, 182(2):419–447, 2010.

[Roh52] Vladimir Rohlin. New results in the theory of four-dimensional manifolds. *Doklady Acad. Nauk. SSSR (N.S.)*, 84:221–224, 1952.

[Rud83] L. Rudolph. Braided surfaces and Seifert ribbons for closed braids. *Comment. Math. Helv.*, 58(1):1–37, 1983.

[Rud90] L. Rudolph. A congruence between link polynomials. *Math. Proc. Cambridge Philos. Soc.*, 107(2):319–327, 1990.

[Rud93] L. Rudolph. Quasipositivity as an obstruction to sliceness. *Bull. Amer. Math. Soc. (N.S.)*, 29(1):51–59, 1993.

[Sav12] Nikolai Saveliev. *An Introduction to the Casson Invariant*. De Gruyter, Berlin, Boston, 2012.

[Sch21] Dirk Schütz. KnotJob. Available at http://www.maths.dur.ac.uk/~dma0ds/knotjob.html, 2021.

[Wei91] Alan Weinstein. Contact surgery and symplectic handlebodies. *Hokkaido Math. J.*, 20(2):241–251, 1991.

[Wol] Wolfram Research, Inc. Mathematica, Version 13.2. Champaign, IL, 2022.

Chapter 3
Severi Strata of Plane Curve Singularities

Duco van Straten and László Koltai

Contents

3.1 The Delta-Invariant 78
 3.1.1 The Definition of the Delta-Invariant 78
 3.1.2 Local Topology of Isolated Hypersurface Singularities 81
 3.1.3 Deformations and the Delta-Invariant 84
3.2 Severi Strata 87
 3.2.1 Miniversal Deformation 87
 3.2.2 Definition of Local Severi Strata 87
 3.2.3 Severi Multiplicities 89
3.3 The Jacobian 92
 3.3.1 Classical Jacobian 92
 3.3.2 Compactified Jacobian 94
 3.3.3 Grassmannian of a Curve 96
3.4 Severi Multiplicities and the HOMFLY Polynomial 97
 3.4.1 Hilbert Scheme of Points 97
 3.4.2 HOMFLY Polynomial 101
References 104

Examples of algebraic curves with singularities are known since antiquity. The cubic curve called the *cissoid of Diocles*, that was used to solve the Delian problem, exhibits a proud cusp at its origin. The *conchoid of Nicomedes*, used for the trisection of an arbitrary angle, contains a node or a cusp, depending on the value of a parameter. A broader approach to curves and singularities was taken by Newton with his classification of cubics and the parametrisation of branches by fractional power series based on the analysis of the *Newton diagram*. In the nineteenth century the subject of curves and singularities matured and the influence of singularities on the

D. van Straten
Johannes Gutenberg University, Mainz, Germany
e-mail: straten@mathematik.uni-mainz.de

L. Koltai (✉)
Eötvös Loránd University, Budapest, Hungary
e-mail: koltail@renyi.hu

© The Author(s), under exclusive license to Springer Nature Switzerland AG 2024

J. Fernández de Bobadilla et al. (eds.), *Singularities and Low Dimensional Topology*,
Bolyai Society Mathematical Studies 30,
https://doi.org/10.1007/978-3-031-56611-0_3

genus of a curve was clearly recognised (after Riemann) by Cayley and Clebsch. The relations between Newton-Puiseux expansions, iterated torus knots and embedded resolutions were well understood around the time of Enriques. For an excellent source of information on curves and singularities, in their historical context, can be found in the book by Brieskorn and Knörrer, [BK86] and the references therein.

In more recent times, the theory singularities of plane curves became part of the theory of isolated hypersurface singularities. This more general theory, with its origins in differential topology on the one hand and classification of critical points at the other hand, led to the introduction of many new invariants (like the Milnor number, the spectrum, etc.) and their connections to old and new problems. We refer to [Mil68, AGV85] and [BG80] background on this.

Nevertheless, the theory of singularities of plane curves is far from being a finished story. Many aspects of *deformations* of curve singularities remain ill understood. Especially the description of possible ways one singularity can be deformed into combinations of other singularities, *adjacencies* between singularities, is very incomplete. The so-called *Severi strata* in the versal deformation space parametrise adjacencies of a singularity to a curve with a prescribed number of nodes. It was discovered by A. Oblomkov and V. Shende [OS12] that, mysteriously, the *multiplicities* of these strata appear as coefficients of the HOMFLY-polynomial of the link of the singularity, and even more, that, conjecturally, the complete HOMFLY-polynomial can be interpreted in terms of the Hilbert scheme of points of the singularity. This conjecture has been proven by D. Maulik [Mau16] using advanced methods of algebraic geometry. These new developments suggests that we are looking at the *tip of an iceberg* and show that the subject of curve singularities is still full of surprises. The purpose of the four lectures given at the winter-school *Singularities and low dimensional Topology* in January 2023, was to introduce the audience to this exciting circle of ideas.

3.1 The Delta-Invariant

3.1.1 The Definition of the Delta-Invariant

Suppose we have a smooth connected projective curve $\widetilde{C}$ in a high dimensional projective space, and assume that we want to study this curve by making it more manageable and project it in the projective plane. This way we get a plane curve C, but via the projection, we might introduce double points, cusps, or other type of singularities. Slightly more formally, suppose we have a map $n : \widetilde{C} \to C$ from a smooth Riemann surface $\widetilde{C}$, to a singular Riemann surface C. We can consider $\widetilde{C}$ as an abstract curve obtained from C by the process of normalisation, that is, we can recover from the singular curve C the original curve $\widetilde{C}$ by undoing the introduced singularities (Fig. 3.1).

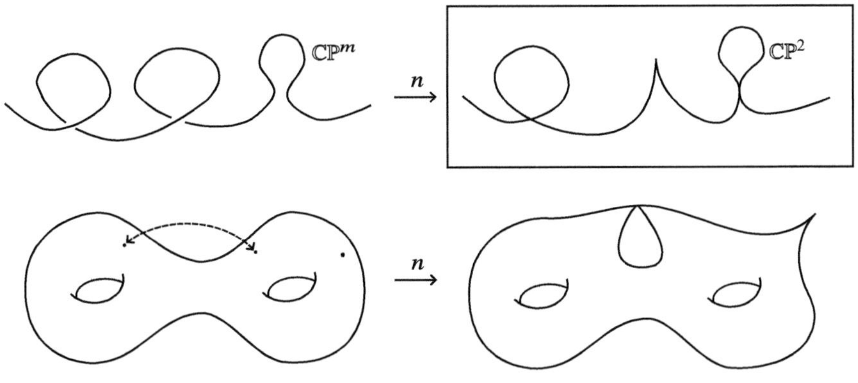

Fig. 3.1 Projecting a curve into $\mathbb{CP}^2$ acquiring singularities

We want to measure the difference between C and $\widetilde{C}$, and to do this, we look at the sheaf of holomorphic functions on C and $\widetilde{C}$, and use the normalisation map n to pull back functions from C to $\widetilde{C}$. In this way we get a map $\mathcal{O}_C \to n_*\mathcal{O}_{\widetilde{C}}$, which can be inserted in an exact sequence:

$$0 \longrightarrow \mathcal{O}_C \longrightarrow n_*\mathcal{O}_{\widetilde{C}} \longrightarrow n_*\mathcal{O}_{\widetilde{C}}/\mathcal{O}_C \longrightarrow 0$$

If we take the Euler characteristics of these sheaves we get the following relation:

$$1 - g_a(C) + \dim_{\mathbb{C}}(n_*\mathcal{O}_{\widetilde{C}}/\mathcal{O}_C) = 1 - g(\widetilde{C}),$$

where $g_a(C) = \dim H^1(\mathcal{O}_C)$ is the arithmetic genus of the curve C, and $g(\widetilde{C})$ is the (geometric) genus of the smooth curve $\widetilde{C}$. The arithmetic genus $g_a(C)$ depends only on the degree of C in the projective plane, so if we assume that $\deg(C) = d$, then $g_a(C) = (d-1)(d-2)/2$. Now if we rearrange this we get the following formula, that goes back to Cayley and Clebsch:

$$g(\widetilde{C}) = \frac{(d-1)(d-2)}{2} - \delta(C),$$

where $\delta(C) = \dim_{\mathbb{C}}(n_*\mathcal{O}_{\widetilde{C}}/\mathcal{O}_C)$. Since $n_*\mathcal{O}_{\widetilde{C}}/\mathcal{O}_C$ is supported on a discrete set, its dimension is just the sum of the local contributions, i.e.

$$\dim_{\mathbb{C}}(n_*\mathcal{O}_{\widetilde{C}}/\mathcal{O}_C) = \sum_{p\in\Sigma} \delta(C, p),$$

where Σ is the set of singular points of C, the ring $(n_*\mathcal{O}_{\widetilde{C}})_p$ is the normalisation of $\mathcal{O}_{C,p}$ and $\delta(C, p) = \dim_{\mathbb{C}}((n_*\mathcal{O}_{\widetilde{C}})_p/\mathcal{O}_{C,p})$ is the *delta-invariant* of the singularity (C, p). Since $\mathcal{O}_{C,p}$ is a curve singularity, its normalisation $(n_*\mathcal{O}_{\widetilde{C}})_p$ is isomorphic to $\oplus_{i=1}^r \mathcal{O}_{\mathbb{C},0}$, where $\mathcal{O}_{\mathbb{C},0} = \mathbb{C}\{s\}$ is the ring of convergent power series in one variable and r is the number of branches of (C, p). We can assume without loss of generality that $p = 0 \in \mathbb{C}^2$, and if we assume that the plane curve singularity $(C, 0)$ is irreducible, i.e. $r = 1$, then we can take an irreducible local equation $f \in \mathbb{C}\{x, y\}$ of C, and then the normalisation map $n^* : \mathcal{O}_{C,0} \to \mathcal{O}_{\mathbb{C},0}$ corresponds to a local parametrisation $s \mapsto (x(s), y(s))$ of $\{f = 0\}$. If $(C, 0)$ has more than one branch, then the normalisation corresponds to parametrising each irreducible component separately.

Let us look at some computation of the delta-invariant using the definition:

1. Let $f(x, y) = y^2 - x^3$. Then its normalisation is given by the parametrisation $s \mapsto (s^2, s^3)$ inducing a map $\mathcal{O}_{C,0} = \mathbb{C}\{x, y\}/(y^2 - x^3) \to (n_*\mathcal{O}_{\tilde{C}})_0 = \mathbb{C}\{s\}$, which sends y to s^3, and x to s^2. Thus we get that $\delta((\{y^2 - x^3 = 0\}, 0)) = 1$. Another way to compute the delta-invariant of an irreducible curve is the following: define the semigroup $\mathcal{S}$ as:

$$\mathcal{S} = \{m \in \mathbb{N} \mid m = \mathrm{ord}(n^* f),\ f \in \mathcal{O}_{C,0}\},$$

that is, we take those natural numbers that arise as the order of functions pulled back from $(C, 0)$. Then it is not hard to see that $\delta = \#(\mathbb{N} \setminus \mathcal{S})$, that is, to compute the delta-invariant of $(C, 0)$ we just have to count the holes of the semigroup $\mathcal{S}$ in $\mathbb{N}$.

0	1	2	3	4	5	6...
×	○	×	×	×	×	×...

2. Let $f(x, y) = y^2 - x^7$. Then we can parametrise $f(x, y) = 0$ by $s \mapsto (s^2, s^7)$, so we get that $\mathcal{O}_{C,0} \subset (n_*\mathcal{O}_{\tilde{C}})_0$ as subring is generated by s^2, s^7, thus $\delta(\{y^2 - x^7\} = 0) = 3$.

0	1	2	3	4	5	6	7	8...
×	○	×	○	×	○	×	×	×...

3. For $f(x, y) = y^3 - x^5$ we do as before, we parametrise the curve by $s \mapsto (s^3, s^5)$, and we get that $\delta = 4$.

0	1	2	3	4	5	6	7	8	9	10...
×	○	○	×	○	×	×	○	×	×	×...

4. Now set $f(x, y) = x^m - y^m$. This singularity is called an ordinary m-fold point since it consists of m smooth branches meeting in one point and any two branches meet transversally, in other words, any two branch has different tangets. In this case, the singularity is not irreducible so we need to use as many parameters as the number of branches (Fig. 3.2).

   We do the computations for $m = 2$. We take the parametrisations $n_1 : s_1 \mapsto (s_1, s_1)$, and $n_2 : s_2 \mapsto (s_2, -s_2)$, which induces a map $\mathbb{C}\{x, y\}/(x^2 - y^2) \to$

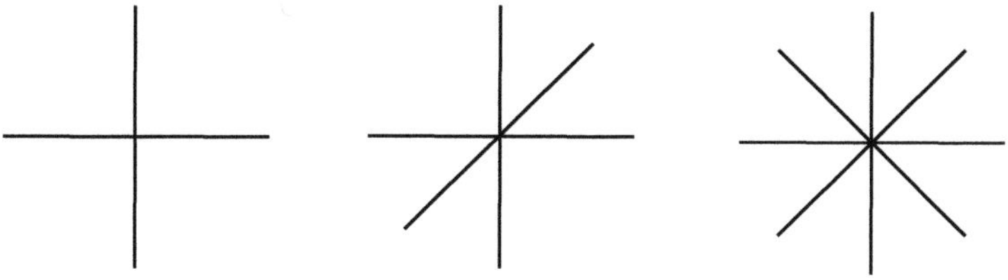

Fig. 3.2 An ordinary m-fold singularity for $m = 2, 3, 4$

$\mathbb{C}\{s_1\} \oplus \mathbb{C}\{s_2\}$. Now $n_1^*(x + y) = 2s_1$, while $n_2^*(x + y) = 0$, thus we have $(s_1, 0) \in \mathcal{O}_{C,0} \subset (n_*\mathcal{O}_{\tilde{C}})_0$, and by pulling back $x - y$ we get that $(0, s_2) \in \mathcal{O}_{C,0} \subset (n_*\mathcal{O}_{\tilde{C}})_0$. We also see that $1 = (1, 1) \in \mathcal{O}_{C,0}$, but there are no functions on $(C, 0)$ that restricts to the constant 1 function on one branch, and to the constant 0 function on the other, so we see that $(1, 0), (0, 1) \notin \mathcal{O}_{C,0}$, thus $\delta = \dim_{\mathbb{C}}((n_*\mathcal{O}_{\tilde{C}})_0/\mathcal{O}_{C,0}) = 1$.

As we see even for $m = 2$ the computation gets a bit cumbersome, and for higher m the computation is becoming even worse. In the situation of reducible curves there are simpler ways to compute the delta-invariant, like

$$\delta(C, 0) = \sum \delta(C_i, 0) + \sum_{i<j} I(C_i, C_j, 0),$$

where $I(C_i, C_j, 0)$ denotes the intersection multiplicity of the i-th and j-th branch of $(C, 0)$. Another way of computing δ is via the formula of Milnor, which relates the delta-invariant of a plane curve with the Milnor number of the curve, which we will discuss later.

Exercise 3.1.1 Take the parametrisation $s \mapsto (s^4, s^6 + s^7)$. Give an equation for the image of this map, and compute its delta-invariant using the definition.

3.1.2 Local Topology of Isolated Hypersurface Singularities

Let $f(x_1, \ldots, x_n)$ be an element of $\mathbb{C}\{x_1, \ldots, x_n\} = \mathcal{O}_n$, the ring of convergent power series in n variables. We want to study f in a small neighborhood of the origin, so we assume that $f(0) = 0$ and we pick $\varepsilon > 0$ small enough, so f is convergent on $B_\varepsilon := \{x \in \mathbb{C}^n \mid |x| < \varepsilon\}$, and let $V(f) := f^{-1}(0) \subset B_\varepsilon$.

Definition 3.1.2 We say that such an f has an *isolated singularity* at 0 if we can shrink ε such that $\{x \in B_\varepsilon \mid \partial_1 f(x) = \cdots = \partial_n f(x) = 0\} = \{0\}$.

Fig. 3.3 Milnor fibration

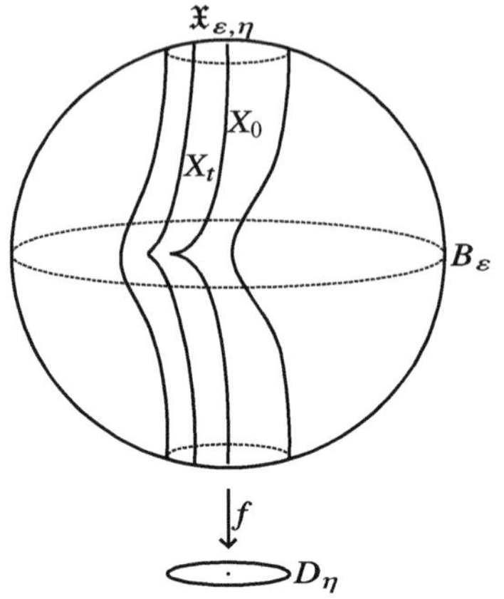

We again shrink ε even further so that for all $0 < \varepsilon' \leq \varepsilon$ the boundary of $B_{\varepsilon'}$ is transversal to $V(f)$.

Definition 3.1.3 In this setup we define the *link* of the singularity L_ε to be the compact $2n-3$ dimensional manifold $V(f) \cap \partial B_\varepsilon$ for small enough ε. We call the embedding $L_\varepsilon \subset \partial B_\varepsilon = S_\varepsilon^{2n-1}$ the *embedded link.*

Now f maps the ball B_ε to $\mathbb{C}$, and we choose a small $0 < \eta \ll 1$ such that for all $t \in D_\eta = \{t \in \mathbb{C} \mid |t| < \eta\}$ the inverse image of t, $f^{-1}(t) = X_t$, is transversal to ∂B_ε. Let $\mathfrak{X}_{\varepsilon,\eta} = \mathfrak{X} = f^{-1}(D_\eta) \cap B_\varepsilon$. Then we have a map $f : \mathfrak{X}_{\varepsilon,\eta} \to D_\eta$ (see Fig. 3.3). For $t \neq 0$, the nearby fiber X_t is called the *Milnor fiber* of f.

Since η is sufficiently small, X_0 is the only fiber that has a singularity. If we remove that singular fiber X_0 and denote the space left by $\mathfrak{X}^*$ then we have the following:

Theorem 3.1.4 *The map $f : \mathfrak{X}^* \to D^* = D \setminus \{0\}$ is a smooth locally trivial fibration. The Milnor fiber X_t is homotopically equivalent to $\bigvee_\mu S^{n-1}$, where $\mu = \dim_{\mathbb{C}}(O_n/(\partial f))$ and $(\partial f) = (\partial_1 f, \ldots, \partial_n f)$ is the Jacobian ideal of f.*

$$\begin{array}{ccccc} X_t & \hookrightarrow & \mathfrak{X}^* & \hookrightarrow & \mathfrak{X} \\ \downarrow & & \downarrow & & \downarrow \\ \{t\} & \longrightarrow & D^* & \hookrightarrow & D \end{array}$$

The algebra $O_n/(\partial f)$ is called the Milnor algebra of the singularity $(\{f = 0\}, 0)$.

Exercise 3.1.5 Show that the Milnor algebra of f is finite dimensional if and only if f defines an isolated singularity.

Example 3.1.6 Let $f_k = x_1^2 + x_2^2 + x_3^2 + x_4^3 + x_5^{6k-1}$. Then Brieskron showed that for $k = 1, \ldots, 28$ the abstract links of the f_k polynomials are exactly the exotic 7 spheres, [Bri66].

Example 3.1.7 In the case of curve singularities, the embedded link consist of a union of knots in the 3-sphere. Below we show some of the simplest local plane curve singularities and the corresponding embedded links (Fig. 3.4).

Let us continue with invariants of local plane curve singularities. In this case we have $f(x, y) \in \mathbb{C}\{x, y\}$ and we can decompose f into irreducible elements. So we write $f = f_1^{\alpha_1} \dots f_r^{\alpha_r}$, where the f_i are irreducible elements in $\mathbb{C}\{x, y\}$, and we notice that f defines an isolated singularity if and only if $\alpha_i = 1$ for $i = 1, \dots, r$. Corresponding to the irreducible decomposition our singularity $(C, 0) = (X_0, 0)$ will be the union of $(C_i, 0) = (V(f_i), 0)$, i.e. $C = C_1 \cup \dots \cup C_r$. Moreover, the Milnor fiber X_t is a Riemann surface with r boundary circles, and the (embedded) link of C will be $L = L_1 \cup \dots \cup L_r \subset S^3$, where the knots L_i are iterated torus knots. With these preparations, we are ready to give Milnor's formula:

Theorem 3.1.8 *Let* $\mu = \dim_{\mathbb{C}}(\mathbb{C}\{x, y\}/(\partial_x f, \partial_y f))$, *and* $r = \#\{$*branches of* $f\}$, *then:*

$$\mu = 2\delta - r + 1.$$

With Milnor's formula it is easy to compute the delta-invariant of $x^m - y^m$. The Milnor number of the ordinary m-fold point is $\dim_{\mathbb{C}}(\mathbb{C}\{x, y\}/(x^{m-1}, y^{m-1})) = (m-1)^2$, and using Milnor's formula we get that $(m-1)^2 = 2\delta - m + 1$ or

$$\delta = \frac{m(m-1)}{2}.$$

Exercise 3.1.9 Compute the delta-invariant of $x^n - y^m$ for arbitrary m and n.

The message is that working directly with the definition is not always the easiest way to do a computation. In fact let us go back to the original construction, where we had the smooth curve $\widetilde{C}$ and we projected it to $\mathbb{CP}^2$ and the image of the projection

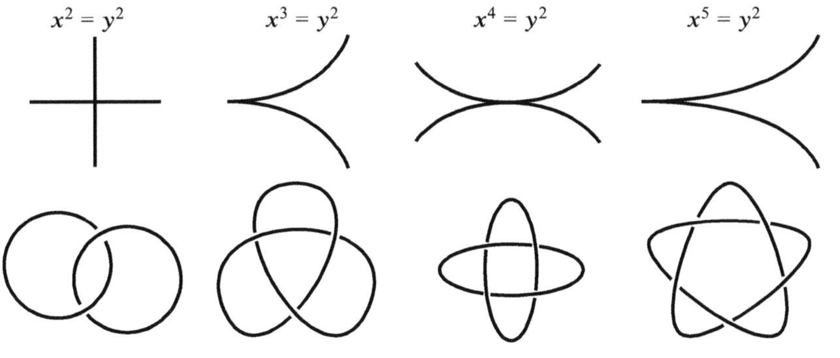

Fig. 3.4 Singularities and their corresponding links

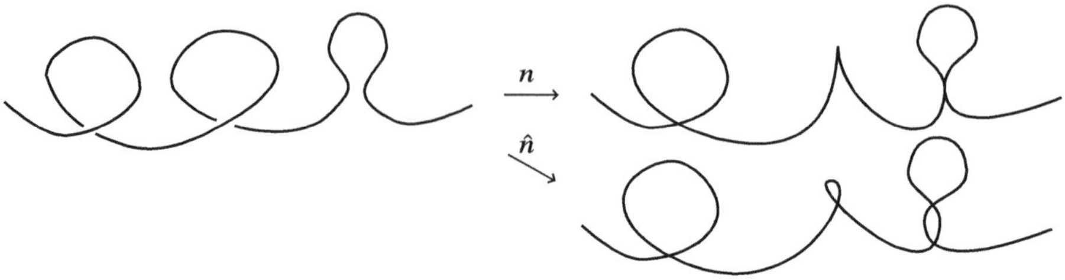

Fig. 3.5 Perturbing the projection

Fig. 3.6 Perturbing the ordinary 4-fold point

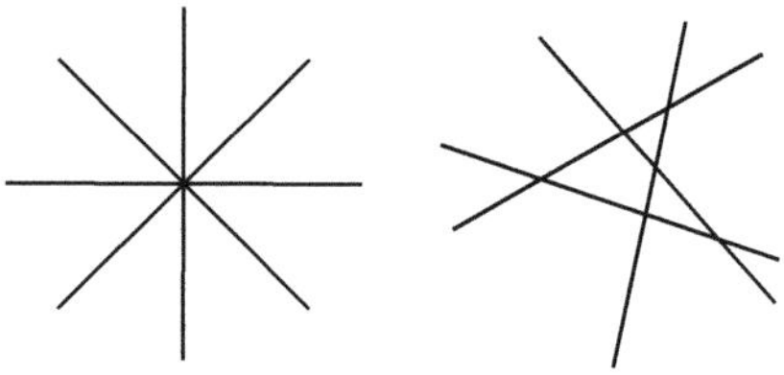

was the singular curve C. Now let us imagine that we change very slightly this projection (Fig. 3.5).

As we see in the picture, the cusp turns into a small loop with an ordinary double point, and the two branches that were tangent turn into a 'double loop' with two ordinary double points. Notice that by changing the projection the delta-invariant remained the same: δ(cusp)=δ(loop with one double point)= 1 and δ(two tangent branches with tangency 2) = δ(double loop with two double points)= 2.

If we look at the ordinary m-fold point again, we see m lines going through the same point, but if we push them a little bit, we will see m lines in generic position (Fig. 3.6). The singularities are then double points and the number of them, which is the delta-invariant of this curve, is just the number of intersections, namely $m(m-1)/2$.

As we can see, it is easier to calculate the delta-invariant in this way, because we only have to count how many double points we get. This is also the reason why the delta-invariant was originally called the "virtual number of double points". However it is not true that for any deformation the delta-invariant stays constant. If we look at the Milnor fibration we see that the central fiber has positive delta-invariant but the Milnor fiber is smooth so it has delta-invariant zero. Nevertheless if we deform the projection, then the delta-invariant remains constant and this is exactly the phenomenon what we will discuss in the next section.

3.1.3 Deformations and the Delta-Invariant

In the last section we saw that if we have a singular curve C, then by deforming it a little bit, we can decompose its singularities into simpler ones, a fact which was very helpful in the computation of the delta-invariant. To make this type of reasoning precise we need the notion of deformations.

Definition 3.1.10 Suppose we have fixed analytic spaces X and T. Then a deformation of X over T is a flat morphism $\phi : \mathfrak{X} \to T$, and an isomorphism between X and the fiber $X_0 = \phi^{-1}(0)$, where 0 is a preferred point of T.

$$\begin{array}{ccccc} X = X_0 & \hookrightarrow & \mathfrak{X} & \hookleftarrow & X_t \\ \downarrow & & \downarrow & & \downarrow \\ \{0\} & \hookrightarrow & T & \hookleftarrow & \{t\} \end{array}$$

In this case we call T the parameter space, $\mathfrak{X}$ the total space and X_0 the special fiber of the deformation.

Example 3.1.11 Let X be a germ defined by $f(x, y) \in \mathbb{C}\{x, y\}$ and let $\mathfrak{X} = \{(x, y, t) \in \mathbb{C}^3 \mid f(x, y) = t\}$, then a deformation of X is given by $\mathfrak{X} \to T$, $(x, y, t) \mapsto t$. Notice that this is just the Milnor fibration.

More generally, if we have a germ X defined by $f(x) \in \mathbb{C}\{x\}$, where $x = (x_1, \ldots, x_n)$, we can give a deformation of it by perturbing its equation, i.e. we take a germ $F(t, x) \in \mathbb{C}\{t, x\}$, where $t = (t_1, \ldots, t_m)$ such that $F(0, x) = f(x)$. Then we can take the germ $\mathfrak{X} = \{F(t, x) = 0\}$. From $\mathfrak{X}$ we have the natural projection to the parameter space given by t. For this projection to be flat, we only have to require that $t_1, \ldots, t_m \in \mathcal{O}_{\mathfrak{X},0}$ is a regular sequence, and for $m = 1$ this only means that $t_1 = t$ is not a zero divisor.

Example 3.1.12

(a) Let $f(x, y) = y^2 - x^3$ be the cusp singularity, then a possible deformation of f is given by $F(t, x, y) = y^2 - x^2(x + t)$. Hence, for $t = 0$ we have the original singularity, and for any $t \neq 0$ the fiber has an ordinary double point (see Fig. 3.7).

(b) Let $f(x.y) = y^2 - x^4$ be the two tangent branches with tangency two, then a possible deformation is given by $F(t, x, y) = y^2 - (x^2 - t)^2$. Hence, for $t = 0$ we

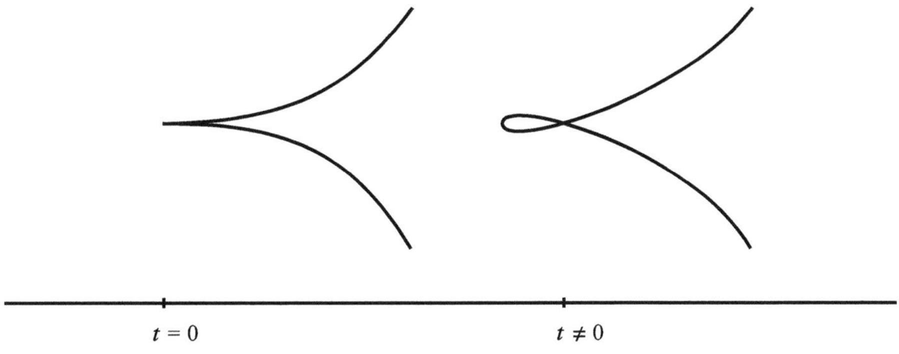

Fig. 3.7 A deformation of $y^2 - x^3$

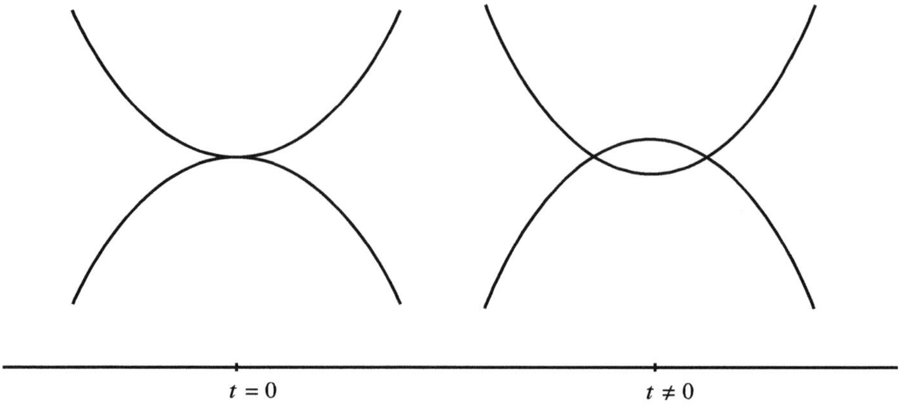

Fig. 3.8 A deformation of $y^2 - x^4$

have the original equation, however for $t \neq 0$ we have two parabolas intersecting in two ordinary double points (see Fig. 3.8).

As we said earlier, not every deformation is delta-constant (e.g. Milnor fibration), but the following theorem of Teissier will give a necessary and sufficient condition for a deformation to be delta-constant.

Theorem 3.1.13 (B. Teissier, 1976, [Tei76]) *Suppose we have a deformation of a curve singularity* $(C, 0)$.

$$\begin{array}{ccc} C & \hookrightarrow & \mathfrak{C} \\ \downarrow & & \downarrow \\ \{0\} & \hookrightarrow & T \end{array}$$

Then we have the following:

1. *δ is upper semi-continuous, which means that* $\delta(C) \geq \sum_{p \in C_t} \delta(C_t, p)$.
2. *We have* $\delta(C) = \sum_{p \in C_t} \delta(C_t, p)$ *if and only if the family has a simultaneous normalisation. This means, that if we take the normalisation of* $n : \tilde{\mathfrak{X}} \to \mathfrak{X}$, *then* $(\tilde{\mathfrak{X}})_t = \widetilde{(C_t)}$.

Corollary 3.1.14 *A deformation of* $\{f(x, y) = 0\}$ *is delta-constant precisely when it arises by deformation of its parametrisation.*

Example 3.1.15

(a) Take the parametrisation of the cusp $s \mapsto (s^2, s^3)$, then a deformation of the parametrisation is $x(s, t) = s^2 - t$, $y(s, t) = (s^2 - t)s$. This parametrisation is particularly nice because we can easily find its equation, which is $F(t, x, y) = y^2 - x^2(x + t)$. By Teissier's theorem this is a delta-constant deformation of the cusp singularity. (Compare with the previous Example (a).)

(b) Let us look at the two tangent branches again. Since it has two branches, we need two variables for the parametrisation. We take $s_1 \mapsto (s_1, s_1^2)$ and $(s_2, -s_2^2)$, and a deformation is given by $(s_1, s_1^2 - t)$, and $(s_2, -(s_2^2 - t))$, and the equation of this deformation is just $y^2 - (x^2 - t)^2$.

3.2 Severi Strata

3.2.1 Miniversal Deformation

How can we deform a singularity? In principle this is very simple. We have an equation and we can deform the equation by some parameters. We do not need too many parameters, in fact a τ-dimensional deformation is enough to induce all the deformations, where

$$\tau = \dim_{\mathbb{C}}(\mathbb{C}\{x, y\}/(f, \partial_x f, \partial_y f)$$

is called the *Tjurina number* of f. Clearly, $\tau \leq \mu$.

To describe this deformation we pick elements $g_1, \dots, g_\tau \in \mathbb{C}\{x, y\}$ such that $\overline{g}_1, \dots, \overline{g}_\tau$ gives a basis of $\mathbb{C}\{x, y\}/(f, \partial_x f, \partial_y f)$, and then we perturbe the original function in a the simplest possible way:

$$F(t_1, \dots, t_\tau, x, y) = f(x, y) + \sum_{i=1}^{\tau} t_i g_i(x, y),$$

where the t_i are new parameters. Now we do something similar as in the case of Milnor fibrations. Let $T = \{t \in \mathbb{C}^\tau \mid |t| < \varrho\}$ for some sufficiently small $\varrho > 0$, and then we look at $C = \{(t, x) \in T \times B \mid F(t, x) = 0\}$. This space has a natural map to T defined by $(t, x) \mapsto t$, and this way we get a family of curves C_t parametrised by τ free parameters. The surprising fact is that if one takes any deformation $G(s, x)$ of the original curve $(C, 0)$, then it is *induced* from this family. This means that one can find power series $t_i(s)$, such that $F(t_1(s), \dots, t_\tau(s), x)$ is, up to a coordinate transformation (that depends on s) equal to G. It is this property that makes F into a so-called *versal family*, and this is the smallest dimensional such versal family, hence it is called *miniversal*. In particular, if one wants to study all deformations of $(C, 0)$, one has to understand this deformation only.

3.2.2 Definition of Local Severi Strata

Definition 3.2.1 We define

$$S_k^\circ = \{t \in T \mid C_t \text{ has exactly } k \text{ nodes as singularities}\}.$$

Then we define S_k as the closure of S_k°, and it called the k-th *Severi stratum* in [She12].

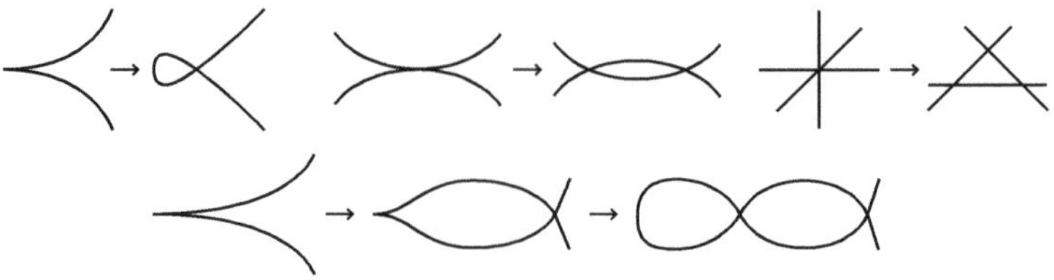

Fig. 3.9 Deformations to curve singularities with exactly δ nodes

From this definition it is not clear whether these sets are empty or not. Indeed, the question is: if you are given a plane curve singularity $(C, 0)$ and a positive integer k, can you always deform the curve $(C, 0)$ into a curve with exactly k nodes? Usually this is not possible, the highest number you can get is δ. In fact we have the following important theorem.

Theorem 3.2.2 (Ch. Scott [Sco92, Sco93], A'Campo [Cam73]) *$S_\delta^\circ \neq \emptyset$, that is, there exists a deformation of $(C, 0)$ such that C_t has exactly δ nodes as singularities.*

Figure 3.9 shows four examples of such deformations. In fact, Scott and A'Campo prove a stronger statement. Let $m_1, m_2, \ldots, m_r, 1, \ldots, 1$ be the multiplicity sequence of an irreducible f (it is the sequence of multiplicities of the local equations obtained by iterated blow ups of the original equation till we obtain a good embedded resolution). If m_r is the last multiplicity which is larger than 1, then it is followed by m_r 1's. In the next discussion we delete these multiplicities and we keep the sequence $m_1, \ldots, m_r$. Scott proved that there exists a delta constant deformation whose generic fiber has r singular points, all ordinary singularities with multiplicities $m_1, \ldots, m_r$. This also shows that

$$\delta = \sum_{i=1}^{r} m_i(m_i - 1)/2.$$

Example 3.2.3 The parametrisation $s \mapsto x(s) = s^4,\ y(s) = s^6 + s^7$ has image with equation $f(x, y) = (y^2 - x^3)^2 - 4x^5y - x^7 = 0$, the (reduced) multiplicity sequence is 4, 2, 2, hence $\delta = 8$, and there exists a deformation with only ordinary singularities of multiplicities 4,2,2.

Now we will list some properties of S_k:

- S_k is nonempty for $0 \leq k \leq \delta$.
- S_k is the closure of the following set: $\{t \in T \mid C_t \text{ has } k \text{ singularities}\}$.
- $S_0 = T$ (closure of the parameter space of fibers without any node singularity).
- $S_1 = \{t \in T \mid C_t \text{ is singular}\} = \Delta$. S_1 is also called the discriminant, and it is a hypersurface in T.
- $\emptyset = S_{\delta+1} \subset S_\delta \subset \cdots \subset S_1 \subset S_0$.
- $\operatorname{codim}(S_k) = k$.

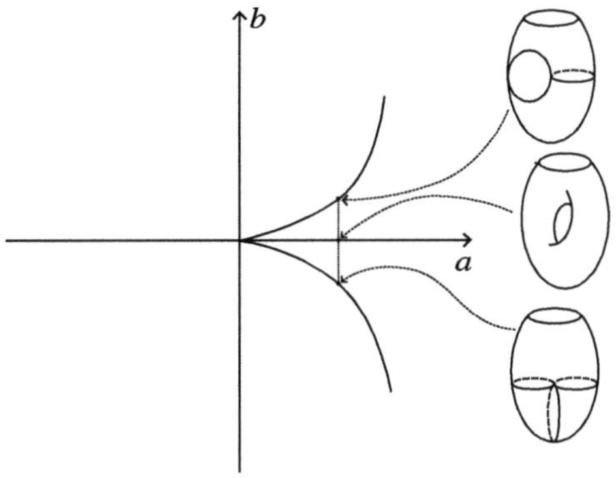

Fig. 3.10 The Severi stratum of $y^2 - x^3$

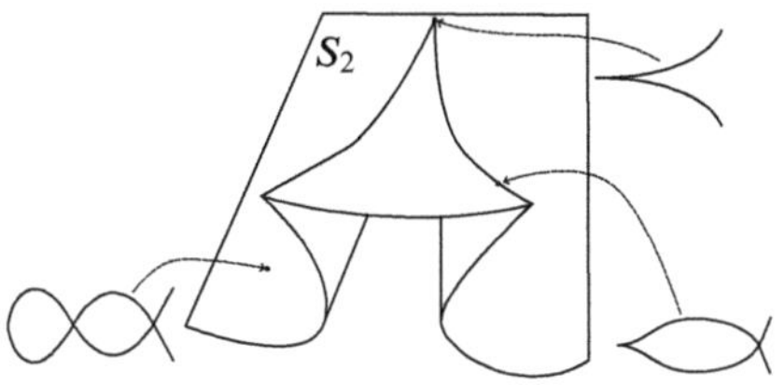

Fig. 3.11 The second Severi stratum of $y^2 - x^5$

Example 3.2.4

(a) Let $f(x, y) = y^2 - x^3$, then $1, x \in \mathbb{C}\{x, y\}$ will map into a basis of the Milnor algebra $\mathbb{C}\{x, y\}/(y, x^2)$. Hence the miniversal deformation is given by the following family of curves $y^2 - x^3 + ax + b = 0$ parameterised by a and b. In this case the Severi strata are $S_1 = \Delta = \{4a^3 - 27b^2 = 0\}$. Figure 3.10 shows how the fibers look over different parts of the parameter space.

(b) Set $f(x, y) = y^2 - x^5$. In this case $1, x, x^2, x^3 \in \mathbb{C}\{x, y\}$ lift a basis of the Milnor algebra, and the deformation is given by $y^2 - x^5 + ax^3 + bx^2 + cx + d = 0$. In this case, the Severi strata live in a 4 dimensional parameter space, and since S_1 is a hypersurface it is quite challenging to draw them. However, Fig. 3.11 gives a schematic picture of S_2, and the fibers over different parts of S_2. The surface S_2 is commonly called the *open swallow tail surface*.

3.2.3 Severi Multiplicities

If we have a plane curve singularity $(C, 0)$, then we define its *Severi multiplicities* as

$$m_k(C, 0) = \text{mult}(S_k),$$

for $k = 0, 1, \ldots, \delta$. Recall that if we have a germ $(X, 0) \in (\mathbb{C}^N, 0)$, then its multiplicity is defined as follows:

$$\text{mult}(X, 0) = \#X \cap L$$

where L is a generic linear subspace near the origin with $\dim(L) = \text{codim}(X, 0)$.

Now let us look at these multiplicities in the most well understood cases, which are the simple singularities listed by Arnold, see e.g. [AGV85]. These are in the hierarchy of singularities the ones that come first, and are classified by Dynkin diagrams of type ADE:

- $A_n : y^2 - x^{n+1} = 0$ 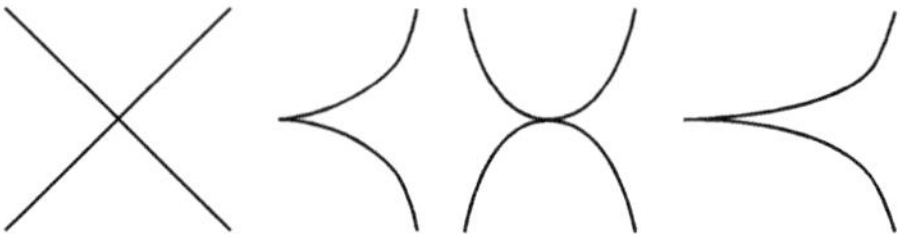
- $D_n : xy^2 - x^{n-1} = 0$ 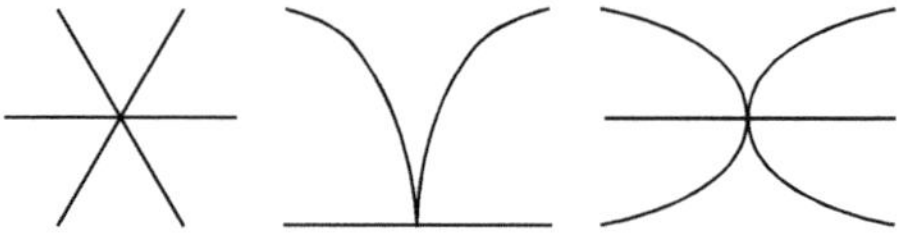
- $E_6 : x^4 - y^3 = 0$ 
- $E_7 : y(y^2 - x^3) = 0$ 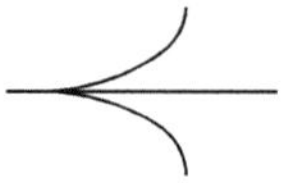
- $E_8 : y^3 - x^5 = 0$ 

Remark 3.2.5 The E_8 singularity looks rather unspectacular, however it is actually the most interesting one. If we just add another variable and look at the surface $z^2 + y^3 + x^5 = 0$, then its link is the Poincaré homology sphere $\Sigma(2, 3, 5)$ and the Milnor fiber is a 4-manifold which bounds this Poincaré sphere, and its intersection lattice is the E_8 lattice, see [Bri66, Mil68].

We can make the ADE diagrams visible by looking at the real part of special deformations, the case of E_6 and D_4 is depicted on Fig. 3.12. It is also a fact that the vertices of the Dynkin diagram correspond to the vanishing cycles in the Milnor fiber, and edges between 2 vertices correspond to the intersection of the corresponding vanishing cycles. Next, we give the Severi multiplicities for these singularities:

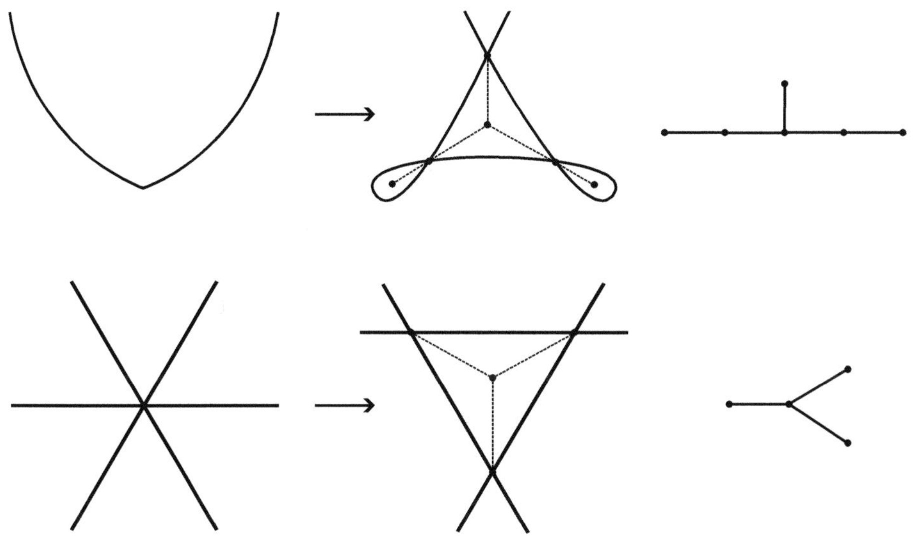

Fig. 3.12 Making the Dynkin diagram visible

- $m_k(A_n) = \binom{n+1-k}{k}$
- $m_k(D_n) = \binom{n-k}{k} + \binom{n-k-1}{k-1} + \binom{n-k}{k-2}$
-

	m_0	m_1	m_2	m_3	m_4
E_6	1	6	10	5	
E_7	1	7	15	11	2
E_8	1	8	21	21	7

Notice that $m_0 = 1$ and $m_1 = \mu$. However, the remaining Severi multiplicities are rather mysterious numbers. Note also that $m_2(A_3) = m_3(D_4) = m_4(D_6) = 1$, that is, if you can create the k double points by shifting the components, then $m_k = 1$. Vivek Shende made the following revealing remark in [She12]:

$$m_k(ADE) = \#\text{ ways to pick } k \text{ disjoint dots from the Dynkin diagram.}$$

Example 3.2.6 Let us look at A_4. In the picture, we can see the Dynkin diagram via a special delta constant deformation, and also via vanishing cycles on the Milnor fiber of A_4. As we can see, picking 2 disjoint points from the Dynkin diagram corresponds to disjoint cycles in the Milnor fiber. Now, a generic elements of S_2 are curves with 2 nodes, and we can achieve these curves by contracting 2 vanishing cycles. Moreover collapsing intersecting vanishing cycles would give worse than node singularities. The statement is that the multiplicity of S_2 is just the number of ways we can pick 2 disjoint vanishing cycles, which is 3 in this case (Fig. 3.13).

Of course, this not a formal proof, but rather a compelling idea that enables one to remember the result.

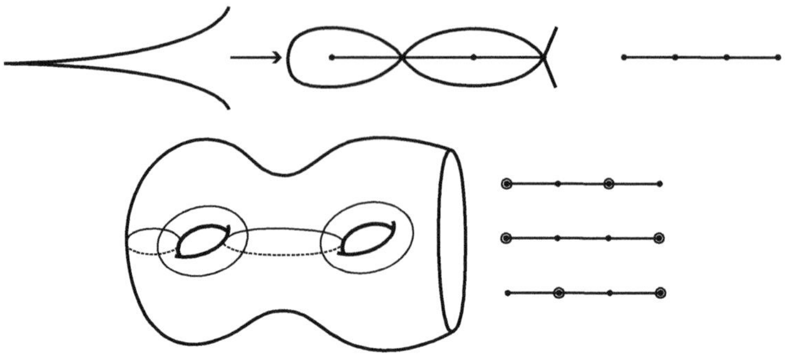

Fig. 3.13 Computing $m_2(A_4)$

3.3 The Jacobian

3.3.1 Classical Jacobian

We start this section with a theorem:

Theorem 3.3.1 (L. Göttsche et al. [GFvS98]) *Let* $(C, 0)$ *be a reduced curve singularity, then*

$$m_\delta(C, 0) = e(\overline{J}(C, 0)),$$

where $\overline{J}(C, 0)$ *is the compactified Jacobian of the plane curve singularity* $(C, 0)$.

Since $\overline{J}(C, 0)$ is a topological space, we can look at $H^*(\overline{J}(C, 0))$, and we get a categorification of m_δ. But what is $\overline{J}(C, 0)$? In order to explain this, first we start with the smooth case.

Let C be a smooth projective curve (compact Riemann surface), then a fundamental fact first proved by Riemann is that the space of holomorphic one forms $H^0(\omega_C)$ has dimension g, where g is the genus of the curve C. This fundamental fact says that the topology of C determines the number of linearly independent holomorphic one forms. Now we can define the Jacobian of C as

$$\mathrm{Jac}(C) = H^0(\omega_C)^*/H_1(C, \mathbb{Z}),$$

where the map $H_1(C, \mathbb{Z}) \to H^0(\omega_C)^*$ is given by integration, i.e. an element $\gamma \in H_1(C, \mathbb{Z})$ maps to integration along γ, that is $\gamma \mapsto (\int_\gamma : \omega \in H^0(\omega_C) \mapsto \int_\gamma \omega \in \mathbb{C})$. Since $H^0(\omega_C) = \mathbb{C}^g$ and $H_1(C, \mathbb{Z}) = \mathbb{Z}^{2g}$, and $\gamma \mapsto \int_\gamma$ is an injection, we obtain that $\mathrm{Jac}(C) \simeq \mathbb{C}^g/\mathbb{Z}^{2g} \simeq (S^1)^{2g}$ is a complex torus. We are talking about the mathematics before the middle the nineteenth century, when Riemann created the theory of Riemann surfaces in order to cope with Abel's theorem on integrals and it was in attempts to find an geometric way to understand Abels theorem, that led Jacobi and Riemann to the above definition, [Mum75]. The construction and the theory of the Jacobian was crucial discovery in the nineteenth century and still plays a big role in algebraic geometry today.

The modern way of introducing the Jacobian, the approach which was pursued by Lefschetz, is a little different. In this approach we use the exponential exact sequence of sheaves:

$$0 \longrightarrow \mathbb{Z}_C \xrightarrow{2\pi i} \mathcal{O}_C \xrightarrow{exp} \mathcal{O}_C^* \longrightarrow 0$$

where $\mathbb{Z}_C$ is the sheaf of locally constant functions on C, $\mathcal{O}_C$ is the sheaf of holomorphic functions on C and $\mathcal{O}_C^*$ is the sheaf of nowhere vanishing holomorphic functions. This sequence of sheaves is exact because locally the logarithm of nowhere zero function is well defined. Now for every sheaf there is sheaf cohomology, and for every short exact sequence of sheaves there is a long exact sequence of sheaf cohomologies. In our case the important part is the following:

$$0 \longrightarrow H^1(C, \mathbb{Z}) \longrightarrow H^1(\mathcal{O}_C) \longrightarrow H^1(\mathcal{O}_C^*) \longrightarrow H^2(C, \mathbb{Z}) \longrightarrow 0.$$

By Serre duality, $H^1(\mathcal{O}_C)$ is the dual space of $H^0(\omega_C)$, this fact was proved by Roch, who was one of the few students Riemann had. If we also use the fact that $H^1(\mathcal{O}_C^*) = \text{Pic}(C)$, the group of isomorphism classes of analytic line bundles, then the above exact sequence reads as the following exact sequence:

$$0 \longrightarrow \text{Jac}(C) \longrightarrow \text{Pic}(C) \xrightarrow{\deg} \mathbb{Z} \longrightarrow 0.$$

Here the map deg, which can also be interpreted as the first Chern class of a line bundle, is the topological aspect of a line bundle. However, its kernel is invisible by topology, it consists of the 'analytic part of the line bundles': it is the group of analytic line bundles, which are topologically trivial. Thus, in this approach we see that different points of the Jacobian correspond to different analytic structures on the topologically trivial line bundle over C.

Another good property of the above exact sequence is that it also can be applied when the curve is not smooth. Even for singular curves, there is a Picard group, and if C is compact then still there is a degree. If the curve has more components the second cohomology might be bigger, in that case we can look at the Chern classes on each component. So we can use this to define the Jacobian of singular curves, usually called *generalised Jacobians*.

Example 3.3.2 Let us take the projective cubic $\{y^2z - x^3 - x^2z = 0\}$ in $\mathbb{CP}^2$. It is a rational curve, and topologically it is a sphere with two points identified. We can normalise this curve by removing the intersection, hence its normalization $\widetilde{C}$ topologically will be a sphere (Fig. 3.14). If we remove the singular point, which is the same as removing the two preimages of the singular point from the sphere, we will get $\mathbb{C}^*$. Its Euler number is 0, but we have to add back the removed point, so the Euler number of the singular curve is 1.

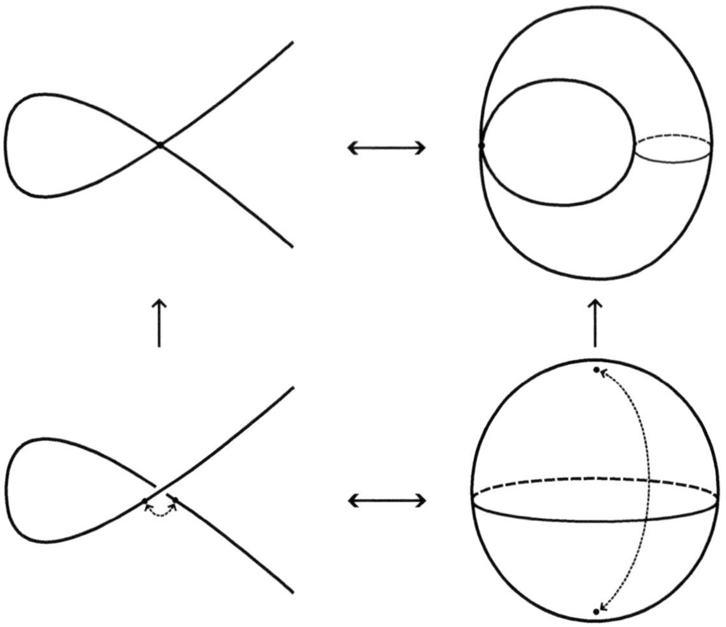

Fig. 3.14 The cubic with a node in $\mathbb{CP}^2$ and its normalisation

Note that the Picard group of the normalization $\widetilde{C}$ is trivial. However, one has the natural question: how do we get a line bundle on C, that is, on a space where we identified two points? We just need to take an ordinary line bundle on $\widetilde{C}$, and over the two points that are glued together we have to linearly identify the two lines. The linear identifications of the two lines is the same as the linear automorphisms of $\mathbb{C}$, which is $\mathbb{C}^*$. In particular, we get that the Jacobian of the singular cubic is $\mathbb{C}^*$. As we see, the Jacobian of the singular cubic is no longer a compact torus, it is $\mathbb{C}^*$.

3.3.2 *Compactified Jacobian*

Definition 3.3.3 For a singular projective curve C the *compactified Jacobian* is defined (following Grothendieck, Altmann and Kleiman, [AK80]) as follows:

$$\overline{J}(C) = \text{Moduli space of torsion free rank 1 sheaves } \mathcal{F} \text{ with } \deg(\mathcal{F}) = 0.$$

Furthermore, for an isolated curve singularity $(C, 0)$, we define the compactified Jacobian of $(C, 0)$ as $\overline{J}(C, 0) = \overline{J}(C)$, where C is a rational curve with $(C, 0)$ its only singularity.

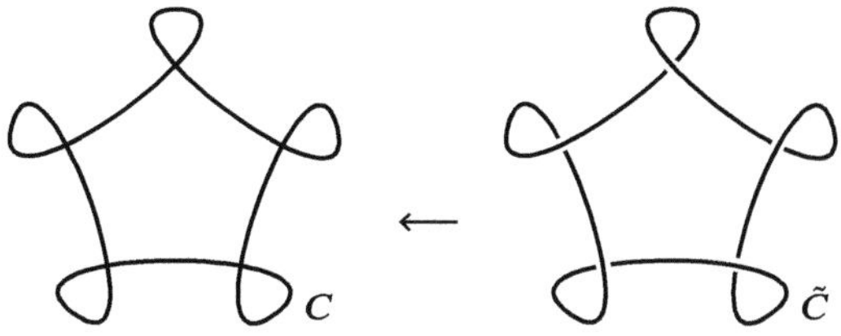

Fig. 3.15 Curve with maximal number of nodes

The construction of $\overline{J}(C)$ is an example where the Grothendieck way of thinking about algebraic geometry makes its appearance. In the smooth case we had line bundles with first Chern class 0. In the present singular case we have something more general, the torsion free rank 1 sheaves with degree 0, which form a compact space. The point is that a line bundle can degenerate to something that is not a line bundle, but a torsion free sheaf. Note that $\overline{J}(C)$ is no longer a group, but it contains the generalised Jacobian $J(C)$ of the singular curve.

The compactified Jacobian of the curve $\{y^2z - x^3 - x^2z = 0\}$, which is the rational curve with an ordinary double point, is just itself, i.e. $\overline{J}(C) = C$. Recall that a smooth cubic curve is isomorphic with its own Jacobian. (Using this isomorphism one can explain the existence of a group structure on the points of a smooth cubic curve.) The interesting new phenomenon is that this isomorphism also holds for singular cubic curves. Hence in this example, the Euler number of $\overline{J}(C)$ is 1. For another example, if we take a cubic rational curve with a cusp singularity, then the compactified Jacobian is again isomorphic to the curve itself: topologically it is just a sphere, hence it has Euler number 2.

Suppose that we have a curve C with maximal number of ordinary double points as singularities, and its normalisation is $\widetilde{C}$ (Fig. 3.15). Since C has maximal number of double points its normalisation $\widetilde{C}$ has genus 0. (Indeed, $g(\widetilde{C}) + \delta(C) = g_a(C)$.) In this case, when $g(\widetilde{C}) = 0$, the compactified Jacobian $\overline{J}(C)$ is homeomorphic to

$$\text{(a rational curve with an ordinary double point)}^{\delta}.$$

In particular, its Euler number is still 1. It is also a general fact, that for non-rational curves C the Euler number of $\overline{J}(C)$ is 0.

We wish to sketch the proof of the theorem stated at the beginning of the section. For this we also need that the compactified Jacobian behaves well in families. That is, if you have a family of curves $C \to T$, it induces a family of compactified Jacobians $\overline{J}(C) \to T$ such that $\overline{J}(C)_t$, the fiber over t, is $\overline{J}(C_t)$.

Proof (Sketch) Choose a general one-parameter family $\{W_t\}_t$ of smooth δ dimensional subspaces of $(\mathbb{C}^\tau, 0)$ such that $W_0 \cap S_\delta = \{0\}$ (Fig. 3.16). Then, for general t, the intersection $W_t \cap S_\delta$ is a set of $m_\delta(C, 0)$ distinct points corresponding to curves with maximal number of ordinary double points. Let $C_t \to W_t$ be the induced family. Then for a fixed t, $\overline{J}(C_t)$, the union of $\overline{J}(C_v)$ for all $v \in W_t$, is a flat family of smooth compact varieties. Therefore $e(\overline{J}(C_t)$ does not depend on t. Notice now that for any element $v \in W_t \setminus S_\delta$ we have $e(\overline{J}(C_v)) = 0$ by the non-rationality of

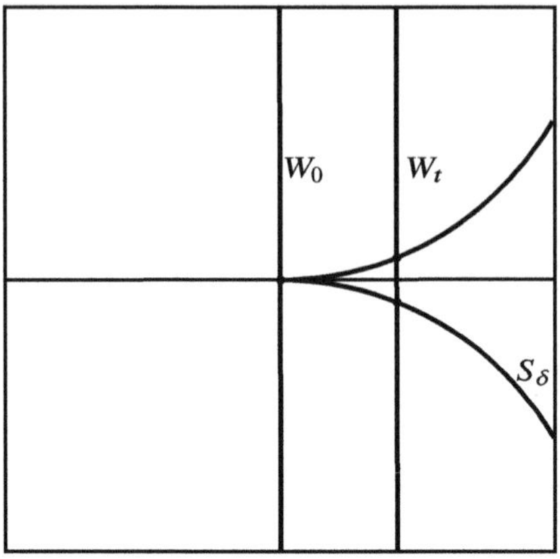

Fig. 3.16 Generic one parameter family of subspaces

the normalization, and for $v \in W_t \cap S_\delta$ we have $e(\overline{J}(C_v)) = 1$. Therefore, using the additivity property of the Euler number we get that

$$e(\overline{J}(C_t)) = \sum_{v \in W_t \cap S_\delta} e(\overline{J}(C_v)).$$

Considering this for $t = 0$ and for a generic t we get

$$e(\overline{J}(C)) = e(\overline{J}(C_0)) = e(\overline{J}(C_t)) = \sum_{v \in W_t \cap S_\delta} e(\overline{J}(C_v)) = m_\delta(C, 0).$$

□

3.3.3 *Grassmannian of a Curve*

Let us assume that $(C, 0)$ is a unibranch singularity. In this subsection we define another space attached to the singularity, $\mathrm{GR}(C, 0)$, the Grassmannian of $(C, 0)$. Denote by R the local ring $\mathcal{O}_{C,0}$ of $(C, 0)$ and by $\widetilde{R}$ the local ring $\mathcal{O}_{\widetilde{C},0}$ of the normalisation $(\tilde{C}, 0)$. The *conductor ideal* c is then defined as $\mathrm{Ann}(\widetilde{R}/R) \subset R$, that is, c is the largest ideal of R which is also an ideal of $\widetilde{R}$.

Definition 3.3.4 With $(C, 0)$ as above, we define the *Grassmannian* of $(C, 0)$ as:

$$\mathrm{GR}(C, 0) = \mathrm{GR}(R) = \{V \in \mathrm{Gr}(\delta, \tilde{R}/c) \mid V + c \text{ is an } R\text{–module}\}.$$

Since $\tilde{R}/c$ as a vectorspace is $\mathbb{C}^{2\delta}$, we see that $\mathrm{Gr}(\delta, \tilde{R}/c)$ is the manifold $\mathrm{Gr}(\delta, 2\delta)$.

Theorem 3.3.5 ([Reg80], [GP93]) *Suppose that* $(C, 0)$ *is a unibranch plane curve singularity. Then* $\overline{J}(C, 0)$ *is homeomorphic to* $GR(C, 0)$.

The point is that any torsion free sheaf of $(C, 0)$ becomes a torsion free module of R, which then necessarily sits in the normalisation $\widetilde{R}$. They have a very precise description: any of them corresponds to a δ-dimensional subspace V of $\tilde{R}/c$, such that $V + c$ is an R-module.

Example 3.3.6 Let us look at the singularity E_6. Its equation is $x^4 - y^3 = 0$, the normalisation map is $s \mapsto (s^3, s^4)$ and the conductor is the ideal generated by s^6 in $\mathbb{C}\{s\}$. The following table shows 5 special elements of $\mathrm{GR}(C, 0)$:

	0	1	2	3	4	5	6	7...
R	×	○	○	×	×	○	×	×...
	○	×	○	○	×	×	×	×...
	○	○	×	○	×	×	×	×...
	○	○	×	×	○	×	×	×...
	○	○	○	×	×	×	×	×...

We have to pick 3 crosses such that translating them by 3 or 4 we get again a cross. These 5 elements of the Grassmannian are the fixed points of a certain $\mathbb{C}^*$ action on $\mathrm{Gr}(C, 0)$. From these 5 fixed points we can determine a cellular decomposition, where each cell corresponds to a fixed point. In this particular case, we have one 0-dimensional cell, one 2-dimensional cell, two 4-dimensional cells and one 6-dimensional cell. Hence the Betti numbers of this space are 1, 1, 2, 1, and the Euler number is 5. This 5 is exactly the multiplicity m_3 of the Severi stratum $S_{\delta=3}$ in the case of E_6.

3.4 Severi Multiplicities and the HOMFLY Polynomial

3.4.1 *Hilbert Scheme of Points*

Let us fix a topological space X, and assume that we have n points of X that move around some way. Mathematically this can be represented via the configuration space of all n-tuples,

$$\underbrace{X \times X \times \cdots \times X}_{n\text{-times}}.$$

This product codifies numbered (or marked) n-tuples of points. However, if we do not wish to mark them, and we want to consider unnumbered collection of points, then we have to factor this product with the action of the symmetric group. In this way we get the symmetric product $\mathrm{Sym}^n(X) = X \times \cdots \times X/\sigma_n$. One disadvantage of the symmetric product is that even if X is a manifold, the symmetric product of X is usually not a manifold. This makes their study difficult.

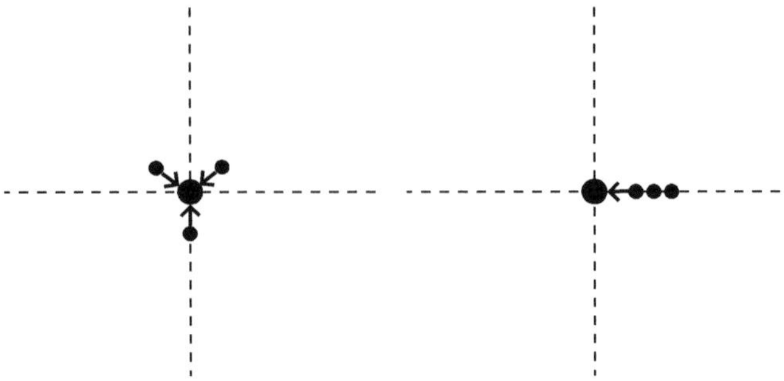

Fig. 3.17 Three points collapsing in two different ways

Definition 3.4.1 Suppose that X is an algebraic variety. The set

$$X^{[n]} = \{\text{0-dimensional subschemes } Z \subset X \text{ such that } \dim_{\mathbb{C}}(\mathcal{O}_Z) = n.\}$$

is called the *Hilbert scheme of points*.

If Z consists of n distinct points of X, then we can evaluate the functions of X at these points. This correspondence provides an isomorphism $\mathcal{O}_Z \simeq \mathbb{C}^n$, so $Z \in X^{[n]}$. Therefore the Hilbert scheme of points perceives the distinct points of X, but it also perceives how these points 'come together', collapse. For example, if $X = \mathbb{C}^2$ and we have 3 points moving in the plane and they all 'come together', then there are different ways in which they can unite (Fig. 3.17). Geometrically we see just one point, but the scheme structure tells us how they collapsed into one point. If we look at the defining equations (y, x^3) and $(x^2, xy, y^2) \subset \mathbb{C}\{x, y\}$, then both of these define geometrically the same point. However, the first contains linear elements the second does not, so they correspond to different elements in $(\mathbb{C}^2)^{[3]}$. The surprising fact about the Hilbert scheme is the following theorem:

Theorem 3.4.2 *If X is a surface, that is, $\dim_{\mathbb{C}}(X) = 2$, then $X^{[n]}$ is smooth!*

For complex surfaces the n-fold symmetric product typically is not a manifold. This is one of the reasons why this Hilbert scheme of points associated with a 2-dimensional space is being studied, and appears in many situations. In higher dimensions, however, the Hilbert scheme is no longer a smooth space, so it is more difficult to work with.

Let us see what happens when we have a smooth curve instead of a surface. In this case the situation is fairly simple, namely $C^{[n]} \simeq \mathrm{Sym}^n(C)$, as the points can 'come together only along one direction' (Fig. 3.18). This is something which appears in classical algebraic geometry too, where people look at the space of effective divisors of given degree n, $\mathrm{Div}^n_+(C)$. It consists of the formal sums of finitely many points of C with positive integral coefficient such that the sum of the coefficients is n:

$$\mathrm{Div}^n_+(C) = \{n_1 P_1 + n_2 P_2 + \cdots + n_k P_k \mid P_i \in C,\ n_i \geq 0,\ n_1 + n_2 + \cdots + n_k = n\}.$$

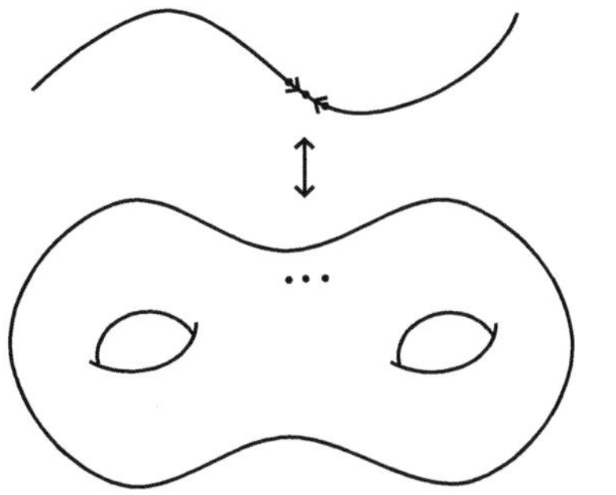

Fig. 3.18 Hilbert scheme of points on a curve

Clearly, $\mathrm{Sym}^n(C) \simeq \mathrm{Div}^n_+(C)$. For smooth curves this Hilbert scheme of points is also a smooth manifold, and it is a fairly well understood. The topology of this space is completely known in terms of the topology of C. However, if C is singular, then $C^{[n]} \not\simeq \mathrm{Sym}^n(C)$, the reason is that at the singular points of the curve there are possibilities for the points to collapse in different ways.

It turns out that understanding the Hilbert scheme of points on singular curves is quite challenging. Nevertheless, Shende was able to determine the structure of the spaces $C^{[n]}$ for singular curves. An important invariant of them is the following generating series

$$Z_C(t) = \sum_{n \geq 0} \chi(C^{[n]}) \cdot t^n.$$

Example 3.4.3 If the curve C is $\mathbb{P}^1$, then to give an effective divisor of degree n is equivalent with the choice of a homogeneous polynomial of degree n in two variables. Indeed, the roots of that polynomial determine n points, where we might have some multiple roots. In particular, we get some points with multiplicities. But by the fundamental theorem of algebra the sum of these multiplicities is n. If we multiply the polynomial with a non-zero constant the set of roots remain unmodified, hence we obtain that $C^{[n]} = \mathbb{P}^n$. Since $\chi(\mathbb{P}^n) = n + 1$, we get that the generating series of $\mathbb{P}^1$ is

$$Z_{\mathbb{P}^1}(t) = \sum_{n \geq 0} \chi(\mathbb{P}^n) t^n = \sum_{n \geq 0} (n+1) t^n = \frac{\partial}{\partial t}\left(\frac{1}{1-t}\right) = \left(\frac{1}{1-t}\right)^2.$$

Now, there is a simple theorem for smooth curves:

Theorem 3.4.4 *Suppose that C is a smooth curve, then*

$$Z_C(t) = \left(\frac{1}{1-t}\right)^{\chi(C)}.$$

In particular, the generating series for smooth curves is very simple, although the cohomology of the variety can be very complicated. The proof of the above formula conceptually is also very simple.

Sketch of proof Let us assume that we decompose C as a disjoint union $C = C_1 \sqcup C_2$. Then we have n points in the union of C_1 and C_2, so we have to distribute them: a points go to C_1 and b points to C_2 such that $a + b = n$. That is, at the level of generating series we are multiplying the two series, $Z_{C_1 \sqcup C_2} = Z_{C_1} Z_{C_2}$. Merely the fact that this series is universally multiplicative is enough to complete the proof. Indeed, at the Euler characteristic level the splitting is also additive. If we split our Riemann surface then we have Mayer-Vietoris (or intersection and union) rule for Euler numbers, which guarantees that the only function which has the universal multiplicative property is indeed the right hand side of the formula. □

Note that we were able to prove the result without knowing the cohomology groups of the Hilbert scheme. In particular we understand what happens when C is smooth. next we look at the case when C has singularities.

Theorem 3.4.5 (V. Shende, [She12]) *Let Σ denote the set of singular points of C and denote by C_{reg} the set of regular points $C \setminus \Sigma$. Then*

$$Z_C(t) = \left(\frac{1}{1-t}\right)^{\chi(C_{reg})} \cdot \prod_{p \in \Sigma} Z_{C,p}(t),$$

where $Z_{C,p}(t)$ depends only on the local singularity (C, p), and these local factors are given by:

$$Z_{C,p}(t) = \sum_{n \geq 0} \chi(C_p^{[n]}) t^n.$$

Here $C_p^{[n]}$ is the part of $C^{[n]}$ which is supported at p. The points of this space are described by ideals which only cut out p and has length n. We can think about the factorization of the generating series as looking at all the subschemes which omit the singular points, and then we have the part which is supported at the singular points. A very surprising and nice result states that for any singular point p the series $Z_{C,p}(t)$ can be computed from the classical invariants of $(C, 0)$.

Theorem 3.4.6 ([She12]) *Let $(C, 0)$ be a curve singularity and set $z = (q - q^{-1})$. Then*

$$\sum_{n \geq 0} \chi((C,0)^{[n]}) q^{2n} = \frac{q^{2\delta}}{(1-q^2)^r} \sum_{d=0}^{\delta} n_d z^{2d},$$

where r is the number of branches, δ is the delta-invariant and $n_d = n_d(C, 0) = m_{\delta - d}(C, 0)$ are the Severi multiplicities.

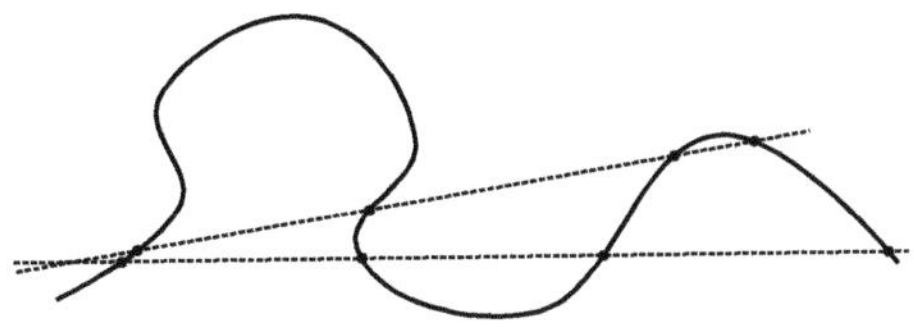

Fig. 3.19 Linear system of a curve embedded in some $\mathbb{P}^n$

The substitution $z = (q - q^{-1})$ is a very specific way of rearranging the terms in the series, and it is a reflection of Serre duality. The proof of the theorem uses the Riemann-Roch theorem, Serre duality and the Abel-Jacobi mapping.

Proof (Idea) If C is a smooth curve, then the Abel-Jacobi mapping is

$$\begin{array}{ccc} C^{[d]} & \longrightarrow & \mathrm{Jac}(C) \\ & \searrow & \downarrow \wr \\ & & \mathrm{Pic}^d(C) \end{array}$$

It maps a degree d effective divisor D to the line bundle $\mathcal{O}_C(D)$. This mapping extends to the singular case $C^{[d]} \longrightarrow \overline{\mathrm{Pic}}^d(C) = \bar{\mathrm{J}}^d(C)$. It maps the subscheme Z to the dual of its ideal sheaf I_Z^*. Here $\bar{\mathrm{J}}^d(C)$ is the space of torsion free rank 1 sheaves of degree d, and this space is isomorphic with $\bar{\mathrm{J}}(C)$. Indeed, if we choose a line bundle L of degree d on C, then tensorization by L provides the needed isomorphism. An important property of the Abel-Jacobi mapping is that the fibers, called linear systems, are projective spaces (Fig. 3.19).

The dimension of these linear systems are controlled by the Riemann-Roch theorem, and the Serre duality gives a certain symmetry and that internal symmetry is expressed by the substitution $z = (q - q^{-1})$. □

In proving these theorems, Shende used the same geometrical setup of the deformation, making the discs transversal to the strata. However, he used the spaces $C^{[n]}$ instead of the compactified Jacobian. Notice that this is a generalization of the theorem which establishes $m_\delta(C) = e(\overline{J}(C))$, however, in this new case all the multiplicities play a role, and in this way it is a much nicer formula.

3.4.2 *HOMFLY Polynomial*

We continue with a purely topological story.

Definition 3.4.7 Suppose we have a link $L \subset S^3$, then we associate to L a Laurent polynomial in two variables $P_L \in \mathbb{Z}[\alpha^{\pm}, q^{\pm}]$ called the *HOMFLY* polynomial. The polynomial is defined by a skein relation:

$$\alpha P_{L_+} - \alpha^{-1} P_{L_-} = z P_{L_0}$$

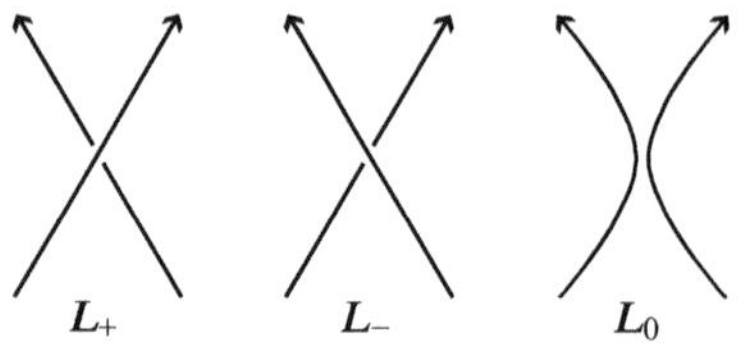

Fig. 3.20 Crossing and smoothing changes

where L_+, L_- and L_0 are links formed by crossing and smoothing changes on a local region of a link diagram, as indicated in Fig. 3.20.

For the $T_{2,n}$ torus knot, the HOMFLY polynomial $P_{T_{2,n}} = P_n$ can be computed recursively by the following rule:

$$P_n = -\alpha z P_{n-1} + \alpha P_{n-2}.$$

If we denote the unknot by U, then its HOMFLY polynomial is:

$$P_U = \frac{\alpha - \alpha^{-1}}{q - q^{-1}} = \frac{\alpha - \alpha^{-1}}{z}.$$

Here we use again the notation $z = (q - q^{-1})$. It turns out that P_n in terms of α and z has the following expression:

$$P_3 = (\alpha^2(z^2 + \underline{2}) - \alpha^4))P_U$$

$$P_5 = (\alpha^4(z^4 + \underline{4}z^2 + \underline{3}) - \alpha^6(z^2 + 2))P_U$$

$$P_7 = (\alpha^6(z^6 + \underline{6}z^4 + \underline{10}z^4 + \underline{4}) - \alpha^8(z^4 + 4z^2 + 3))P_U$$

Remark 3.4.8 Notice, that in the terms with the lowest order in α the underlined values in P_3,P_5 and P_7 are precisely the Severi multiplicities of the singularities A_2,A_4 and A_6, which have links $T_{2,3}$, $T_{2,5}$ and $T_{2,7}$. We can check that the same happens for the torus knot $T_{3,4}$. We have also seen this sort of expression appearing in the Euler numbers of the Hilbert schemes, the generating series of the Hilbert scheme organises exactly in this way.

Noticing this, Shende and Oblomkov formulated a conjecture. First the conjecture was only about the lowest term in α, regarding the part which is expressible in terms of the multiplicities. Then, looking at higher order terms, it appeared that there is only a minimal change what is needed, and this is the conjecture of Oblomkov and Shende, which is now a theorem of Maulik. It is important to note, that in many topology books, the HOMFLY polynomial is written in terms of q, and in the resulting expressions it is much harder to recognize the appearance of the Severi multiplicities m_k!

Theorem 3.4.9 (Oblomkov-Shende Conjecture [OS12], Maulik's Theorem, [Mau16]) *Let $(C, 0) \subset (\mathbb{C}^2, 0)$ be a plane curve singularity with link L. Then*

$$P_L(\alpha, q) = \left(\frac{\alpha}{q}\right)^{\mu(C)-1} \cdot Z_{(C,0)},$$

where

$$Z_{(C,0)} = \sum_{n,m} \chi((C, 0)_m^{[n]}) q^{2n} (1 - \alpha^2)^m$$

and

$$(C, 0)_m^{[n]} = \{I \in C^{[n]} \mid I \text{ has a set of minimal generator of cardinality } m\}.$$

The proof of this theorem is so complicated, that we can not even give a sketch or idea of the proof. It uses all sorts of algebraic geometry, equivalences of derived categories, Hilbert schemes and spaces of stable pairs, a long story which can not be included in these lecture notes.

Example 3.4.10 Let us look at the A_2 singularity. The normalisation map is $s \mapsto (s^3, s^2)$ and the semigroup is the following:

0	1	2	3	4	5	6...
×	○	×	×	×	×	×...

Hence the local ring of A_2 is isomorphic with $\mathbb{C}\{s^2, s^3\} = \mathbb{C}\{x, y\}/(x^3 - y^2)$, and the following list gives all the ideals of this ring:

(i) (1) with $\dim(\mathbb{C}\{s^2, s^3\}/(1)) = 0$.
(ii) (s^k), for $k \geq 2$, and $\dim(\mathbb{C}\{s^2, s^3\}/(s^k)) = k$. For example the ideal (s^2) looks like

0	1	2	3	4	5	6...
×	○	⊗	×	⊗	⊗	⊗...

(iii) (s^k, s^{k+1}), for $k \geq 2$, and $\dim(\mathbb{C}\{s^2, s^3\}/(s^k, s^{k+1})) = k - 1$. For example the ideal (s^3, s^4) looks like

0	1	2	3	4	5	6...
×	○	×	⊗	⊗	⊗	⊗...

In both examples, the dimension of the factor is 2, but there is a distinction, because the first has one generator while the second one has two. The generating series

$\left(\frac{\alpha}{q}\right)^{\mu(A_2)-1} \cdot Z_{A_2}$ equals:

$$\left(\frac{\alpha}{q}\right)^{2-1} \cdot \Big[(1-\alpha^2)(\underbrace{1}_{(i)} + \underbrace{q^4+q^6+q^8+\ldots}_{(ii)})$$

$$+ (1-\alpha^2)^2\underbrace{(q^2+q^4+q^6+\ldots)}_{(iii)}\Big]$$

$$= \frac{\alpha-\alpha^3}{q} \cdot \Big[1 + \frac{q^4}{1-q^2} + \frac{(1-\alpha^2)q^2}{1-q^2}\Big] = \frac{\alpha-\alpha^3}{q-q^3}(1+q^4-\alpha^2 q^2)$$

$$= \frac{\alpha-\alpha^{-1}}{q-q^{-1}} \cdot [\alpha^2(q^2+q^{-2})-\alpha^4] = \frac{\alpha-\alpha^{-1}}{q-q^{-1}} \cdot [\alpha^2(z^2+2)-\alpha^4]$$

$$= P_U \cdot [\alpha^2(z^2+2)-\alpha^4]$$

$$= P_{T_{2,3}}.$$

As mentioned in the introduction, we can not go into the proof of D. Maulik. There are further conjectures of Oblomkov et al. [ORS18] that relate the complete Khovanov homology of the link to invariants of the Hilbert scheme of a plane curve singularity. These further conjectures remain unproven, as far as I know.

References

[Cam73] N. A'Campo, *Sur la monodromie des singularités isolées d'hypersurfaces complexes*, Inventiones mathematicae volume **20**, pp. 147–169 (1973)

[AK80] A. Altman and S. Kleiman, *Compactifying the Picard scheme*, Adv. Math. 35, 50–112, (1980).

[AGV85] Arnold, Gusein-Zade, Varchenko, *Singularities of differentiable maps. Vol. I and II*, Monograph Math. 82/83, Birkhäuser, 1985.

[Bri66] E. Brieskorn, *Beispiele zur Differentialtopologie von Singularitäten*, Inventiones math. **2**, 1–14 (1966).

[BK86] E. Brieskorn, H. Knörrer, *Plane algebraic curves*, Springer, (1986).

[GFvS98] L. Göttsche, B. Fantechi, D. van Straten, *Euler number of the compactified Jacobian and multiplicity of rational curves*, J. of Alg. Geometry, Vol. **8** No. 1, 115–133 (1998).

[BG80] R. Buchweitz, G.-M. Greuel, *The Milnor Number and Deformations of Complex Curve Singularities.*, Inv. Math **58**, 241–282, (1980).

[GP93] G.-M. Greuel, G. Pfister, *Moduli spaces for torsion free modules on curve singularities I*, Journ. Alg. Geom. **2**, 81–135, (1993).

[Mil68] J. Milnor, *Singular points of Complex Hypersurfaces*, Princeton University Press 1968.

[Mau16] D. Maulik, *Stable pairs and the HOMFLY polynomial*, Invent. math. 204, 787–831 (2016).

[Mum75] D. Mumford, *Curves and their Jacobians*, University of Michigan Press, (1975).

[Sco92] A. Scott, *On the Higher Singularities of Plane Curves*, American Journal of Mathematics, Vol. **14**, No. 4, pp. 301–325, (1892).

[Sco93] A. Scott, *The Nature and Effect of Singularities of Plane Algebraic Curves*, American Journal of Mathematics, Vol. **15**, No. 3, pp. 221–243, (1893).

[She12] V. Shende, *Hilbert schemes of points on a locally planar curve and the Severi strata of its versal deformation*, Compositio Math. **148** 531–547, (2012).

[OS12] A. Oblomkov, V. Shende, *The Hilbert scheme of a plane curve singularity and the HOMFLY homology of its link*, Duke Math. J. **161** (7), 1277–1303, (2012).

[ORS18] A. Oblomkov, J. Rasmussen, V. Shende, *The Hilbert scheme of a plane curve singularity and the HOMFLY homology of its link*, Geom. Topol. **22**, 645–691, (2018).

[Reg80] C. J. Rego, *The compactified Jacobian*, Ann. Sci. Ec. Norm. Sup. **13**, 211–223, (1980).

[Tei76] B. Teissier, *The hunting of Invariants in the geometry of Discriminants*, in: *Real and Complex Singularities*, Nordic Summer School, Symposium in Mathematics, Oslo, 1976.

Chapter 4
Lattice Cohomology

András Némethi and Tamás Ágoston

Contents

4.1 Isolated Singular Germs 108
4.1.1 Preliminary 108
4.1.2 Curves in $(\mathbb{C}^2, 0)$ 111
4.2 Lattice Cohomology 119
4.2.1 Combinatorial Setup 119
4.2.2 Overview and Motivation 124
4.2.3 The Case of Isolated Curve Singularities 126
4.2.4 Surface Singularities: The Topological Cohomology 128
References 137

This note is a gentle introduction to the lattice cohomology of isolated complex analytic germs. The analytic version is sensitive to the analytic type of the germs. It can be defined in any positive dimension, in particular for reduced curve singularities as well. Here we will treat mostly the curve case. The topological version is defined for the topological type of an isolated surface singularity with a rational homology sphere link.

These cohomology theories are categorifications of famous numerical invariants. E.g., the Euler characteristic of the lattice cohomology of a reduced curve singularity is its delta invariant.

A. Némethi (✉)
Alfréd Rényi Institute of Mathematics, Budapest, Hungary

Eötvös Loránd University, Budapest, Hungary

Babeş-Bolyai University, Cluj-Napoca, Romania

Basque Center for Applied Math, Bilbao, Spain
e-mail: nemethi.andras@renyi.hu

T. Ágoston
Eötvös Loránd University, Budapest, Hungary
e-mail: tamas.agoston@ttk.elte.hu

© The Author(s), under exclusive license to Springer Nature Switzerland AG 2024
J. Fernández de Bobadilla et al. (eds.), *Singularities and Low Dimensional Topology*, Bolyai Society Mathematical Studies 30,
https://doi.org/10.1007/978-3-031-56611-0_4

In Sect. 4.1 we review some notations and elementary properties of singular analytic germs. In the case of isolated plane curve singularities we compare numerical invariants read from the embedded topological type with invariants read from the abstract analytic type.

In Sect. 4.2 we treat the lattice cohomology. We provide some combinatorial statements and also several examples both in the curve and surface cases.

4.1 Isolated Singular Germs

4.1.1 Preliminary

Definition 4.1.1 Let $f_i : (\mathbb{C}^n, 0) \to (\mathbb{C}, 0)$ be holomorphic germs ($1 \leq i \leq k$), and set $f = (f_1, \ldots, f_k)\colon (\mathbb{C}^n, 0) \to (\mathbb{C}^k, 0)$. Then if $(V(f), 0)$ is defined as

$$(V(f), 0) := \big\{\mathbf{x} \in (\mathbb{C}^n, 0) | f(\mathbf{x}) = 0\big\} \subset (\mathbb{C}^n, 0)$$

we call $(V(f), 0)$ the germ of an analytic set. If $k = 1$ then we call $(V(f), 0)$ a *hypersurface* germ.

At each point $\mathbf{x} \in V(f)$, we can consider the Jacobian

$$\mathrm{d}f(\mathbf{x}) = \left(\frac{\mathrm{d}f_i}{\mathrm{d}x_j}\right)_{i,j} \in \mathbb{C}^{k\times n}.$$

The *dimension* of $(V(f), 0)$ is defined as

$$\dim(V(f), 0) = n - \max_{\mathbf{x}\in V(f)} \mathrm{rk}\,\mathrm{d}f(\mathbf{x}) = n - \mathrm{rk}\,\mathrm{d}f(\mathbf{x}_{gen}),$$

where the second equality also shows that for a generic choice of $\mathbf{x}$, the rank of the Jacobian will be maximal. By definition, points where the rank is *not* maximal form the *singular locus* of $(V(f), 0)$:

$$\mathrm{Sing}(V(f), 0) = \{\mathbf{x} \in (V(f), 0) | \,\mathrm{rk}\,\mathrm{d}f(\mathbf{x}) < n - \dim(V(f), 0)\}.$$

Remark 4.1.2 In the above setting, $\mathrm{codim}\big((V(f), 0) \subset (\mathbb{C}^n, 0)\big) = n - \dim V(f) \leq k$. In particular, for the hypersurface case—i.e. when $k = 1$ and $f \neq 0$—the codimension is 1.

Remark 4.1.3 $\mathrm{Sing}(V(f), 0)$ is also the germ of an analytic set: it is described by the simultaneous vanishing of f and the determinants of all $(\mathrm{codim}\ V) \times (\mathrm{codim}\ V)$ minors of $\mathrm{d}f$.

For hypersurfaces, i.e. $f: (\mathbb{C}^n, 0) \to (\mathbb{C}, 0)$, the point $\mathbf{x} \in (V(f), 0)$ is singular if and only if $\mathrm{d}f(\mathbf{x}) = 0$.

Example 4.1.4 If $f = (xy, xz, yz): \mathbb{C}^3 \to \mathbb{C}^3$, then $V(f) \subset \mathbb{C}^3$ is the union of the coordinate axes. The Jacobian is

$$\mathrm{d}f = \begin{pmatrix} y & x & 0 \\ z & 0 & x \\ 0 & z & y \end{pmatrix},$$

and one can verify that the rank of $\mathrm{d}f$ is 2 for all $(x, y, z) \in V(f)$ except 0 ($\det(\mathrm{d}f) = -2xyz \equiv 0$).

Definition 4.1.5 The analytic germ $(f, 0)$ is said to define an *isolated singularity* when 0 is an isolated point in $\mathrm{Sing}(V(f), 0)$.

Example 4.1.6

- If $f = x_1^2 + \cdots + x_n^2$ then $\mathrm{Sing}(V(f), 0) = \{0\}$, hence $(f, 0)$ is an isolated singularity;
- if $f = x_1^2(x_1^2 + x_2^3)$ then $\mathrm{Sing}(V(f), 0) = \{x_1 = 0\}$, hence $(f, 0)$ is not an isolated singularity.

For a given germ $(V(f), 0) \subset (\mathbb{C}^n, 0)$ with $f = (f_1, \ldots, f_k)$, we can consider its topological and analytic types. The analytic type is characterized by the $\mathbb{C}$-algebra isomorphism type of its coordinate ring $\mathcal{O}_{V,0} = \mathcal{O}_{\mathbb{C}^n,0} \,/\, (f_1, \ldots, f_k)$, where $\mathcal{O}_{\mathbb{C}^n,0} = \mathbb{C}\{z_1, \ldots, z_n\}$ is the (local) ring of holomorphic functions defined near 0 (its unique maximal ideal consists of those germs which vanish at 0) and $(f_1, \ldots, f_k)$ denotes the ideal generated by $f_1, \ldots, f_k$.

Example 4.1.7 Let $f = x^2 - y^3$, $f' = (x^2 - y^3, z - x^{100})$ and consider them at the point 0. Then the two local algebras, $\mathbb{C}\{x, y\}/(x^2 - y^3)$ and $\mathbb{C}\{x, y, z\}/(x^2 - y^3, z - x^{100})$ are isomorphic, hence f and f' define analytically equivalent curve singularities at 0.

Regarding the topological type, we will distinguish two notions. The abstract topological type is the topology of $V(f)$ in some sufficiently small neighborhood of 0, while the embedded topology is the topology of the pair of $V(f) \subset \mathbb{C}^n$ within some sufficiently small neighborhood of 0. In order to formulate these statements more precisely, the following facts are crucial:

Theorem 4.1.8 (See [Mil69, Nem22] and the References Therein) *Let $(f, 0)$ be an analytic germ and assume that $(V, 0) = (V(f), 0) \subset (\mathbb{C}^n, 0)$ defines an isolated singularity. Let also S_r^{2n-1} and B_r^{2n} denote the sphere and ball in $\mathbb{C}^n$ centered at 0* *with radius r. Then for some sufficiently small* ε_0*:*

(1) $S^{2n-1}_{\varepsilon} \pitchfork V$ *for every* $0 < \varepsilon \leq \varepsilon_0$.
(2) There is a homeomorphism $\Phi\colon (B^{2n}_{\varepsilon_0}, V \cap B^{2n}_{\varepsilon_0}) \to \big(C(S^{2n-1}_{\varepsilon_0}), C(V \cap S^{2n-1}_{\varepsilon_0})\big)$ *where* $C(X)$ *denotes the real cone over X. This map is a diffeomorphism away from 0, and takes the sections* $V \cap S^{2n-1}_{\varepsilon}$ *parametrized by* ε *to sections of the cone* $C(V \cap S^{2n-1}_{\varepsilon_0})$ *parallel to the base.*

Definition 4.1.9 In the above setting, the intersections $V \cap S^{2n-1}_{\varepsilon}$ are all diffeomorphic for sufficiently small ε. We call this the *(abstract) link* L_V of $(V, 0)$.

Likewise, the diffeomorphism type of the pair $(S^{2n-1}_{\varepsilon}, V \cap S^{2n-1}_{\varepsilon})$ is constant for sufficiently small ε. We call this the *embedded link* of $(V, 0)$ into S^{2n-1}_{ε}.

Remark 4.1.10 Because of the cone structure, the local topological type of $(V, 0)$ is given by the homeomorphism type of the abstract link, while the embedded topological type of $(\mathbb{C}^n, V, 0)$ is equivalent to the topological type of the embedded link respectively. In particular, when talking about the (abstract) topological and embedded topological types, in fact we might consider the respective links.

The analytic type (or the embedded topological type) determines the abstract topological type.

Remark 4.1.11 If $\dim(V(f), 0) = d$, then V has dimension $\dim_{\mathbb{R}}(V(f), 0) = 2d$ over $\mathbb{R}$. Hence, by the transversality property from Theorem 4.1.8, the link L_V is a compact, oriented, smooth real manifold of dimension $2d - 1$, embedded in the sphere S^{2n-1}.

Example 4.1.12 Let $(V, 0)$ be an isolated plane curve singularity, that is, $\dim(V, 0) = 1$ and $(V, 0) \subset (\mathbb{C}^2, 0)$. Then the abstract link L_V is a compact real 1-manifold, i.e. a disjoint union of circles. The number of such circles agrees with the number of irreducible components of the curve germ $(V, 0)$ (for more details see below), and we obtain the embedded link as well:

$$L_V = \bigsqcup_{i=1}^{r} S^1 \hookrightarrow S^3.$$

Example 4.1.13 If $\dim(V, 0) = 2$ then $(V, 0)$ is called a surface singularity. Then L_V is a compact, oriented real 3-manifold. In fact, it is a plumbed 3-manifold associated with a plumbing graph whose intersection form is negative definite. If $(V, 0)$ is irreducible then the graph is connected.

Theorem 4.1.14 (Grauert [Gra71]) *Every oriented plumbed 3-manifold with a negative definite intersection form is the link of a surface singularity.*

Example 4.1.15 For the following singularities, the plumbing graphs are the following:

- $\{x^2 + y^2 + z^2 = 0\}$ $\longrightarrow$ 

- $\{x^2 + y^2 + z^{n+1} = 0\}$ $\longrightarrow$ 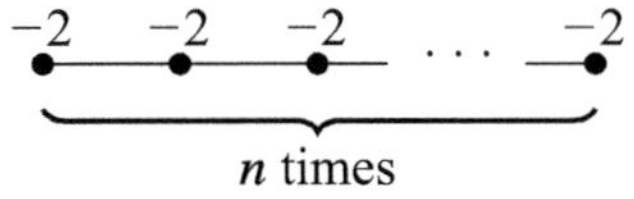

- $\{x^2 + y^3 + z^5 = 0\}$ $\longrightarrow$ 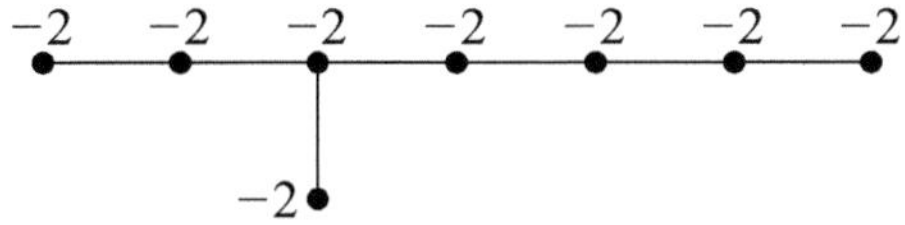

4.1.2 Curves in $(\mathbb{C}^2, 0)$

In this section, we assume $f : (\mathbb{C}^2, 0) \to (\mathbb{C}, 0)$. For such an $f \in \mathcal{O}_{\mathbb{C}^2,0} = \mathbb{C}\{x, y\}$, we have a decomposition $f_1^{\alpha_1} \cdots f_r^{\alpha_r}$, where each f_i is an irreducible germ and $f_i \neq f_j$ $(i \neq j)$ up to invertible elements of $\mathbb{C}\{x, y\}$. This induces the irreducible decomposition $(V(f), 0) = \bigcup_{i=1}^{r} (V(f_i), 0)$. Correspondingly, we get the embedded link

$$L_f = \bigsqcup_{i=1}^{r} L_{f_i} \approx \bigsqcup_{i=1}^{r} S^1 \hookrightarrow S^3,$$

where L_f and L_{f_i} denote the links of $(V(f), 0)$ and $(V(f_i), 0)$ respectively. The germ defines an isolated singularity if and only if $\alpha_i = 1$ for all i. Such isolated singularities are also called *plane curve singularities*.

Fact 4.1.16 ([Zar71]) *The embedded topological type of an isolated plane curve singularity is completely classified by the following data:*

- *The embedded topological types of each component L_{f_i} of the link. This is an iterative torus knot, encoded by the Puiseux pairs of f_i.*
- *The linking numbers between the components L_{f_i} and L_{f_j} $(i \neq j)$. Algebraically, the linking number of L_{f_i} and L_{f_j} $(i \neq j)$ equals the (local) intersection multiplicity at the origin of f_i and f_j.*

A notable property of the embedded topological type is the following:

Theorem 4.1.17 (Milnor [Mil69]) *For a sufficiently small $\varepsilon > 0$, the map $\Phi\colon S^3_\varepsilon \setminus V(f) \to S^1$, $\Phi(z) = \frac{f(z)}{|f(z)|}$ ($z = (x, y)$) defines a C^∞ locally trivial fibration (called the Milnor fibration):*

$$\begin{array}{ccc} F & \longrightarrow & S^3_\varepsilon \setminus L_f \\ & & \downarrow \Phi \\ & & S^1 \end{array}$$

In particular, each $L_f \subset S^3_\varepsilon$ is a fibered link. The Milnor fiber F is homotopically equivalent to a bouquet of S^1's, and its first Betti number is equal to the Milnor number $\mu(f) = \mu(f, 0)$:

$$F \cong \bigvee_{i=1}^{\mu(f,0)} S^1 \quad \textit{where} \quad \mu(f,0) = \dim_{\mathbb{C}} \frac{\mathcal{O}_{\mathbb{C}^2,0}}{\left(\frac{\mathrm{d}f}{\mathrm{d}x}, \frac{\mathrm{d}f}{\mathrm{d}y}\right)}.$$

Remark 4.1.18 A similar statement holds for isolated hypersurface singularities in any dimension ($F = \vee S^{n-1}$, $b_{n-1}(F) = \mu(f, 0) = \dim_{\mathbb{C}}(\mathcal{O}_{\mathbb{C}^n,0}/(\mathrm{d}f))$), see [Mil69].

As for understanding certain numerical invariants of the analytic type of a plane curve singularity, consider first the irreducible case.

4.1.2.1 Irreducible Plane Curve Singularities

Proposition 4.1.19 (See e.g. [BK86, Wal04]) *Let f define an irreducible plane curve singularity $(V(f), 0) \hookrightarrow (\mathbb{C}^2, 0)$. Then there exists an analytic homeomorphism $n\colon (\mathbb{C}, 0) \to (V(f), 0)$, also called the Puiseux parametrization, or the normalization of $\big(V(f), 0\big)$. Furthermore, this parametrization $n\colon t \mapsto \big(x(t), y(t)\big)$ can be written in the form*

$$\begin{aligned} x(t) &= t^m, \\ y(t) &= a_n t^n + a_{n+1} t^{n+1} + \cdots, \quad (a_i \in \mathbb{C},\ a_n \neq 0) \end{aligned}$$

such that $f\big(x(t), y(t)\big) \equiv 0$.

Example 4.1.20 For $f = x^a - y^b$ with a, b coprime (which implies that f is irreducible), we have

$$t \mapsto \begin{pmatrix} x(t) = t^b \\ y(t) = t^a \end{pmatrix}.$$

For any analytic map $g \in \mathcal{O}_{\mathbb{C}^2,0}$, we can then take the pullback $n^*g \in \mathcal{O}_{\mathbb{C},0} = \mathbb{C}\{t\}$: the bijection n induces $n^*\colon \mathcal{O}_{V,0} \hookrightarrow \mathcal{O}_{\mathbb{C},0} = \mathbb{C}\{t\}$, and we can apply that to $g|_V$. Using diagrams:

$$\begin{array}{ccccc} (\mathbb{C},0) & \xrightarrow{n} & \big(V(f),0\big) & \hookrightarrow & (\mathbb{C}^2,0) \\ & \searrow^{n^*g} & \downarrow{\scriptstyle g|_V} & \swarrow_{g} & \\ & & (\mathbb{C},0) & & \end{array}$$

Definition 4.1.21 Define the *δ-invariant* of $(f,0)$ as

$$\dim_{\mathbb{C}} \mathcal{O}_{\mathbb{C},0}/\operatorname{Im} n^* = \dim_{\mathbb{C}} \mathbb{C}\{t\}/\mathcal{O}_{V,0} < \infty,$$

and its *semigroup* (or *monoid*) *of values* as

$$\mathcal{S} = \left\{\operatorname{ord}_t n^*g | g \in \mathcal{O}_{V,0}\right\} \subset \mathbb{Z}_{\geq 0}.$$

Note that this satisfies

$$\delta = \dim \frac{\mathbb{C}\{t\}}{\mathcal{O}_{V,0}} = \left|\mathbb{Z}_{\geq 0} \setminus \mathcal{S}\right| < \infty.$$

In particular, there exists a smallest element $c \in \mathbb{Z}_{\geq 0}$ such that $c + \mathbb{Z}_{\geq 0} \subset \mathcal{S}$, which is called the *conductor* element of the semigroup $\mathcal{S}$.

Example 4.1.22 Let $f = x^3 - y^4$. Then, via Example 4.1.20, we get

$$n^*g(t) = g\big(x(t), y(t)\big) = g(t^4, t^3)$$

for any $g \in \mathbb{C}\{x, y\}$. In particular, for $g = x^k y^\ell$, we get $n^*g(t) = t^{4k+3\ell}$, so $\mathcal{S} \supseteq \langle 3, 4\rangle$. One can also verify that this is in fact an equality.

Example 4.1.23 More generally, for $f = x^a - y^b$ with a, b coprime, we get

$$\mathcal{S} = \langle a, b\rangle\,, \quad \delta = \frac{(a-1)(b-1)}{2}, \quad c = (a-1)(b-1).$$

As for the topological type, in this case the abstract link is S^1 (since f is irreducible), and it is embedded in S^3 as the torus knot $T_{a,b}$. The Milnor number is $\mu = (a-1)(b-1)$ too.

Example 4.1.24 Consider the singularity parametrised as $n\colon t \mapsto (t^4, t^6 + t^7)$. Then $n^*g(t) = g(t^4, t^6 + t^7)$, and one can show that

$$\mathcal{S} = \langle 4, 6, 13\rangle\,, \quad \delta = 8, \quad c = 16.$$

Exercise 4.1.25 Find a polynomial f that corresponds to this parametrization, and compute the Milnor number as well.

One can observe that in all these examples, we always have $c = 2\delta = \mu$. This is indeed the case when f is an irreducible plane curve germ. For the general statement see Theorem 4.1.39.

It is a notable fact that the semigroup can be computed from the embedded topological type. In fact, the semigroup is a complete embedded topological invariant, i.e. it always distinguishes singularities with different embedded topological types (i.e. different embedded algebraic links).

Another complete invariant of irreducible isolated plane curve singularities is the Alexander polynomial $\Delta(t)$ of the embedded link $S^1 \subset S^3_\varepsilon$. The connection between the two can be expressed as follows:

Proposition 4.1.26 (Campillo et al. [CDG99]) *For an irreducible plane curve singularity* $(V, 0) \subset (\mathbb{C}^2, 0)$ *with an associated semigroup* $\mathcal{S}$, *we have*

$$\sum_{s \in \mathcal{S}} t^s = \frac{\Delta(t)}{1-t}.$$

Remark 4.1.27 The Alexander polynomial is not a complete invariant of knots in S^3 in general (for not necessarily algebraic knots). It does, however distinguish between all links of irreducible plane curve singularities.

It can be immediately seen then that $\mathcal{S}$ and Δ can indeed be computed from each other. Another complete invariant is the Hilbert series $H(t)$. Its equivalence with the semigroup can be seen from its very definition.

Definition 4.1.28 For an irreducible plane curve singularity $(V, 0)$ with semigroup $\mathcal{S}$, we can define its *Hilbert function* $\ell \mapsto \mathfrak{h}(\ell)$ and *Hilbert series* $H(t)$ respectively as

$$\mathfrak{h}(\ell) = |\{s \in \mathcal{S} | s < \ell\}| \quad (\ell \in \mathbb{Z}_{\geq 0}) \quad \text{and} \quad H(t) = \sum_{\ell \geq 0} \mathfrak{h}(\ell) t^\ell.$$

Proposition 4.1.29 *With the above notations,*

$$H(t) = \frac{t}{1-t} \sum_{s \in \mathcal{S}} t^s = \frac{t\Delta(t)}{(1-t)^2}.$$

Remark 4.1.30 The function $\mathfrak{h}$ *is* actually the Hilbert function of a filtration. We will talk about that in detail later in this section.

Example 4.1.31 Consider the plane curve singularity defined by $f = x^2 - y^3$. Recall from Example 4.1.23 that its embedded link is the torus knot $T_{2,3}$ (the trefoil), and the semigroup is $\mathcal{S} = \langle 2, 3\rangle = \mathbb{Z}_{\geq 0} \setminus \{1\}$ (with $\delta = 1$ and $c = 2$). Then

$$\sum_{s\in\mathcal{S}} t^s = 1 + t^2 + t^3 + t^4 + \cdots = 1 + \frac{t^2}{1-t} = \frac{1-t+t^2}{1-t},$$

and $1 - t + t^2$ is indeed the Alexander polynomial of the knot $T_{2,3}$. Also, we get

$$\begin{aligned} H(t) &= \sum_{\ell\geq 0} \mathfrak{h}(\ell)t^\ell = t + t^2 + 2t^3 + 3t^4 + \cdots \\ &= t + \frac{t^2}{(1-t)^2} = \frac{t - t^2 + t^3}{(1-t)^2} = \frac{t\Delta(t)}{(1-t)^2}. \end{aligned}$$

4.1.2.2 General Curves

Next, we define the delta invariant and the Hilbert function for non-irreducible curve germs as well. In fact, in this subsection we will consider an arbitrary isolated curve singularity (which is not necessarily embedded in $(\mathbb{C}^2, 0)$). In particular the next definition will not use any embedded data. (On the contrary, for the definition of the Milnor number μ one needs an embedding in $(\mathbb{C}^2, 0)$), or at least a smoothing.)

Now, write $(V, 0) = \bigcup_{i=1}^{r} (V_i, 0)$, where $(V_i, 0) \subset (\mathbb{C}^n, 0)$ are irreducible curve germs.

Consider the normalizations $n_i : (\mathbb{C}_{t_i}, 0) \to (V_i, 0)$ to get the mapping $\mathcal{O}_{\mathbb{C}^n,0} \ni g \mapsto n_i^* g \in \mathbb{C}\{t_i\}$ induced via the diagram

$$\begin{array}{ccccccc} (\mathbb{C}_{t_i}, 0) & \xrightarrow{n_i} & (V_i, 0) & \hookrightarrow & (V, 0) & \hookrightarrow & (\mathbb{C}^n, 0) \\ & \searrow^{n_i^* g} & & & \downarrow{\scriptstyle g|_V} & \swarrow_{g} & \\ & & & & (\mathbb{C}, 0) & & \end{array}$$

and the valuations $(\mathrm{ord}_{t_1} n_1^* g, \ldots, \mathrm{ord}_{t_r} n_r^* g)$ on $\mathcal{O}_{\mathbb{C}^n,0}$. Since here $n_i^* g$ only depends on $g|_V$, we might as well equivalently define $n_i^* g$ on $\mathcal{O}_{V,0}$ via the left part of the diagram. Group the valuations as

$$\mathfrak{v}(g) = (\mathrm{ord}_{t_1} n_1^* g, \ldots, \mathrm{ord}_{t_r} n_r^* g) \quad (g \in \mathcal{O}_{V,0}).$$

This induces a $(\mathbb{Z}_{\geq 0})^r$-filtration on $\mathcal{O}_{V,0}$, which in turn gives rise to a Hilbert function (and series):

$$\mathcal{F}(\ell) = \left\{ g \in \mathcal{O}_{V,0} | \forall i : \mathrm{ord}_{t_i} n_i^* g \geq \ell_i \right\} \subset \mathcal{O}_{V,0}, \quad (\ell = (\ell_1, \ldots, \ell_r)),$$

$$\mathfrak{h}(\ell) = \dim \mathcal{O}_{V,0}/\mathcal{F}(\ell),$$

$$H(\mathbf{t}) = \sum_{\ell \geq 0} \mathfrak{h}(\ell)\mathbf{t}^{\ell} \quad (\mathbf{t}^{\ell} = t_1^{\ell_1} \cdots t_r^{\ell_r}),$$

and provides an injective morphism $n^* : \mathcal{O}_{V,0} \hookrightarrow \bigoplus_{i=1}^{r} \mathbb{C}\{t_i\}$.

Remark 4.1.32 Observe also that the definition of $\ell \mapsto \mathfrak{h}(\ell)$ does not depend at all on the embedding (realization) of $(V, 0)$ in some $(\mathbb{C}^n, 0)$, it depends only the local algebra $\mathcal{O}_{V,0}$.

Definition 4.1.33 We define the *δ-invariant* of $(V, 0)$ as

$$\dim_{\mathbb{C}} \bigoplus_{i=1}^{r} \mathbb{C}\{t_i\}/\mathrm{Im}\, n^* < \infty,$$

and its *semigroup of values* as

$$\mathcal{S} = \left\{\mathfrak{v}(g) | g \in \mathcal{O}_{V,0}\right\} = \left\{(\mathrm{ord}_{t_1} n_1^* g, \ldots, \mathrm{ord}_{t_r} n_r^* g) | g \in \mathcal{O}_{V,0}\right\} \subset (\mathbb{Z}_{\geq 0})^r.$$

There exists a unique smallest element $c \in (\mathbb{Z}_{\geq 0})^r$ such that $c + (\mathbb{Z}_{\geq 0})^r \subset \mathcal{S}$, which is called the *conductor* element of $\mathcal{S}$.

Example 4.1.34 Let $(V, 0) = \{xy = 0\} \subset (\mathbb{C}^2, 0)$. Then $n_1(t_1) = (t_1, 0)$ and $n_2(t_2) = (0, t_2)$, so for some $g(x, y)$, we get

$$n_1^* g = g(t_1, 0) \in \mathbb{C}\{t_1\}, \quad n_2^* g = g(0, t_2) \in \mathbb{C}\{t_2\}.$$

One can then verify that $\delta = 1$, $\mu = 1$, $c = (1, 1)$, and $\mathcal{S} = (0, 0) \cup ((1, 1) + (\mathbb{Z}_{\geq 0})^2)$.

Example 4.1.35 Let $(V, 0) = \{xy = yz = zx = 0\} \subset (\mathbb{C}^3, 0)$, i.e. the union of the 3 coordinate axes. Then we get the maps $n_1(t_1) = (t_1, 0, 0)$, $n_2(t_2) = (0, t_2, 0)$, $n_3(t_3) = (0, 0, t_3)$. We can then compute $\delta = 2$.

Example 4.1.36 Let $(V, 0) = \{xy(x - y) = 0\} \subset (\mathbb{C}^2, 0)$. We leave as an exercise to the reader to show that $\delta = 3$ in this case.

Note however, that while this singularity has the same abstract topologically type as the germ from the previous example (the link is $S^1 \sqcup S^1 \sqcup S^1$) and even 'geometrically' they are very similar (3 lines, each pair of them intersecting in the same point), the δ-invariants differs. Indeed the two are *not* analytically isomorphic. E.g., $\{xy = yz = zx = 0\}$ cannot be embedded into $(\mathbb{C}^2, 0)$.

Example 4.1.37 Let $(V, 0) = \left\{x^2 + y^4 = 0\right\} \subset (\mathbb{C}^2, 0)$. Again, we leave as an exercise to check that $\mu = 3$ and $\delta = 2$ and $r = 2$.

The semigroup $\mathcal{S}$ and the Hilbert function determine each other:

Lemma 4.1.38 (See e.g. [GN15]) *The semigroup can be deduced from the Hilbert function as follows:*

$$\mathcal{S} = \{\ell \in (\mathbb{Z}_{\geq 0})^r \mid \mathfrak{h}(\ell + E_i) > \mathfrak{h}(\ell) \text{ for every } i = 1, \ldots, r\}.$$

On the other hand, $\mathfrak{h}(\ell + E_i) - \mathfrak{h}(\ell) \in \{0, 1\}$ *for any* $\ell \geq 0$ *and* $i \in \{1, \ldots, r\}$. *Moreover,* $\mathfrak{h}(\ell + E_i) = \mathfrak{h}(\ell) + 1$ *if there is an element* $s \in \mathcal{S}$ *such that* $s_i = \ell_i$ *and* $s_j \geq \ell_j$ *for* $j \neq i$. *Otherwise* $\mathfrak{h}(\ell + E_i) = \mathfrak{h}(\ell)$. *(Here* $\{E_i\}_{i=1}^r$ *denotes the standard basis of* $\mathbb{Z}^r$.*)*

4.1.2.3 The Case of Plane Curve Singularities Revisited

We have the following two statements valid specifically for plane curve singularities $(V(f), 0) \subset (\mathbb{C}^2, 0)$. The first one relates the delta invariant with the Milnor number.

Theorem 4.1.39 (Jung, Milnor, See [Mil69]) *For a plane curve singularity* $(f, 0)$, *we have* $\mu(f) = 2\delta(f) - (r - 1)$ *where* r *is the number of irreducible components of* $(V(f), 0)$.

The second group of statements relates the Hilbert function (or, equivalently, the semigroup) with the multivariable Alexander polynomial $\Delta(\mathbf{t})$ of the link (here $(\mathbf{t} = (t_1, \ldots, t_r))$).

The key intermediate bridge is the multivariable Poincaré series. First define the *extended* Hilbert series as

$$H^{ext}(\mathbf{t}) = \sum_{\ell \in \mathbb{Z}^r} \mathfrak{h}(\ell)\mathbf{t}^\ell \in \mathbb{Z}[[t_1, t_1^{-1}, \ldots, t_r, t_r^{-1}]],$$

where for an arbitrary $\ell \in \mathbb{Z}^r$ we set $\mathfrak{h}(l) = \mathfrak{h}(\max\{l, 0\})$. Note that $H^{ext}|_{(\mathbb{Z}_{\geq 0})^r} = H$.

Then we define the multivariable Poincaré series $P(\mathbf{t})$ as follows. If $r = 1$, then the Poincaré series of the graded ring $\oplus_l \mathcal{F}(l)/\mathcal{F}(l + E_1)$ is $P(t) = -H(t)(1 - t^{-1})$. For general r, one defines the Poincaré series by

$$P(t_1, \ldots, t_r) := -H^{ext}(t_1, \ldots, t_r) \cdot \prod_i (1 - t_i^{-1}).$$

This means that the coefficient $\mathfrak{p}(\ell)$ of $P(\mathbf{t}) = \sum_\ell \mathfrak{p}(\ell) \cdot t_1^{\ell_1} \ldots t_r^{\ell_r}$ satisfies

$$\mathfrak{p}(\ell) = \sum_{I \subset \{1, \ldots, r\}} (-1)^{|I|-1} \mathfrak{h}(\ell + E_I), \quad \Big(E_I = \sum_{i \in I} E_i\Big).$$

$\mathbb{Z}[[t_1, t_1^{-1}, \ldots, t_r, t_r^{-1}]]$ is a module over the ring of Laurent power series, hence the multiplication by $\prod_i (1-t_i^{-1})$ is well-defined. One can check (using the property $\mathfrak{h}(l) = \mathfrak{h}(\max\{l, 0\})$) that $P(\mathbf{t})$ is a power series involving only nonnegative powers of t_i. In fact, the support of $P(\mathbf{t})$ is included in $\mathcal{S}$, that is, $\mathfrak{p}(\ell) = 0$ whenever $\ell \notin \mathcal{S}$.

If $r = 1$, then by Lemma 4.1.38 $P(t) = \sum_{s\in\mathcal{S}} t^s = -\sum_{s\notin\mathcal{S}} t^s + 1/(1-t)$, where $-\sum_{s\notin\mathcal{S}} t^s$ is a polynomial. Furthermore, $P(\mathbf{t})$ is a polynomial for $r > 1$ (see Theorem 4.1.40 below).

The following result motivated the introduction of $P(\mathbf{t})$, it creates the bridge between the abstract analytic invariant $H(\mathbf{t})$ and the embedded topological invariant $\Delta(\mathbf{t})$.

Theorem 4.1.40 (Campillo, Delgado, Gusein-Zade, See [CDG02, CDG03, CDG07]) *The following identity holds:*

$$P(\mathbf{t}) = \begin{cases} \Delta(\mathbf{t}), & \text{if } r > 1, \\ \Delta(\mathbf{t})/(1-t), & \text{if } r = 1. \end{cases}$$

Multiplication by $\prod_i (1 - t_i^{-1})$ of series with $\mathfrak{h}(0) = 0$ is injective if $r = 1$, however it is not if $r > 1$. In particular, in such cases it can happen that for two different series $H(\mathbf{t})$ we obtain the very same $P(\mathbf{t})$. (This really can happen for concrete isolated curve singularities, which have higher embedded dimensions, that is for non-plane curve singularities, see [CDG07].) Nevertheless, for plane curve singularities even for $r > 1$ one can recover $H(\mathbf{t})$ as follows.

Using this, we write the multivariate analogue of Proposition 4.1.29:

Theorem 4.1.41 (Gorsky–Némethi [GN15]) *For a plane curve singularity* $(V, 0) = \bigcup_{i=1}^{r} (V_i, 0)$, *with the above notations we have*

$$H(\mathbf{t}) = \frac{1}{\prod_{i=1}^{r}(1-t_i)} \Big(\sum_{\emptyset \neq I \subseteq \{1,\ldots,r\}} (-1)^{|I|+1} \Big(\prod_{i\in I} t_i\Big) P_{V_I}(\mathbf{t}_I) \Big)$$

where $\mathbf{t} = (t_1, \ldots, t_r)$ *and for* $\emptyset \neq I = \{i_1, \ldots, i_s\} \subseteq \{1, \ldots, r\}$, *we denote* $(V_I, 0) = \bigcup_{i\in I}(V_i, 0)$, $\mathbf{t}_I = (t_{i_1}, \ldots, t_{i_s})$, *and* P_{V_I} *is the Poincaré series of* $(V_I, 0)$ *in the variables* $t_{i_1}, \ldots, t_{i_s}$.

The multivariable Alexander polynomials (hence the multivariable Poincaré series/polynomials) can be directly computed from the splice diagrams of the pair $L_V \subset S^3$ (or from the embedded resolution graph of $V, 0) \subset (\mathbb{C}^2, 0))$, see [EN85]. In this way we can compute $H(\mathbf{t})$ as well. We emphasize again:

Proposition 4.1.42 *For any isolated plane curve isolated singularity* $(V, 0)$, *its semigroup, Hilbert and Poincaré series, as well as the multivariate Alexander* *polynomial of the link are all complete embedded topological invariants. For multivariable Poincaré series see also [Moy08] and [Nem08b].*

4.2 Lattice Cohomology

4.2.1 Combinatorial Setup

The lattice cohomology was originally introduced as a way to construct a particular topological invariant for normal surface singularities. In recent years, however, the method was generalized and used to study the analytic structure of singularities in any dimension as well. Let us take a look at the general, abstract setup first. We follow [Nem08a, Nem22], see also [AN23].

Goal 4.2.1 *Given some sort of combinatorial data—that will, ultimately, come from some topological or analytic information—we aim to construct a bigraded* $\mathbb{Z}[U]$*-module* $\mathbb{H}^* = \bigoplus_{q\geq 0} \mathbb{H}^q$ *where*

- $\mathbb{H}^q$ *is a* $\mathbb{Z}$*-graded* $\mathbb{Z}[U]$*-module,*
- *for every* $x \in \mathbb{H}^q$*, there is some* $n(x) \in \mathbb{Z}_{\geq 0}$ *such that* $U^{n(x)} \cdot x = 0$.

Here $\mathbb{Z}[U]$ *is the polynomial ring, with* U *denoting the free variable.*

Example 4.2.2 Some particular $\mathbb{Z}$-graded $\mathbb{Z}[U]$-modules:

- $\mathcal{T}_0^+ = \mathbb{Z}[U, U^{-1}] \,/\, U\mathbb{Z}[U] = \mathbb{Z}\langle 1, U^{-1}, U^{-2}, \ldots\rangle$ with $\deg U^{-k} = 2k$,
- $\mathcal{T}_{2n}^+ = \mathbb{Z}[U, U^{-1}] \,/\, U^{-n+1}\mathbb{Z}[U] = \mathbb{Z}\langle U^{-n}, U^{-n-1}, U^{-n-2}, \ldots\rangle$ (with the same grading),
- $\mathcal{T}_{2n}^+(m) = U^{-n-m+1}\mathbb{Z}[U] \,/\, U^{-n+1}\mathbb{Z}[U] = \mathbb{Z}\langle U^{-n}, U^{-n-1}, \ldots, U^{-n-m+1}\rangle$ (with the same grading and with $\mathbb{Z}$-rank m).

In each of these cases, multiplication with U is a degree-(-2) morphism, which together with the grading being bounded from below implies in particular that the second condition above is satisfied.

We should also introduce some further notations that we will use extensively. Assume that $\{E_i\}_{i=1}^r$ is a basis of $\mathbb{R}^r$.

Definition 4.2.3

(a) For a set $I \subseteq \{1, \ldots, r\}$, let $E_I = \sum_{i\in I} E_i$.
(b) For $a, b \in \mathbb{R}^r$ with $a \leq b$, let $R(a, b) = \{x \in \mathbb{R}^r | a \leq x \leq b\}$ denote the (hyper)rectangle spanned by a and b with edges parallel to the coordinate axes. We will also allow $a, b \in (\mathbb{R} \cup \{\pm\infty\})^r$.

Next, we present the construction.

Definition 4.2.4 Consider the lattice $L = \mathbb{Z}^r = \mathbb{Z}\langle E_i\rangle_{i=1}^r \hookrightarrow \mathbb{Z}^r \otimes_{\mathbb{Z}} \mathbb{R} = \mathbb{R}^r$, which induces a cubical cellular decomposition of $\mathbb{R}^r$:

- 0-cubes: $\ell \in L$ lattice points (technically the singleton sets $\{\ell\}$, but for convenience we identify these with the lattice points themselves when it causes no confusion),
- 1-cubes: $R(\ell, \ell + E_i)$ segments, where $\ell \in L$ and $i \in \{1, \ldots, r\}$,
- q-cubes: $R(\ell, \ell + E_I)$ cubes, where $l \in L$ and $I \subset \{1, \ldots, r\}$, $|I| = q$.

We denote the set of q-cubes by Q_q.

On this cubical complex we introduce a set of *compatible weight functions*:

- $w_0 \colon Q_0 = \mathbb{Z}^r \to \mathbb{Z}$ such that $\forall n : \left|w_0^{-1}\big((-\infty, n)\big)\right| < \infty$ (in particular there exists some $m_w = \min w_0 \in \mathbb{Z}$),
- $w_q \colon Q_q \to \mathbb{Z}$ for any q-cell $\square$, where $w_q(\square) = \max_{\ell \in L \cap \square} w_0(\ell)$,
- $w = \bigcup_{q \geq 0} w_q$.

(We could allow more general configurations here, but this will suffice for now; the interested reader is invited to check e.g. [Nem22].)

For every $n \in \mathbb{Z}$, consider the 'level set'

$$S_n = \bigcup_{w(\square) \leq n} \square,$$

which gives us a chain of inclusions of finite cubical spaces

$$\emptyset \hookrightarrow S_{m_w} \hookrightarrow S_{m_w+1} \hookrightarrow \cdots .$$

Due to the way the weights were defined, we can view the weights as a $\mathbb{Z}$-filtration on the cubical complex: cubes $\square$ with $w(\square) \leq n$ form the subcomplex S_n.

Definition 4.2.5 Given the above setup, we define the graded $\mathbb{Z}$-module

$$\mathbb{H}^q(L, w) = \bigoplus_{n \geq m_w} H^q(S_n, \mathbb{Z}),$$

where $\mathbb{H}^q(L, w)$ is $\mathbb{Z}$-garded in such a way that $H^q(S_n, \mathbb{Z})$ is the homogeneous component of degree $2n$. Together with a U-action

$$0 \xleftarrow{\cdot U} H^q(S_{m_w}, \mathbb{Z}) \xleftarrow{\cdot U} H^q(S_{m_w+1}, \mathbb{Z}) \xleftarrow{\cdot U} \cdots,$$

induced by the inclusion maps, we get a graded $\mathbb{Z}[U]$-module structure as well, just as we wanted. Finally, set $\mathbb{H}^*(L, w) = \bigoplus_{q \geq 0} \mathbb{H}^q(L, w)$. It is called the *lattice cohomology* associated with the pair (L, w). (We will often simply write $\mathbb{H}^q$ and $\mathbb{H}^*$ if the meaning is clear from the context.)

Definition 4.2.6 We define the *reduced lattice cohomology* as well. For $q = 0$ we set

$$\bigoplus_{n \geq m_w} H^0(S_n, \mathbb{Z}) = \bigoplus_{n \geq m_w} \left(\mathbb{Z} \oplus \widetilde{H}^0(S_n, \mathbb{Z})\right)$$

with its natural $\mathbb{Z}$-graded $\mathbb{Z}[U]$-module structure. This gives us a splitting of graded $\mathbb{Z}[U]$-modules

$$\mathbb{H}^0 = \mathcal{T}^+_{2m_w} \oplus \mathbb{H}^0_{red}, \quad \text{where } \mathbb{H}^0_{red} := \bigoplus_{n \geq m_w} \widetilde{H}^0(S_n, \mathbb{Z}).$$

We then set $\mathbb{H}^*_{red} = \mathbb{H}^0_{red} \oplus \bigoplus_{q>0} \mathbb{H}^q$.

Remark 4.2.7 For all $q \geq r$, we have $\mathbb{H}^q = 0$.

Definition 4.2.8 The same construction can be done for just a smaller part of the lattice, indeed, for any subcomplex of the cubical complex. The most oft-occurring cases are the restrictions to a (possibly infinite) rectangle $R(a, b)$, in particular to $(\mathbb{R}_{\geq 0})^r$.

If we restrict an already given weight function w to a smaller subcomplex R, we will use the notation $\mathbb{H}^*(L, w)|_R = \mathbb{H}^*(L \cap R, w|_R)$. In fact, there also always exists a graded $\mathbb{Z}[U]$-module morphism $\mathbb{H}^*(L, w) \to \mathbb{H}^*(L, w)|_R$.

Example 4.2.9 Let $r = 1$, and consider just $L_{\geq 0} = \mathbb{Z}_{\geq 0} \subset \mathbb{R}_{\geq 0}$. If $w_0 \colon \mathbb{Z}_{\geq 0} \to \mathbb{Z}$ is an increasing function then all S_n are of the form $R(0, a)$, in particular contractible. Hence, $\mathbb{H}^*_{red} = 0$.

Example 4.2.10 Again, let $L_{\geq 0} = \mathbb{Z}_{\geq 0} \subset \mathbb{R}_{\geq 0}$. Now consider a monoid $\mathcal{S} \subset \mathbb{Z}_{\geq 0}$ (a semigroup with $0 \in \mathcal{S}$) such that $\delta = \left|\mathbb{Z}_{\geq 0} \setminus \mathcal{S}\right| < \infty$ (cf. Definition 4.1.21), and define

$$\mathfrak{h}(\ell) = |\{a \in \mathcal{S} | a < \ell\}|,$$

$$\overline{\mathfrak{h}}(\ell) = \left|\{a \in \mathbb{Z}_{\geq 0} \setminus \mathcal{S} | a < \ell\}\right| = \ell - \mathfrak{h}(\ell),$$

$$w_0(\ell) = \mathfrak{h}(\ell) - \overline{\mathfrak{h}}(\ell) = 2\mathfrak{h}(\ell) - \ell.$$

Note that for every ℓ, we have $w_0(\ell + 1) - w_0(\ell) = \pm 1$ depending on whether $\ell \in \mathcal{S}$ or not, and due to $\left|\mathbb{Z}_{\geq 0} \setminus \mathcal{S}\right| < \infty$, the weight function w_0 becomes strictly increasing after a point. Hence in particular w_0 satisfies the conditions from Definition 4.2.4, and it induces a lattice cohomology $\mathbb{H}^* = \mathbb{H}^0$.

Now, set $\mathcal{S} = \langle 3, 4 \rangle$ as in Example 4.1.22. We have $\delta = \left|\mathbb{Z}_{\geq 0} \setminus \mathcal{S}\right| = 3$, and $c = 6$ is the conductor, the element above which every integer is in $\mathcal{S}$: this is the point where w_0 becomes increasing. Computing the functions in question (filled dots in the top row denote elements of $\mathcal{S}$):

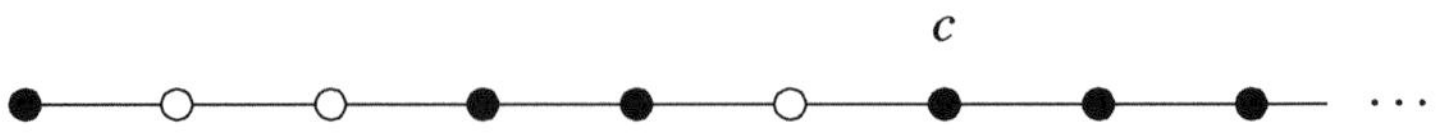

ℓ	0	1	2	3	4	5	6	7	8
$\mathfrak{h}(\ell)$	0	1	1	1	2	3	3	4	5
$\overline{\mathfrak{h}}(\ell)$	0	0	1	2	2	2	3	3	3
$w_0(\ell)$	0	1	0	-1	0	1	0	1	2

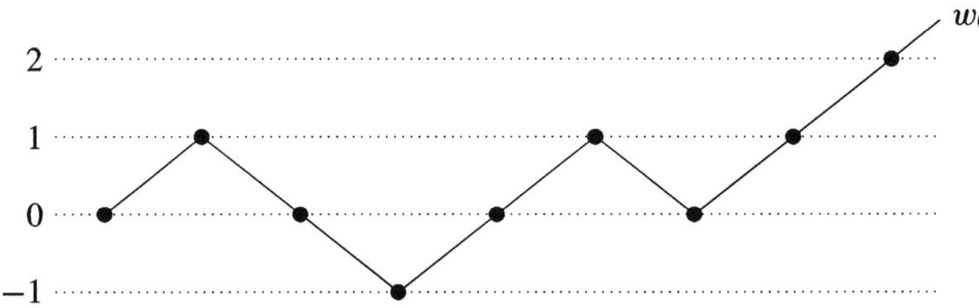

We see that in the induced $\mathbb{H}^0$, there is one generator at level -1, three at level 0, and for all $n > 0$, the set S_n is contractible. (The part of the lattice above $c = 6$ does not matter since there w_0 is increasing, so for any such n the subset $S_n \cap R(0, 6)$ is a strong deformation retract of S_n.) Drawing the so-called *graded root* which encodes this information:

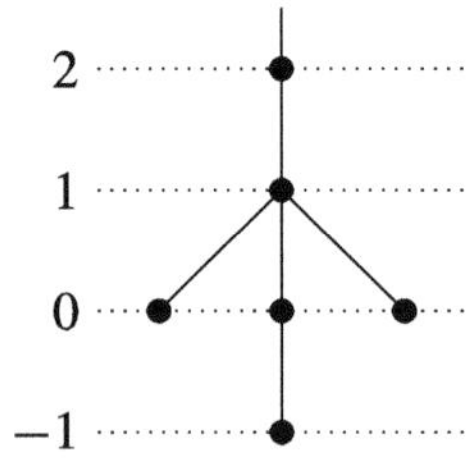

Here the integers on the left are the w_0-weights, the black dots at weight/level k correspond to the connected components of S_k, with the edges denoting the inclusion relations. Algebraically, we can write $\mathbb{H}^0 = \mathcal{T}^+_{-2} \oplus \mathcal{T}_0(1) \oplus \mathcal{T}_0(1)$, and $\mathbb{H}^{\geq 1} = 0$. For more on graded roots see e.g. [Nem07].

Example 4.2.11 For $\mathcal{S} = \langle 2, 7\rangle$ we get $\delta = 3$, $c = 6$, and

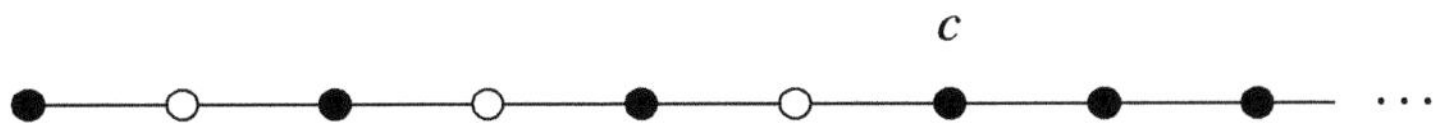

ℓ	0	1	2	3	4	5	6	7	8
$\mathfrak{h}(\ell)$	0	1	1	2	2	3	3	4	5
$\overline{\mathfrak{h}}(\ell)$	0	0	1	1	2	2	3	3	3
$w_0(\ell)$	0	1	0	1	0	1	0	1	2

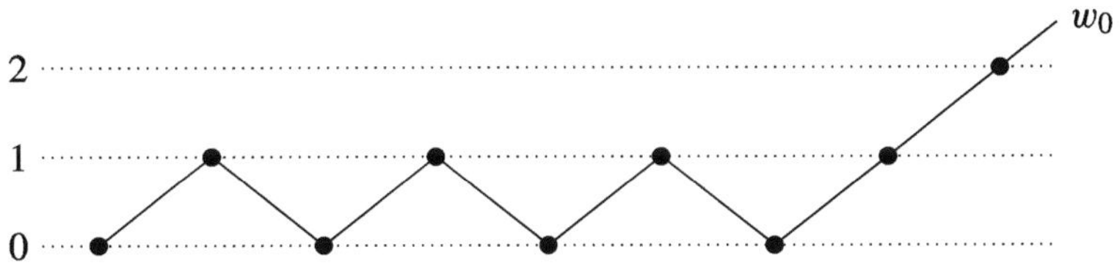

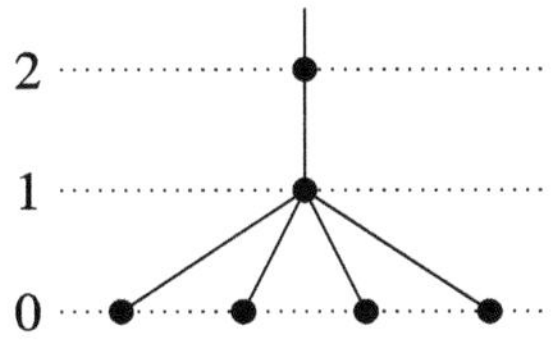

Before we proceed further, we introduce one more general concept for lattice cohomology:

Definition 4.2.12 Assume that $\mathrm{rank}_{\mathbb{Z}}\mathbb{H}^*_{red} < \infty$. Then let the *Euler characteristic* of $\mathbb{H}^*$ be

$$\mathrm{eu}(\mathbb{H}^*) = -m_w + \sum_{q\geq 0}(-1)^q\,\mathrm{rank}_{\mathbb{Z}}\mathbb{H}^q_{red}.$$

For a finite rectangle this can be expressed in the following way as well:

Proposition 4.2.13 *In the case of $R = R(a, b)$ for finite $a, b \in \mathbb{Z}^r$:*

$$\mathrm{eu}(\mathbb{H}^*)(R(a,b), w) = \sum_{q\geq 0}\ \sum_{\square\in Q_q\cap R}(-1)^{q+1}w(\square).$$

Looking back on the previous examples, we notice that the Euler characteristic is always equal to δ. This is not a coincidence, though it does rely on some conditions being satisfied:

Theorem 4.2.14 (Ágoston–Némethi [AN23]) *Consider the finite rectangle* $R(0, c)$ *for some* $c \in (\mathbb{Z}_{>0})^r$*, and let the weight function* w_0 *be obtained in the following way:*

- $\mathfrak{h}\colon L \to \mathbb{Z}$ *is increasing (with respect to the partial ordering) and* $\mathfrak{h}(0) = 0$,
- $\mathfrak{h}^\circ\colon L \to \mathbb{Z}$ *is decreasing (with respect to the partial ordering),*
- $\mathfrak{h}$ *satisfies the following 'matroid inequality' (MAT):*

$$\mathfrak{h}(\ell_1) + \mathfrak{h}(\ell_2) \geq \mathfrak{h}\big(\max(\ell_1, \ell_2)\big) + \mathfrak{h}\big(\min(\ell_1, \ell_2)\big),$$

- *the pair* $(\mathfrak{h}, \mathfrak{h}^\circ)$ *satisfies the following 'combinatorial duality property' (CDP):*

$$\mathfrak{h}(\ell + E_i) - \mathfrak{h}(\ell) \text{ and } \mathfrak{h}^\circ(\ell + E_i) - \mathfrak{h}^\circ(\ell) \text{ cannot be simultaneously } \neq 0,$$

- $w_0(\ell) = \mathfrak{h}(\ell) + \mathfrak{h}^\circ(\ell) - \mathfrak{h}^\circ(0)$.

Then for the induced lattice cohomology $\mathbb{H}^*(R(0, c), w)$*, we have* $\mathrm{eu}\,\mathbb{H}^* = \mathfrak{h}^\circ(0) - \mathfrak{h}^\circ(c)$.

One can easily verify that for the examples before (when $r = 1$), setting $\mathfrak{h}^\circ(\ell) = \mathfrak{h}(\ell) - \ell$ gives us the weight function we were computing, and also that $\mathfrak{h}$ and $\mathfrak{h}^\circ$ satisfy both the MAT and CDP properties.

This is the method we will use to construct the weight function for $\mathbb{H}^*$ in various contexts (but in some cases the weight function can be more complicated).

Remark 4.2.15 The MAT property is automatically satisfied whenever $\mathfrak{h}$ is obtained as the Hilbert function of a filtration induced by a valuation: given a decreasing $\mathbb{Z}^r$-filtration $\mathscr{F}$ of a vector space M where $\mathscr{F}(0) = M$ and $\mathscr{F}(c) = 0$ for some $c \in (\mathbb{Z}_{\geq 0})^r$, we get an increasing function $\mathfrak{h}(\ell) = \dim M/\mathscr{F}(\ell)$. If $\mathscr{F}$ is obtained from some valuation $\mathfrak{v}\colon M \to (\mathbb{Z}_{\geq 0} \cup \{\infty\})^r$ via $\mathscr{F}(\ell) = \{f \in M | \mathfrak{v}(f) \geq \ell\}$ then MAT holds.

As a result, there is a plethora of situations in which the above theorem may be applicable. The CDP condition is more particular, but certain duality properties can often be used to ensure it being satisfied, like it was in the previous examples—and the general construction for curve singularities, to be discussed later.

4.2.2 Overview and Motivation

The introduction and study of lattice cohomology associated with various situations of singularity theory is motivated by two packages of results and principles. Firstly, by the appearances of different (co)homology theories in low-dimensional topology (Heegaard Floer, Khovanov, contact, etc.), see e.g. [OS04a, OS04b, Ras05, Ras10]. The other motivation is based on several results of surface singularities, which aims to understand certain analytic invariants in terms of the topology of the link, and in this way compares invariants of singularities—both topological and analytical—and studies their connections to each other. While going over a part of this larger picture, we will also give a bit of a historical overview on the topic of lattice cohomologies.

As mentioned at the beginning of Sect. 4.2, lattice cohomology was first introduced in [Nem08a] as a topological invariant of normal surface singularities. We will talk about it in detail in Sect. 4.2.4. The construction of a bigraded $\mathbb{Z}[U]$-module is a cohomology paralleled of the *Heegaard–Floer homology* HF^- of the link (a smooth, compact 3-manifold)—and in fact they were proved to be equivalent by Zemke in 2021 [Zem]. This cohomology $\mathbb{H}^*_{top}$ also has connections to other notable topological properties, e.g. its Euler characteristic is the *Seiberg–Witten invariant* of the link corresponding to the canonical $spin^c$-structure. (This means that $\mathbb{H}^*_{top}$ is a categorification of the Seiberg–Witten invariant.) Its weight function is constructed from the Riemann-Roch expression χ.

Analogously, it seemed natural to try and construct an analytic version: one whose Euler characteristic is the so-called *geometric genus* of a surface singularity $(V, 0)$, which can be considered an analytic analogue to the Seiberg–Witten invariant (though we will not go into details now). For detailed discussions and parallel statements between the topological and analytical invariants (in particular, between the Seiberg-Witten invariant and the geometric genus) see [Nem22]. In this sense, $\mathbb{H}^*_{an}(V, 0)$ appears as a 'categorification' of the geometric genus.

Such an invariant was constructed by Ágoston and Némethi in [AN21a] (see also [AN21b] for the equivariant case which covers the cohomology modules corresponding to all the $spin^c$-structures of the link). The construction (and the weight function) relies an the divisorial filtration associated with a good resolution. The analytic lattice cohomology $\mathbb{H}^*_{an}(V, 0)$ also has a direct connection to $\mathbb{H}^*_{top}$ in the form of graded $\mathbb{Z}[U]$-module morphism $\mathbb{H}^*_{an}(V, 0) \to \mathbb{H}^*_{top}(V, 0)$.

This particular construction valid for surfaces inspired the more general combinatorial setup described e.g. in Theorem 4.2.14 and used to construct other lattice cohomologies: not just for surfaces, but for isolated singularities of any dimension $n \geq 2$ [AN21c] (using again the divisorial filtration of a resolution), as well as for curves [AN23] (using the Hilbert function associated with the valuative filtration of the normalization). The curve case—which covers the case of all reduced curves, with arbitrary embedded dimension, i.e. not necessarily plane curves—will be discussed in Sect. 4.2.3.

Unfortunately, at this moment, no direct topological analogue is known in the cases $n > 2$ yet. (Conjecturally we have to find connection with some version of Embedded Contact Homology.) The construction of $\mathbb{H}^*_{top}$ for surfaces does not easily lend itself to generalizations in the same manner. In fact, for $n > 2$, the abstract link itself holds too little information to be sufficient. (This is why we need to introduce some additional information, e.g. the contact structure.) The same is true in the case $n = 1$, in that case too the abstract link has very little information (it gives only the number of local irreducible components). However, in the case of plane curve singularities $(V, 0)$ specifically, the knot/link Floer homology HFL^- of the embedded link (and other equivalent embedded invariants like the multivariate Alexander polynomial Δ, or the motivic multivariable Poincaré series) can be

Table 4.1 Cohomology theories

Dim	Topological	Analytic
$\dim_{\mathbb{C}}(V,0)=2$ $\dim_{\mathbb{R}} L_V=3$	$HF^- = \mathbb{H}^*_{top}$ (from χ of a resolution) $eu(\mathbb{H}^*_{top})=$ Seiberg-Witten invariant	$\mathbb{H}^*_{an}$ (div. filtration of a resolution) $eu(\mathbb{H}^*_{an})=$ geometric genus
$\dim_{\mathbb{C}}(V,0)=d>2$ $\dim_{\mathbb{R}} L_V=2d-1$	?	$\mathbb{H}^*_{an}$ (div. filtration of a resolution) $eu(\mathbb{H}^*_{an})=$ geometric genus
$\dim_{\mathbb{C}}(V,0)=1$ $L_V=\sqcup_{i=1}^r S^1$	?	$\mathbb{H}^*_{an}$ (val. filtration of normalization) $eu(\mathbb{H}^*_{an})=$ delta invariant
$(V,0)\subset(\mathbb{C}^2,0)$ $L_V=\sqcup S^1\subset S^3$	(Embedded topological) HFL^-	(Filtered version) spectral sequence $\Rightarrow \mathrm{Gr}^F_* \mathbb{H}_{*,an}$

related with a filtered version of the analytic lattice (co)homology, see [Nema]. (For the filtered version in the surface case see also [Nemb].) Some of these cohomology theories are summarized in Table 4.1.

4.2.3 The Case of Isolated Curve Singularities [AN23]

Fix an isolated curve singularity $(V,0)$ with irreducible decomposition $\bigcup_{i=1}^{r}(V_i,0)$ (which is not necessarily a plane curve singularity). We consider the lattice $\mathbb{Z}^r$ and the first quadrant of $\mathbb{Z}^r\otimes_{\mathbb{Z}}\mathbb{R}$, namely $(\mathbb{R}_{\geq 0})^r$, together with its cubical decomposition. Additionally we define the weight function as follows.

Definition 4.2.16

$$w_0(\ell)=\mathfrak{h}(\ell)+\mathfrak{h}^\circ(\ell)-\mathfrak{h}^\circ(0)=2\mathfrak{h}(\ell)-|\ell|$$

where

$$|\ell|=\sum_{i=1}^{r}\ell_i,\quad \mathfrak{h}^\circ(\ell)=\mathfrak{h}(\ell)-|\ell|\,.$$

This then induces the lattice cohomology $\mathbb{H}^*((\mathbb{R}_{\geq 0})^r,w)$.

Proposition 4.2.17

(1) *For any $c'\geq c$ (where c is the conductor), one has a graded $\mathbb{Z}[U]$-module isomorphism $\mathbb{H}^*((\mathbb{R}_{\geq 0})^r,w)=\mathbb{H}^*(R(0,c'),w)$.*

(2) For any $c' \geq c$ the pair $(\mathfrak{h}, \mathfrak{h}^\circ)$ satisfies the conditions in Theorem 4.2.14. In particular,

$$\mathrm{eu}\,\mathbb{H}^*((R(0,c), w) = \mathfrak{h}(c) - |c| = \delta.$$

The graded $\mathbb{Z}[U]$-module $\mathbb{H}^*((\mathbb{R}_{\geq 0})^r, w) = \mathbb{H}^*(R(0,c'), w)$ is called the lattice cohomology of $(V, 0)$, and it is denoted by $\mathbb{H}^*(V, 0)$. It is a categorification of the delta invariant of $(V, 0)$.

$\mathbb{H}^*_{red}(V, 0) = 0$ if and only if $(V, 0)$ is smooth. Examples 4.2.10 and 4.2.11 provide the lattice cohomology for irreducible plane curves $\{x^3 + y^4 = 0\}$ and $\{x^2 + y^7 = 0\}$ respectively. The next examples has two irreducible components.

Example 4.2.18 Let $(V, 0) = \{x^2 - y^4 = 0\}$. Here $x^2 - y^4 = (x - y^2)(x + y^2)$, so $(V, 0)$ has two components (two isomorphic components, in fact, so we can expect the Hilbert function etc. to have an additional symmetry).

Both components $(V_1, 0)$ and $(V_2, 0)$ are regular, so their respective links are the unknots; hence, their individual Alexander polynomials are trivial. Their linking number is the intersection multiplicity (cf. Fact 4.1.16), which is 2. From this, one can get the Poincaré series (cf. Theorems 4.1.40 and 4.1.41) for all V_I where $\emptyset \neq I \subset \{1, 2\}$:

$$P_{V_1}(t_1) = \frac{1}{1 - t_1}, \qquad P_{V_2}(t_2) = \frac{1}{1 - t_2}, \qquad P_{V_{1,2}} = 1 + t_1 t_2.$$

Finally, using Theorem 4.1.41, we get

$$H(t_1, t_2) = \frac{1}{(1 - t_1)(1 - t_2)} \Big(\frac{t_1}{1 - t_1} + \frac{t_2}{1 - t_2} - t_1 t_2 (1 + t_1 t_2) \Big).$$

Computing the coefficients $\mathfrak{h}(\ell)$ and the weight function:

$\mathfrak{h}$

4	4	4	5	6
3	3	3	4	5
2	2	2	3	4
1	1	2	3	4
0	1	2	3	4

w

4	3	2	3	4
3	2	1	2	3
2	1	0	1	2
1	0	1	2	3
0	1	2	3	4

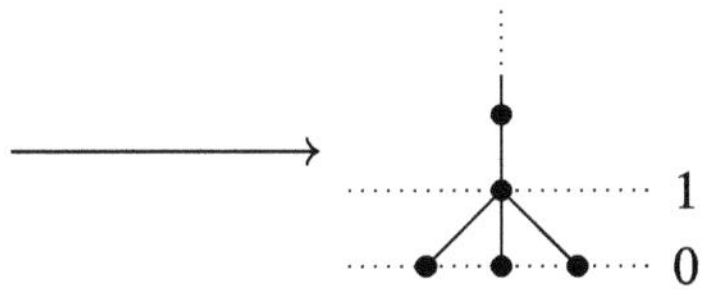

Note that, compatibly with the theory, $\mathrm{eu}\,\mathbb{H}^*(V, 0) = \delta = 2$.

4.2.4 Surface Singularities: The Topological Cohomology [Nem08a, Nem22]

Moving on to another context in which we can apply the lattice cohomology construction, consider a normal surface singularity $(V, 0) \hookrightarrow (\mathbb{C}^n, 0)$. We will aim to understand the topological lattice cohomology: the one that was originally defined, in 2000.

4.2.4.1 Plumbed 3-Manifolds and Topological Lattice Cohomology

Now we will want to rely on solely (abstract) topological invariants for obtaining a suitable weight function on some lattice. As it so happens, both a lattice and some combinatorial data on that lattice naturally arises from the topology.

Fact 4.2.19

(1) For a normal (in particular, isolated) surface singularity $(V, 0)$, *the link* L_V *is a connected, smooth, oriented real 3-manifold.*
(2) L_V *is plumbed 3-manifold associated with a connected negative definite plumbing graph* Γ.

Assuming the plumbing construction as known, let us just quickly recall the basic notations and concepts and how they apply in our current context:

- A *plumbing graph* is a graph $\Gamma = \big(\mathcal{V}, \mathcal{E}, \{e_v\}_{v\in\mathcal{V}}, \{g_v\}_{v\in\mathcal{V}}\big)$ where the decorations e_v and g_v are integers assigned to each vertex $v \in \mathcal{V}$. We will not allow loops now (edges from a vertex to itself), but multiple edges are possible. For simplicity we will assume that Γ is connected, cf. Fact 4.2.19(2).
- Through the *plumbing construction*, we get the plumbed 4-manifold $P = P(\Gamma)$, with its boundary being the plumbed 3-manifold $M = M(\Gamma)$. Here we have $H_2(P, \mathbb{Z}) \simeq \mathbb{Z}\langle E_v\rangle_{v\in\mathcal{V}}$: to each vertex corresponds a compact oriented submanifold E_v of real dimension two, all of which together freely generate $H_2(P, \mathbb{Z})$. Then the genus of E_v is the decoration $g_v \geq 0$.
- The intersection form $\big(\bullet, \bullet\big)$ on $\mathbb{Z}\langle E_v\rangle_{v\in\mathcal{V}}$ is also given by the plumbing graph:

$$(E_u, E_v) = \begin{cases} e_v, & \text{if } u = v, \\ \#\{\text{edges between } u \text{ and } v\}, & \text{if } u \neq v. \end{cases}$$

  (Hence, e_v is the Euler number of the normal bundle of $E_v \subset P(\Gamma)$.) A plumbing graph being *negative definite* means that this intersection form is such.

In particular, we indeed have a lattice $L = \mathbb{Z}\langle E_v\rangle_{v\in\mathcal{V}} = H_2(P, \mathbb{Z})$ and the intersection form—both of which are topological. Well, almost: the plumbing graph Γ of a 3-manifold is not unique, so if we rely on it to construct a lattice cohomology, we will need to show that different choices of Γ result in the same $\mathbb{H}^*$. The *set* of plumbing graphs for the link is a topological invariant, so if the previous statement is true then $\mathbb{H}^*$ would indeed be topological as well.

We will further restrict our attention to singularities $(V, 0)$ whose link is a $\mathbb{Q}HS^3$ (rational homology sphere), i.e. all homologies over $\mathbb{Q}$ are isomorphic to that of S^3. This means that the Betti numbers satisfy $b_0(L_V) = b_3(L_V) = 1$ and $b_1(L_V) = b_2(L_V) = 0$. In terms of the plumbing graph Γ, this means:

Proposition 4.2.20 *A negative definite connected plumbed* 3*-manifold with plumbing graph* $\Gamma = \big(\mathcal{V}, \mathcal{E}, \{e_v\}, \{g_v\}\big)$ *is a* $\mathbb{Q}HS^3$ *if and only if all* $g_v = 0$*, and* Γ *is a tree.*

As a result, we will omit the genus decorations altogether from now on and just assume $g_v = 0$, and also no multiple edges (and circles in the graph) will be present.

Aside from these restrictions though, it is not immediately obvious what the graph being negative definite (ND) means, how we can identify such a graph.

Example 4.2.21 Some examples and non-examples of negative definite graphs.

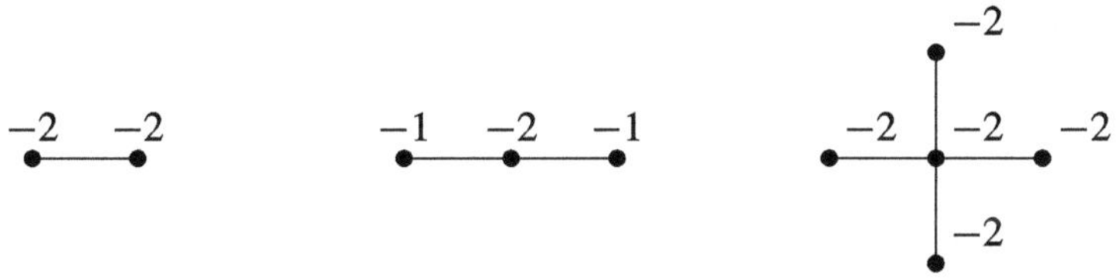

ND not ND not ND

In the following graph

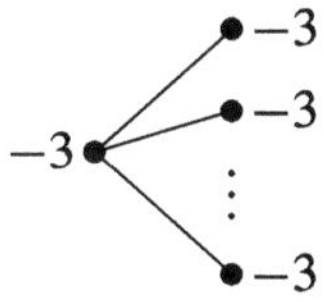

if the number of (-3) vertices on the right (i.e. the number of end vertices) is < 9 then the graph is negative definite, otherwise it is not.

In general, negative definiteness can be tested by Sylvester's Criterion. Another criterion is the following:

Proposition 4.2.22 *Assume that there exists* $D = \sum_{v \in \mathcal{V}} d_v E_v$ *such that:*

- $\forall v : d_v > 0$,
- $\forall u : (D, E_u) \leq 0$,
- $\exists u : (D, E_u) < 0$.

Then $\big(\bullet, \bullet\big)$ *is negative definite.*

Example 4.2.23 Let us test the above property for the effective cycle $D = \sum_{v \in \mathcal{V}} E_v$. Let κ_v denote the degree of the vertex $v \in \mathcal{V}$. Then $D = \sum_{v \in \mathcal{V}} E_v$ satisfies the properties from Proposition 4.2.22 if and only if

- $\forall v : e_v + \kappa_v \leq 0$,
- $\exists v : e_v + \kappa_v < 0$.

In particular, these two conditions imply negative definiteness. Graphs with such properties are very special, they are called *minimal rational*.

Example 4.2.24 The graph

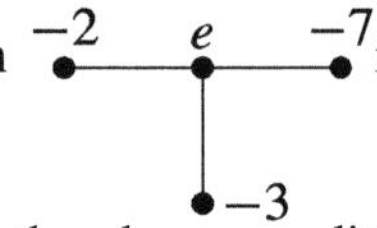

is *minimal rational* (hence negative definite) for $e \leq -3$ as per the above condition. However, this is not a *necessary* condition: both $e = -2$ and $e = -1$ result in negative definite graphs as well.

Also important to note:

Proposition 4.2.25

- *If Γ is negative definite and $\Gamma' \subseteq \Gamma$ is a (full) subgraph then Γ' is negative definite too.*
- *If $\Gamma = (\mathcal{V}, \mathcal{E}, \{e_v\})$ is negative definite and $e'_v \leq e_v$ for all v then $\Gamma' = (\mathcal{V}, \mathcal{E}, \{e'_v\})$ is negative definite too.*

Getting back to constructing the lattice cohomology $\mathbb{H}^*_{top}(V, 0)$, the question is what the weight function w should be on the lattice $L = \mathbb{Z}\langle E_v \rangle_{v \in \mathcal{V}}$.

Definition 4.2.26 Let $Z_K \in L \otimes \mathbb{Q}$ denote the anticanonical cycle.

Remark 4.2.27 The adjunction formula implies $(Z_K, E_v) = (E_v, E_v) + 2$ for all v, so Z_K can be computed from the intersection form (which is nondegenerate). In particular, Z_K is topological.

Definition 4.2.28 Let us define

$$\chi : L \to \mathbb{Z}, \quad \chi(\ell) = \frac{1}{2}(\ell, Z_K - \ell).$$

From the adjunction formula, we get that indeed $\chi(\ell) \in \mathbb{Z}$, in particular $\chi(0) = 0$ and $\chi(E_v) = 1$ for all $v \in \mathcal{V}$.

Remark 4.2.29 Note that

$$\chi(\ell) = \frac{1}{8}(Z_K, Z_K) - \frac{1}{2}(\ell - Z_K/2, \ell - Z_K/2).$$

Given that $(\bullet, \bullet)$ is negative definite, this implies that χ is bounded from below, and even more, for any $k \in \mathbb{Z}$, the set $\chi^{-1}((-\infty, k]))$ is finite. (Also, χ is symmetric with respect to the involution $\ell \mapsto Z_K - \ell$ if $Z_K \in L$.)

As per the above observation, we can set χ as the weight function:

Definition 4.2.30 For a negative definite plumbing graph Γ (when Γ is a tree and with vanishing g_v's), define the weight function

$$w_0\colon L = \mathbb{Z}\langle E_v\rangle_{v\in\mathcal{V}} \to \mathbb{Z}, \quad w_0(\ell) = \chi(\ell)$$

on the lattice L. We denote the resulting cohomology by $\mathbb{H}^*(\Gamma)$.

The following statement shows that it is independent of the choice of the plumbing graph of the (fixed) link.

Proposition 4.2.31 *If two connected negative definite plumbing graphs Γ and Γ' give rise to diffeomorphic 3-manifolds $M(\Gamma)$ and $M(\Gamma')$ through the plumbing construction then $\mathbb{H}^*(\Gamma) \simeq \mathbb{H}^*(\Gamma')$.*

Definition 4.2.32 Let $(V, 0)$ be a normal surface singularity with $\mathbb{Q}HS^3$ link. We can define $\mathbb{H}^*_{top}(V, 0) = \mathbb{H}^*(L_V)$ as $\mathbb{H}^*(\Gamma)$ for a connected negative definite plumbing graph of the link L_V, and the resulting object is a topological invariant of $(V, 0)$.

As for the Euler characteristic, one can show that

Proposition 4.2.33 ([Nem11])

(1) If the plumbing graph Γ gives rise to a $\mathbb{Q}HS^3$ link, then $\mathrm{rank}_{\mathbb{Z}}\mathbb{H}^*_{red}(\Gamma) < \infty$. *Hence in this case the Euler characteristic is well-defined.*

(2) For a negative definite plumbed $\mathbb{Q}HS^3$ M, we have $\mathrm{eu}\,\mathbb{H}^*(M) = \mathfrak{sw}_{can}(M)$, *the normalised Seiberg-Witten invariant for the canonical $spin^c$-structure of M.*

It has also been recently proved that:

Theorem 4.2.34 (Zemke [Zem]) *The topological invariant $\mathbb{H}^*(M)$ is equivalent to the Heegaard–Floer homology $HF^-(M)$. For more topological connections of the lattice cohomology see also [OS03a, OS03b, OSS14, OSS16].*

4.2.4.2 The Topological Lattice Cohomology and Other Singularity Invariants

Now, we want to relate $\mathbb{H}^*_{top}(V, 0)$ to some other properties of the normal surface singularity $(V, 0)$.

Definition 4.2.35 The *geometric genus* of a germ $(V, 0)$ is

$$p_g = \dim_{\mathbb{C}} \frac{\{\text{holomorphic 2-forms on } V \setminus \{0\}\}}{\{\text{holomorphic 2-forms on } V \setminus \{0\} \text{ ``that can be extended to 0''}\}}.$$

Technically this means the following. Let $\pi : \widetilde{V} \to V$ be a resolution of the singularity of $(V, 0)$, that is, $\widetilde{V}$ is smooth, π is proper and surjective. Set $E := \pi^{-1}(0) \subset \widetilde{V}$, the exceptional set of π. Then

$$p_g = \dim_{\mathbb{C}} \frac{\{\text{holomorphic 2-forms on } \widetilde{V} \setminus E\}}{\{\text{holomorphic 2-forms on } \widetilde{V} \setminus E \text{ that can be extended to } \widetilde{V}\}}.$$

(The right hand side of the identity is independent of the choice of the resolution π.)

Definition 4.2.36 The germ $(V, 0)$ is called *rational* if $p_g(V, 0) = 0$.

It turns out that rationality is expressible in terms of the plumbing graph of the link. (Below all the plumbing graphs are considered connected and negative definite.)

Theorem 4.2.37 (Artin [Art62, Art66]) *A normal surface singularity $(V, 0)$ has $p_g = 0$ if and only if the plumbing graph Γ of the link satisfies $\chi(\ell) > 0$ for all nonzero integral effective cycles $\ell \gneq 0$.*

In particular, Artin's Criterion replaces the a priori analytic property of the vanishing of the geometric genus p_g with a combinatorial property of the plumbing graph Γ of the link. This motivated to formulate this property for plumbing graphs as well.

Definition 4.2.38 The plumbing graph Γ is *rational* if $\chi(\ell) > 0$ for all nonzero integral effective cycles $\ell \gneq 0$.

Proposition 4.2.39 *The following facts hold:*

(1) if Γ is rational then any full connected subgraph $\Gamma' \subset \Gamma$ is rational,
(2) if Γ is rational and $M(\Gamma) = M(\Gamma')$ then Γ' is rational too,
(3) if $\Gamma = (\mathcal{V}, \mathcal{E}, \{e_v\})$ is rational, and $e'_v \leq e_v$, then $\Gamma' = (\mathcal{V}, \mathcal{E}, \{e'_v\})$ is rational too,
(4) if Γ is negative definite and $E = \sum_{v \in \mathcal{V}} E_v$ satisfies $(E, E_v) \leq 0$ for all $v \in \mathcal{V}$ then Γ is rational (that is, minimal rational graphs are rational),
(5) for fixed $(\mathcal{V}, \mathcal{E})$, the plumbing graph $\Gamma = (\mathcal{V}, \mathcal{E}, \{e_v\})$ is always rational if all $\{e_v\}$ are sufficiently small.

Example 4.2.40 The graph 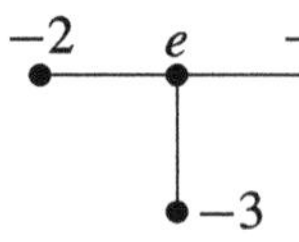

is negative definite for all $e \leq -1$ but rational only for $e \leq -2$.

The condition in Theorem 4.2.37 translates directly to the lattice cohomology of the graph:

Theorem 4.2.41 ([Nem05, Nem08a, Nem22]) *Γ is rational if and only if for the associated lattice cohomology $\mathbb{H}^*$,*

- $\mathbb{H}^0 = \mathcal{T}_0^+$ *and* $\mathbb{H}^q = 0$ *for* $q > 1$, *i.e.* $\mathbb{H}^*_{red} = 0$;
- *or equivalently:* $S_n = \emptyset$ *for all* $n < 0$ *and* S_n *is contractible for all* $n \geq 0$.

To prove this statement, first we observe that when computing $\mathbb{H}^*$, one can restrict the lattice to $(\mathbb{R}_{\geq 0})^{|\mathcal{V}|}$ and not lose anything:

Theorem 4.2.42 *Let S_n be the level sets corresponding to $\mathbb{H}^*(\Gamma) = \mathbb{H}^*(\mathbb{R}^{|\mathcal{V}|}, w)$. Then S_n has the same homotopy type as $S_n \cap (\mathbb{R}_{\geq 0})^{|\mathcal{V}|}$. In fact, $S_n \cap (\mathbb{R}_{\geq 0})^{|\mathcal{V}|}$ is a strong deformation retract of S_n.*

Proof Sketch First we can verify directly that

Lemma 4.2.43 $\chi(A + B) = \chi(A) + \chi(B) - (A, B)$ *for all* $A, B \in L$.

Then, let us fix some $x = \sum_v n_v E_v \leq 0$, $x \neq 0$. We intend to show that there is an $v \in \{1, \ldots, r\}$ with $n_v > 0$ such that $\chi(x + E_v) \leq \chi(x)$. This would essentially mean that the 'negative quadrant' can be retracted to the origin using several inductive steps by a χ-nonincreasing combinatorial flow.

Note that for any $y \geq 0$, $y \neq 0$, it must exists some E_v in the support of y with $(y, E_v) < 0$, because otherwise we would get $(y, y) \geq 0$, which would contradict the negative definiteness. Multiplying y with -1, we get that for the above x, there exists a vertex $v \in \mathcal{V}$ with $(x, E_v) > 0$. Then, by the above lemma,

$$\chi(x + E_v) = \chi(x) + \chi(E_v) - (x, E_v) = \chi(x) + 1 - (x, E_v) \leq \chi(x).$$

Next, if $x \not\geq 0$, then we write x as $x_1 - x_2$, where $x_1 \geq 0$, $x_2 > 0$ and they have disjoint supports. Then we can repeat the above argument for x_2: we find E_v in the support of x_2 such that $\chi(x + E_v) \leq \chi(x)$. This allows us to contract the parts of S_n outside the positive quadrant back inside it. (Technically, one would need to check that the choice of v can be made consistently to give us a retraction on the entire S_n, but we omit that now. For details see e.g. [Nem22].) □

Corollary 4.2.44 *For a plumbing graph Γ, we have an isomorphism $\mathbb{H}^*(\Gamma) \simeq \mathbb{H}^*((\mathbb{R}_{\geq 0})^{|\mathcal{V}|}, w)$.*

In this way we are a little bit closer to Artin's Criterion, Theorem 4.2.37, since both are tests in the first quadrant. Hence, let us return back to the rationality of Γ:

Proof of Theorem 4.2.41 By the above discussion it is enough to consider the cubical decomposition of $(\mathbb{R}_{\geq 0})^{|\mathcal{V}|}$. On the other hand, By Artin's Criterion 4.2.37 we can replace rationality with the positivity condition on χ for the lattice points in the positive quadrant (aside from 0).

Assume then that $\chi(\ell) > 0$ for $\ell \gneq 0$. We of course automatically get $S_n = \emptyset$ for $n < 0$, so we need to check that S_n is contractible for $n \geq 0$. We apply the same method as when proving Theorem 4.2.42: we intend to show that

for any $x \gneqq 0$ there exists E_v in the support of x such that $\chi(x - E_v) \le \chi(x)$.

Assume by contradiction that this is not the case, i.e. $\chi(x - E_v) \ge \chi(x) + 1$ for all E_v in the support of x. Then

$$\begin{aligned}
\chi(x - E_v) - 1 \ge \chi(x) &= \chi(x - E_v + E_v) \\
&= \chi(x - E_v) + \chi(E_v) - (x - E_v, E_v) = \\
&= \chi(x - E_v) + 1 - (x - E_v, E_v), \\
(x - E_v, E_v) &\ge 2, \\
(x, E_v) &\ge (E_v, E_v) + 2 = (Z_K, E_v)
\end{aligned}$$

for all E_v in the support of x. Summing up we obtain that for $x \gneqq 0$ we have $(x, x) \ge (Z_K, x)$, i.e.

$$\chi(x) = \frac{1}{2}(x, Z_K - x) \le 0,$$

which contradicts Artin's Criterion of rationality.

Now once again, we can use this to retract $S_n \cap (\mathbb{R}_{\ge 0})^{|\mathcal{V}|}$ by a χ-nonincreasing flow this time down to a single lattice point 0.

Conversely, assume S_n is contractible for $n \ge 0$; in particular, S_0 is connected. Since $\chi(0) = 0$ and $\chi(E_v) = 1$ for all v, this means S_0 can not contain any point $x \in (\mathbb{Z}_{\ge 0})^{|\mathcal{V}|}$ other than 0 as otherwise we would have at least 2 connected components in S_0. This finishes the proof. □

Example 4.2.45 ([Nem22]) Though $p_g = 0$ is determined by the link, p_g itself is *not* topological. Indeed, e.g. for the link defined by the graph

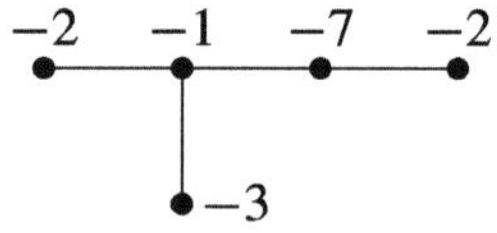

one can find an analytic structure with $p_g = 2$ (let $V = \{x^2 + y^3 + z^{13} = 0\}$), while the generic analytic structure has $p_g = 1$.

In general, for any Γ we have

Proposition 4.2.46 ([Nem05, Nem22]) *S_n is connected for $n \ge 1$.*

Also, we can analyze a slightly weaker condition than Artin's Rationality Criterion:

Definition 4.2.47 ([Nem05, Nem22]) The graph Γ is *elliptic* if $\chi(\ell) \geq 0$ for all $\ell \geq 0$, but Γ is not rational.

Proposition 4.2.48 *Γ is elliptic if and only if $\mathbb{H}^0_{red}(\Gamma) = \mathcal{T}_0(1)^k$ for some integer $k > 0$. For elliptic graphs $\mathbb{H}^{>0} = 0$.*

Example 4.2.49 In case of an elliptic graph Γ, the graded root looks like this:

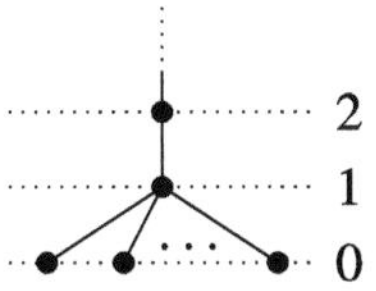

Note that if $Z_K \in L$ then χ is symmetric (cf. Remark 4.2.29), and in particular $\chi(Z_K) = 0$, so 0 and Z_K correspond to two different "legs" of the above graded root.

Example 4.2.50 Let $(V, 0) = \{x^5 + y^5 + z^6 = 0\} \subset (\mathbb{C}^3, 0)$. Then the graded root is

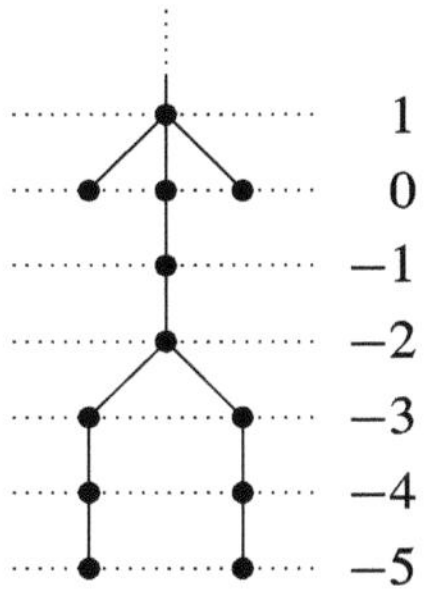

Problem 4.2.51 Characterize the graded $\mathbb{Z}[U]$-modules $\mathbb{H}^0$ (or the graded roots) that can be realized analytically, i.e. as the $\mathbb{H}^0_{top}$ of some surface singularity $(V, 0)$.

4.2.4.3 Bad Vertices and the Reduction Theorem

As a closing note, one can observe that in regards to $\mathbb{H}^*(\Gamma)$, some vertices of Γ 'matter more than others'. We can identify a set of 'bad vertices' in a graph, and focus only on those in a sense. More concretely,

Definition 4.2.52 ([Nem05, Nem22]) Assume $(\mathcal{V}, \mathcal{E}, \{e_v\}_v)$ is a fixed plumbing graph Γ with Euler numbers $\{e_v\}_v$. We say that $\{v_1, \ldots, v_k\} \subseteq \mathcal{V}$ is a *set of bad*

vertices of Γ if there exist $e'_{v_j} \leq e_{v_j}$ such that by replacing all e_{v_j} with e'_{v_j}, we get a rational graph $\Gamma' = (\mathcal{V}, \mathcal{E}, \{e'_v\}_v)$.

Remark 4.2.53 By Proposition 4.2.39, we know that at least $\mathcal{V}$ itself is such a set—usually a much smaller set also suffices though.

Example 4.2.54 When $M(\Gamma)$ is a $\mathbb{Q}HS^3$ then the set of vertices with degree > 2 (the 'nodes') is a suitable set of bad vertices.

Example 4.2.55 In particular, given a 'star-shaped' graph Γ, i.e. a tree with a single node (and $M(\Gamma)$ a $\mathbb{Q}HS^3$), then the singleton set containing that node is a set of bad vertices.

Definition 4.2.56 A graph Γ with $M(\Gamma)$ a $\mathbb{Q}HS^3$ is called *almost rational* if admits a set of bad vertices of cardinality one.

As for how to use these 'bad vertices' exactly:

Theorem 4.2.57 (Reduction Theorem; László–Némethi [LN15]) *Let* Γ *be a plumbing graph for a* $\mathbb{Q}HS^3$*, with a set of bad vertices* $\overline{\mathcal{V}} = \{v_1, \ldots, v_k\}$*. Then*

$$\mathbb{H}^*(\Gamma) \simeq \mathbb{H}^*((\mathbb{R}_{\geq 0})^{|\overline{\mathcal{V}}|}, \overline{w})$$

for a suitable choice of weight function $\overline{w}\colon (\mathbb{Z}_{\geq 0})^{|\overline{\mathcal{V}}|} = (\mathbb{Z}_{\geq 0})^k \to \mathbb{Z}$ *(which we will not explicitly construct here).*

This means that with a modification of the weight function (that is combinatorially and algorithmically computable from χ), we can often drastically reduce the dimension of the lattice we need to consider. This is, of course, very important for any kind of computation, but also, as an immediate consequence we obtain:

Corollary 4.2.58 *If* $\overline{\mathcal{V}}$ *is a set of bad vertices in* Γ *then* $\mathbb{H}^q_{red}(\Gamma) = 0$ *for all* $q \geq |\overline{\mathcal{V}}|$. *(In particular,* $\mathbb{H}^* = \mathbb{H}^0$ *for almost rational graphs.)*

Example 4.2.59 Consider the following graph:

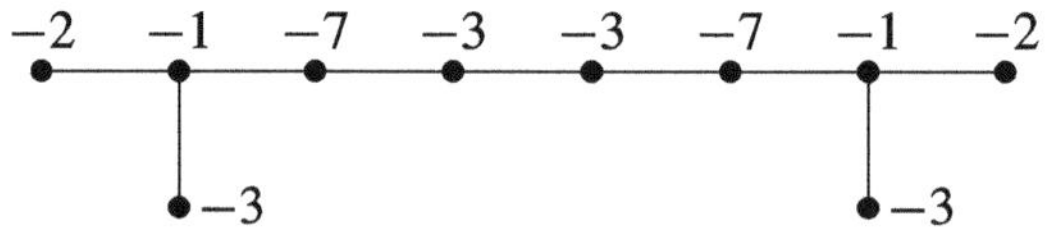

Here we have 2 nodes so $\mathbb{H}^q = 0$ for all $q \geq 2$. By a computation $\text{rank}_{\mathbb{Z}}\mathbb{H}^1 = 1$ (supported in degree zero), and the graded root is the following (which provides $\mathbb{H}^0$ as well):

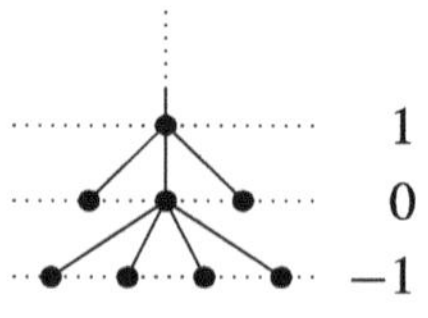

Remark 4.2.60 For any particular $q > 0$ one can also construct examples for graphs with non-vanishing $\mathbb{H}^q$. (Of course, such a graph needs to have at least $q+1$ nodes.)

References

[AN23] Ágoston, T. and Némethi, A.: The analytic lattice cohomology of isolated curve singularities, arXiv 2301.08981, 2023.

[AN21a] Ágoston, T. and Némethi, A.: The analytic lattice cohomology of surface singularities, arXiv 2108.12294, 2021.

[AN21b] Ágoston, T. and Némethi, A.: Analytic lattice cohomology of surface singularities, II (the equivariant case), arXiv 2108.12429, 2021.

[AN21c] Ágoston, T. and Némethi, A.: The analytic lattice cohomology of isolated singularities, arXiv 2109.11266, 2021.

[Art62] Artin, M.: Some numerical criteria for contractibility of curves on algebraic surfaces. *Amer. J. of Math.*, **84**, 485–496, 1962.

[Art66] Artin, M.: On isolated rational singularities of surfaces. *Amer. J. of Math.*, **88**, 129–136, 1966.

[BK86] Brieskorn, E., Knörrer, H.: Plane Algebraic Curves, Birkhäuser, Boston, 1986.

[CDG99] Campillo, A., Delgado de la Mata, F. and Gusein-Zade, S. M.: On the monodromy of a plane curve singularity and the Poincaré series of the ring of functions on the curve, Functional Analysis and Its Applications 33, 56–57 (1999).

[CDG02] Campillo, A., Delgado de la Mata, F. and Gusein-Zade, S. M.: Integrals with respect to the Euler characteristic over spaces of functions and the Alexander polynomial, Proc. of the Steklov Inst. of Math. 238, 134–147 (2002).

[CDG03] Campillo, A., Delgado de la Mata, F. and Gusein-Zade, S. M.: The Alexander polynomial of a plane curve singularity via the ring of functions on it, Duke Math. Journal 117 (1), 125–156 (2003).

[CDG07] Campillo, A., Delgado de la Mata, F. and Gusein-Zade, S. M.: Multi-index Filtrations and Generalized Poincaré Series, Monatshefte für Mathematik 150, 193–209 (2007).

[EN85] Eisenbud, D. and Neumann. W. D.: Three-Dimensional Link Theory and Invariants of Plane Curve Singularities, Annals of Math. Studies 110, Princeton University Press 1985.

[GN15] Gorsky, E. and Némethi, A.: Lattice and Heegaard Floer homologies of algebraic links, Int. Math. Research Notices 2015 (23), 12737–12780 (2015).

[Gra71] Grauert, H.: Über die Deformation isolierter Singularitäten analytischer Mengen, Inventiones mathematicae, 15 (3), 171–198 (1971).

[LN15] László, T. and Némethi, A.: Reduction theorem for lattice cohomology, Int. Math. Research Notices **2015** (1) (2015), 2938–2985.

[Mil69] Milnor, J.: Singular Points of Complex Hypersurfaces, Annals of Math. Studies 61, 1969, Princeton Univ. Press.

[Moy08] Moyano-Ferdnández, J. J.: Poincaré series associated with curves defined over perfect fields, PhD Thesis, University of Valladolid, 2008.

[Nem05] Némethi, A.: On the Ozsváth–Szabó invariant of negative definite plumbed 3-manifolds, Geometry & Topology 9(2), 991–1042 (2005).

[Nem07] Némethi, A.: Graded roots and singularities, in Singularities in Geometry and Topology, World Scientific 394–463 (2007).

[Nem08a] Némethi, A.: Lattice cohomology of normal surface singularities, Publ. of the Res. Inst. for Math. Sci. 44(2), 507–543 (2008).

[Nem08b] Némethi, A.: Poincaré series associated with surface singularities, in Singularities I: Algebraic and Analytic Aspects, Contemp. Math. 474, AMS 271–297 (2008).
[Nem11] Némethi, A.: The Seiberg–Witten invariants of negative definite plumbed 3-manifolds, J. Eur. Math. Soc. 13(4), 959–974 (2011).
[Nem22] Némethi, A.: Normal surface singularities, Ergebnisse der Math. und ihrer Grenzgebiete 74, Springer 2022.
[Nema] Némethi, A.: Filtered lattice homology of curve singularities, arXiv:2306.13889 .
[Nemb] Némethi, A.: Filtered lattice homology of surface singularities, arXiv:2307.16581.
[OS03a] Ozsváth, P. S. and Szabó, Z.: On the Floer homology of plumbed three-manifolds, Geometry & Topology 7(1), 185–224 (2003).
[OS03b] Ozsváth, P. S. and Szabó, Z.: Knot Floer homology and the four-ball genus, Geometry & Topology 7, 615–639 (2003).
[OS04a] Ozsváth, P. and Szabó, Z.: Holomorphic disks and three-manifold invariants: properties and applications, Ann. of Math. (2), 159(2004), no. 3, 1159–1245.
[OS04b] Ozsváth, P. and Szabó, Z.: Holomorphic discs and topological invariants for closed three-manifolds. Ann. of Math. (2) 159 (2004), no. 3, 1027–1158.
[OSS14] Ozsváth, P., Szabó, Z. and Stipsicz, A.: A spectral sequence on lattice homology, Quantum Topol. 5 (2014), 487–521.
[OSS16] Ozsváth, P., Szabó, Z. and Stipsicz, A.: Knot lattice homology in L-spaces, Journal of Knot Theory and Its Ramifications, Vol. 25, No. 01, 1650003 (2016).
[Ras05] J. Rasmussen: Knot polynomials and knot homologies, in *Geometry and Topology of Manifolds*, Boden et. al. eds., Fields Institute Communications 47, 261–280 (2005).
[Ras10] J. Rasmussen: Khovanov homology and the slice genus, Invent. math. 182, 419–447 (2010).
[Wal04] Wall, C.T.C.: Singular points of plane curves, London Math. Soc. Student Texts **63**, Cambridge University Press 2004.
[Zar71] Zariski, O.: General theory of saturation and of saturated local rings II: Saturated local rings of dimension 1, Am. J. Math 93 (4), 872–964 (1971).
[Zem] Zemke, I.: The equivalence of lattice and Heegaard Floer homology, arXiv:2111.14962.

Chapter 5
A Tour of Monopole Floer Spectra

Matthew Stoffregen

Contents

5.1 Seiberg-Witten Theory on Closed 4-Manifolds 141
  5.1.1 The Seiberg-Witten Equations 141
5.2 The Bauer-Furuta Construction 148
5.3 Equivariant Topology 153
5.4 Proof of the 10/8-Theorem 157
  5.4.1 G-Equivariant K-Theory 158
5.5 Floer Spectra for 3-Manifolds 162
  5.5.1 The Seiberg-Witten Floer Spectrum $SWF(Y)$ for $b_1(Y) = 0$ 163
  5.5.2 Spaces From Flows 167
  5.5.3 $SWF(Y)$ for $b_1(Y) > 0$ 172
5.6 Lattice Homology and Spectra 178
  5.6.1 Lattice Homology 178
  5.6.2 Relationship with 3-Manifolds 180
  5.6.3 Lattice Spectrum: The General Construction 181
  5.6.4 Path Lattice Homotopy 183
References 185

Goals

There are a few goals for these lectures:

1. First, to introduce the *Bauer-Furuta invariant* [Fur01, BF04] of closed 4-manifolds, and the *monopole Floer spectra* defined by Manolescu [Man03], with various generalizations over time [KM02, Man16, KLS18a, KLS18b, SS21].
2. To sketch a proof of the $\frac{10}{8}$-Theorem of Furuta.
3. In keeping with the theme of the program, to define *Lattice spectra*, and, for a small family, to show equivalence of monopole Floer and lattice spectra.

M. Stoffregen (✉)
Michigan State University, East Lansing, MI, USA
e-mail: stoffre1@msu.edu

© The Author(s), under exclusive license to Springer Nature Switzerland AG 2024

J. Fernández de Bobadilla et al. (eds.), *Singularities and Low Dimensional Topology*, Bolyai Society Mathematical Studies 30,
https://doi.org/10.1007/978-3-031-56611-0_5

Monopole Floer spectra $SWF(Y, \mathfrak{s})$ are a partial generalization of the *monopole Floer homology groups* $HM(Y, \mathfrak{s})$ defined by Kronheimer-Mrowka [KM07]. As of this writing, the theory of monopole Floer spectra is not well-developed, but the hope is that these lectures may provide a road map so that if one encountered a situation where the spectra may be useful, one would be able to find the right sources to put them to use.

The constructions above take some time to set up, but we hope to point to some topological applications of the Bauer-Furuta invariant and monopole Floer spectra along the way. Generally, we follow a minimalist approach to the constructions, outsourcing many important parts. The star of the show will be Furuta's $\frac{10}{8}$-Theorem, which remains one of the principal highlights of these ideas. We remark that it also has a relative version, due much later to Manolescu [Man14], which provides further motivation for studying the spectra, apart from the closed manifold invariant.

Outline

The lectures aim to be somewhat "modular" in that there are a few competing strands that run throughout, each mostly independent:

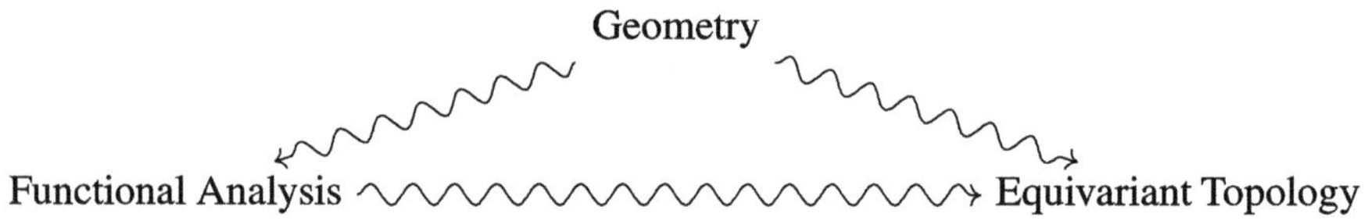

In the first lecture (Sect. 5.1), we use the geometry of particular 4-manifolds to reach a certain functional analytic setup—that of *very Fredholm* maps between Hilbert spaces. Section 5.2 gives a translation of very Fredholm maps into equivariant stable homotopy classes. Sections 5.3 and 5.4 then introduce what we will need from equivariant stable homotopy theory to study the equivariant stable homotopy classes that result.

The lectures are not quite linear in the schematic above, in that we will occasionally go backwards to collect additional properties of the objects constructed at each stage.

We refer the reader interested in more of the analytical details to the survey [Lin16] and the book [KM07].

A key sin of omission of these lectures is that we do not consider the Bauer-Furuta invariant in *families*, which has seen a vast amount of activity recently, see [LM21, KM20, BK20] for just a few examples. We hope that these sources may be relatively accessible, given the material from these lectures.

5.1 Seiberg-Witten Theory on Closed 4-Manifolds

We will introduce the Seiberg-Witten equations [Wit94] on a closed 4-manifold, and interpret them, following Furuta [Fur01] and Bauer-Furuta [BF04], in terms of homotopy theory. Our presentation is closest to Furuta's [Fur01].

5.1.1 The Seiberg-Witten Equations

Let us start with the definition of a spin manifold M. Fix (M, g) an oriented Riemannian manifold. Then the set of orthonormal frames of TM defines a principal $SO(n)$-bundle $P_{SO(n)}$. To be specific, $P_{SO(n)}$ is the set of pairs (x, f) where $x \in M$ and $f : \mathbb{R}^n \to T_x M$ is an isomorphism preserving the norm and the orientation.

For all n, the Lie group $SO(n)$ has a nontrivial double cover, which inherits the structure of a Lie group, $\mathrm{Spin}(n)$. A principal $\mathrm{Spin}(n)$-bundle P over a space X induces a quotient $SO(n)$-bundle, using the covering map fiberwise.

A spin structure on M is then a diagram of the form:

$$\begin{array}{ccc} P_{\mathrm{Spin}(n)} & \longrightarrow & P_{SO(n)} \\ & \searrow & \downarrow \\ & & M \end{array}$$

Here $P_{\mathrm{Spin}(n)}$ is a $\mathrm{Spin}(n)$-principal bundle over M, and, for each $x \in M$, the map $P_{\mathrm{Spin}(n),x} \to P_{SO(n),x}$ is a nontrivial double cover.

For us, the benefit of a spin structure on a 4-manifold M is a rank 2 quaternionic vector bundle S (which will turn out to come with a splitting $S^+ \oplus S^-$ over the quaternions), equipped with a Hermitian form, together with a morphism:

$$\varrho \colon TM \to \mathrm{Hom}(S, S), \tag{5.1}$$

where ϱ lands in morphisms preserving the Hermitian form. All this data is subject to the following:

At each $x \in M$ we can find an orthonormal frame $\{e_0, e_1, e_2, e_3\}$ of $T_x M$ so that, for some basis of S_x,

$$\varrho(e_0) = \begin{pmatrix} 0 & -I_2 \\ I_2 & 0 \end{pmatrix}, \varrho(e_i) = \begin{pmatrix} 0 & -\sigma_i^* \\ \sigma_i & 0 \end{pmatrix} \tag{5.2}$$

where σ_i are the Pauli matrices:

$$\sigma_1 = \begin{pmatrix} i & 0 \\ 0 & -i \end{pmatrix}, \sigma_2 = \begin{pmatrix} 0 & -1 \\ 1 & 0 \end{pmatrix}, \sigma_3 = \begin{pmatrix} 0 & i \\ i & 0 \end{pmatrix}.$$

It will also be convenient for us to view a spin structure in terms of the *associated bundle* construction, recalled here. Fix a principal G-bundle $P \to M$, and a G-representation V, where we write the action as $\varrho\colon G \to \mathrm{End}(V)$. Then the associated vector bundle is defined by:

$$P \times_\varrho V := (P \times V)/((pg, v) \sim (p, \varrho(g)v))$$

One should check that $P \times_\varrho V$ is indeed a vector bundle, with fiber isomorphic to V. The most important example is that, for any differentiable manifold M, the associated vector bundle of the frame bundle is the tangent bundle.

The morphisms of vector bundles in Eq. (5.2) can be considered in terms of associated bundles. Write $\mathbb{H}$ for the quaternions. The symmetry group of the quaternions is exactly $Sp(1) = SU(2)$, i.e. the unit quaternions. In 4 dimensions, we have the coincidence $\mathrm{Spin}(4) = SU(2) \times SU(2)$, with generic element written (p, q), and we can consider two Spin(4)-representations $_+\mathbb{H}$ and $_-\mathbb{H}$, whose underlying vector spaces are both $\mathbb{H}$, given by

$$(p, q)v = qv \text{ for } v \in {}_+\mathbb{H}$$
$$(p, q)v = pv \text{ for } v \in {}_-\mathbb{H}$$

along with some less natural-looking ones:

$$(p, q)v = pvq^{-1} \text{ for } v \in {}_-\mathbb{H}_+$$
$$(p, q)v = qvq^{-1} \text{ for } v \in {}_+\mathbb{H}_+.$$

One can check that the action on $_-\mathbb{H}_+$ satsfies that $(p, q) = (-1, -1)$ acts trivially on $_-\mathbb{H}_+$, so that the representation factors through $\mathrm{Spin}(4)/\mathbb{Z}_2 = SO(4)$, and it is the canonical representation of $SO(4)$.

Then, a spin structure on a manifold yields associated vector bundles:

$$\mathbb{S}^\pm := P_{\mathrm{Spin}(4)} \times_{\mathrm{Spin}(4)} \pm\mathbb{H} \tag{5.3}$$
$$TM \cong P_{\mathrm{Spin}(4)} \times_{\mathrm{Spin}(4)} {}_-\mathbb{H}_+ \tag{5.4}$$
$$\Lambda := \mathbb{R} \oplus \Lambda^2_+ T^*M \cong P_{\mathrm{Spin}(4)} \times_{\mathrm{Spin}(4)} {}_+\mathbb{H}_+. \tag{5.5}$$

The second isomorphism follows from the fact that $_-\mathbb{H}_+$ is the standard representation of $SO(4)$, together with the fact mentioned above, that the associated vector bundle of the frame bundle is the tangent bundle. The last bundle in that list may not seem so natural at the moment, but it will appear later (when we introduce $\Lambda^2_+ T^*M$).

The above equations should be interpreted as going from our first perspective on spin structures (as a principal bundle) to our second perspective, producing a pair of quaternionic vector bundles. To make this precise, we observe that there is a map:

$$\varrho' \colon {}_{-}\mathbb{H}_{+} \times {}_{+}\mathbb{H} \to {}_{-}\mathbb{H}.$$

which is Spin(4)-linear. Indeed, set $\varrho'(a \times \phi) = a\phi$ by multiplication in $\mathbb{H}$. Then $paq^{-1}q\phi = pa\phi$ of course, giving that ϱ' is Spin(4)-linear. Rearranging, we obtain a map (still call it ϱ' for convenience):

$$\varrho' \colon {}_{-}\mathbb{H}_{+} \to \mathrm{Hom}({}_{+}\mathbb{H}, {}_{-}\mathbb{H}).$$

Analyzing in coordinates gives us Eq. (5.2). The associated bundle construction is compatible with morphisms, and associated to ϱ' we obtain a morphism of bundles:

$$\varrho' \colon TM \to \mathrm{Hom}(\mathbb{S}^{+}, \mathbb{S}^{-}).$$

To compare with the description in Eq. (5.2), we have a bundle $\mathbb{S}$ defined by $\mathbb{S}^{+} \oplus \mathbb{S}^{-}$. The map ϱ' may then be viewed as a morphism $\mathbb{S} = \mathbb{S}^{+} \oplus \mathbb{S}^{-} \to \mathbb{S}$ using the matrix description:

$$\varrho = \begin{pmatrix} 0 & -(\varrho')^{*} \\ \varrho' & 0 \end{pmatrix}$$

Expressed thus, ϱ and ϱ' carry the same information.

In Eq. (5.1), the right-hand side has a composition law, and so it makes sense to ask if the morphism extends to some kind of morphism of algebras. Let $F(TM)$ be the free algebra on the bundle TM; i.e. $F(TM) = \bigoplus_{i=0}^{\infty} (TM)^{\otimes i}$, with multiplication given by tensor product. Indeed, categorical nonsense says that we have a morphism of algebras:

$$F(\varrho) \colon F(TM) \to \mathrm{Hom}(\mathbb{S}, \mathbb{S}).$$

The morphism $F(\varrho)$ satisfies additional relations, derivable from the description in terms of the Pauli matrices above, given by:

$$e_i^2 = -1, e_i e_j + e_j e_i = 0,$$

where e_i is an orthonormal basis for TM. The free (tensor) algebra $F(TM)$ of TM modulo these relations is called the *Clifford algebra* $Cl(TM)$, and so we have a morphism of algebras $Cl(TM) \to \mathrm{Hom}(\mathbb{S}, \mathbb{S})$. That is to say, $\mathbb{S}$ is a *bundle of modules* over the algebra-bundle $Cl(TM)$; such a bundle of modules will be called a *Clifford module*. Further, we also note that the action of the Clifford algebra is skew-adjoint, in that:

$$\langle v \cdot s_1, s_2 \rangle = -\langle s_1, v \cdot s_2 \rangle.$$

A *spin connection* on $\mathbb{S}$ can be viewed in several ways as well. If one likes connections on principal bundles, a spin connection is a connection on $P_{\mathrm{Spin}(n)}$ compatible with the Levi-Civita connection on $P_{SO(n)}$. More practical for our purposes, we will use connections on vector bundles.

We do not assume familiarity with the notion of connections, although we do assume familiarity with the Levi-Civita connection. A connection is a mechanism for considering derivatives of sections of bundles, instead of only being able to differentiate functions.

Definition 5.1.1 Let $E \to M$ be a smooth complex vector bundle over M. Let $\Gamma(E)$ be the space of smooth sections of E. A *covariant derivative* on $E \to M$ is a $\mathbb{C}$-linear map

$$\nabla : \Gamma(E) \to \Gamma(T^*M \otimes E)$$

so that

$$\nabla(fs) = df \otimes s + f \nabla s$$

holds for all smooth functions on M and smooth sections $s \in \Gamma(E)$.

The collection of elements $(\nabla_v s)_x := \iota_v(\nabla s)_x \in E_x$, as v varies, should be viewed as the "derivatives of the section s" in the direction of v.

Example 5.1.2 Let $\underline{\mathbb{R}}^m$ be the trivial $\mathbb{R}^m$ bundle over $\mathbb{R}^n$. The "trivial connection" on $\underline{\mathbb{R}}^m$ is given by sending a section $f \in \Gamma(\mathbb{R}^m) = C^\infty(\mathbb{R}^n, \mathbb{R}^m)$ to $df \in T^*\mathbb{R}^n \otimes \underline{\mathbb{R}}^m$. The element df might require some explanation, in that f is not a $\mathbb{R}$-valued function; really we should write $\nabla f = (df_1, \ldots, df_n)$ where f_i are the components of f.

In this case, it is pretty clear that covariant differentiation 'looks like a derivative'. Note that if we had chosen a *different* trivialization of $\underline{\mathbb{R}}^m$, we would get a different connection. To wit, let $g : \underline{\mathbb{R}}^m \to \underline{\mathbb{R}}^m$ be a bundle automorphism. Then define

$$\nabla^g = g \cdot \nabla \cdot g^{-1}.$$

Let us see that this expression makes sense. Indeed,

$$\Gamma(\underline{\mathbb{R}}^m) \xrightarrow{g_*^{-1}} \Gamma(\underline{\mathbb{R}}^m) \xrightarrow{\Gamma} (T^*\mathbb{R}^n \otimes \underline{\mathbb{R}}^m) \xrightarrow{(\mathrm{id} \otimes g)_*} \Gamma(T^*\mathbb{R}^n \otimes \underline{\mathbb{R}}^m).$$

In the last line, we have used that for bundles E, F with endomorphisms $\alpha : E \to E$ and $\beta : F \to F$, we obtain a bundle map $\alpha \otimes \beta : E \otimes F \to E \otimes F$. A short calculation establishes $\nabla^g(f) = g(dg^{-1})f + df$, itself equal to $-(dg)g^{-1}f + df$. (This expression requires some interpretation to make sure it's valid, which you should do!)

A connection can be orthogonal, unitary, etc. Indeed, a connection ∇_E is orthogonal if for sections σ_1, σ_2 of E, and vector fields v, we have

$$v\langle\sigma_1, \sigma_2\rangle = \langle\nabla_v\sigma_1, \sigma_2\rangle + \langle\sigma_1, \nabla_v\sigma_2\rangle.$$

A section σ is called *parallel* for ∇ if $\nabla\sigma = 0$.

Exercise 5.1.3 Let E, F be vector bundles equipped with connections ∇_E, ∇_F. Show that E^* has a naturally defined connection, and define a connection on $E \otimes F$ that satisfies the Leibniz rule.

We can rephrase spin connections in terms of associated bundles:

Definition 5.1.4 A spin connection is a quaternionic connection on $\mathbb{S}$ so that ϱ is parallel.

A fundamental fact is:

Proposition 5.1.5 *Every spin manifold (M, g) admits a unique spin connection.*

At this point we observe that we have an operator which takes sections of our bundle $\mathbb{S}$ to sections with a T^*X factor, and we also have an operator going the other way: a tangent (or cotangent) vector and a spinor produce a spinor. Having an object with an endomorphism is usually much better than having two objects with a morphism between them, which is the case here as well, as follows.

A Clifford module structure and spin connection gives us a first-order operator on the bundle itself:

$$\not{\partial}\colon C^\infty(\mathbb{S}) \xrightarrow{\nabla} C^\infty(T^*M \otimes \mathbb{S}) \xrightarrow{\langle\cdot,\cdot\rangle} C^\infty(TM \otimes \mathbb{S}) \to C^\infty(\mathbb{S}).$$

We will call this the Dirac operator $\not{\partial}$. It is first-order, and elliptic. We will not define elliptic operators, but suffice it to say that the $\overline{\partial}$ operator on $\mathbb{C}$-valued functions is an elliptic operator, and that solutions of $\not{\partial}s = 0$ share many properties with holomorphic functions.

The splitting of Spin(4) has one more important ramification. Recall that the Hodge star operator $*\colon \Omega^2(M) \to \Omega^2(M)$ squares to the identity. In a local coordinate patch, it is easy to see that $*$ has a three-dimensional $(+1)$-eigenspace $\Omega^+(M)$ and a three-dimensional (-1)-eigenspace $\Omega^-(M)$, referred to as the *self-dual* and *anti-self-dual* 2-forms, respectively. By a similar token, the volume form ω on M squares to the identity in the Clifford algebra $Cl(M)$, and induces a splitting $\mathbb{S} = \mathbb{S}^+ \oplus \mathbb{S}^-$ as the $\pm$ eigenspaces of $Cl(\omega)$ (this splitting has already appeared several times, but this is the intrinsic way to give the splitting).

Exercise 5.1.6 Check that Clifford multiplication gives isomorphisms (the morphisms on the right are the traceless skew-adjoint $\mathbb{C}$-linear morphisms):

$$\Omega^+ = \mathrm{Hom}(\mathbb{S}^+, \mathbb{S}^+) \text{ and } \Omega^- = \mathrm{Hom}(\mathbb{S}^-, \mathbb{S}^-)$$

Then, in addition to having the usual exterior derivative operator $d\colon \Omega^1 \to \Omega^2$, we also have

$$d^+ \colon \Omega^1 \to \Omega^+$$

obtained by $\pi_+ \circ d$, where $\pi_+ \colon \Omega^2(M) \to \Omega^+(M)$ is projection to $\Omega^+(M)$. (Note that π_+ is a morphism of bundles.)

The Seiberg-Witten equations are not too terrible to write down at this point. We will write them down only for a spin manifold (as that is the case we will be most interested in). The *ungauged* Seiberg-Witten equations are as follows. Let $a \in \Omega^1(M)$ and φ be a spinor. We write $\partial\!\!\!/$ for the Dirac operator on M (with respect to our choice of spin structure). Then the equations are:

$$\partial\!\!\!/\varphi + \varrho(a)\varphi = 0 \tag{5.6}$$

$$\varrho(d^+a) - \varphi \otimes \varphi^* + \frac{1}{2}|\varphi|^2\mathrm{id} = 0. \tag{5.7}$$

A word on the term $\varphi\otimes\varphi^*$. This line takes place as an equation of endomorphisms of $\mathbb{S}^+$. The expression φ^* is the function $\mathbb{S}^+ \to \mathbb{C}$ by pairing $\langle\varphi, \cdot\rangle$.

Exercise 5.1.7 The morphism

$$\varphi \otimes \varphi^* - \frac{1}{2}|\varphi|^2\mathrm{id}$$

is traceless.

It turns out that Eqs. (5.6), (5.7) have an excess of symmetry, and would have too large a solution space. That symmetry can be removed from the picture by considering the *Seiberg-Witten equations in Coulomb gauge*:

$$\begin{aligned} \partial\!\!\!/\varphi + \varrho(a)\varphi &= 0, \\ \varrho(d^+a) - \varphi \otimes \varphi^* + \frac{1}{2}|\varphi|^2\mathrm{id} &= 0. \\ d^*a &= 0. \end{aligned} \tag{5.8}$$

Just staring at the equations yields a couple more formal observations, before we try to think about solutions of the equations. First, recall that the spinors φ live in quaternionic, in particular complex, bundles. So spinors φ have an action by $U(1) = S^1 \subset \mathbb{C}$. To be more precise, we have an action:

$$U(1) \times \Gamma(\mathbb{S}) \to \Gamma(\mathbb{S}).$$

Let us write this action in terms of our associated bundle construction from before. Now, we will think of each of ${}_+\mathbb{H}$, ${}_-\mathbb{H}$, ${}_-\mathbb{H}_+$, ${}_+\mathbb{H}_+$ as a $\mathrm{Spin}(4) \times S^1$ space, where we set:

$$(p, q, q_0)v = qvq_0^{-1} \text{ for } v \in {}_+\mathbb{H}$$
$$(p, q, q_0)v = pvq_0^{-1} \text{ for } v \in {}_-\mathbb{H}$$
$$(p, q, q_0)v = pvq^{-1} \text{ for } v \in {}_-\mathbb{H}_+$$
$$(p, q, q_0)v = qvq^{-1} \text{ for } v \in {}_+\mathbb{H}_+.$$

Since Clifford multiplication is induced by quaternionic multiplication $\varrho\colon {}_-\mathbb{H}_+ \times {}_+\mathbb{H} \to \mathbb{H}$, we see that $\varrho(a)e^{i\theta}\varphi = e^{i\theta}\varrho(a)\varphi$, by a quick calculation. The Dirac operator is also complex linear (though it does not, of course, commute with multiplication on φ by nonconstant elements of $\Gamma(M; \mathbb{C})$). The term $|\varphi|^2$ is unaffected by scalar multiplication, as is $\varphi \otimes \varphi^*$ (note that competing factors of $e^{i\theta}$ cancel here). That is, the Seiberg-Witten equations have an S^1-symmetry.

Finally, to put things in the abstract framework: we take the perspective that the Seiberg-Witten equations should be viewed instead as the Seiberg-Witten *map*:

$$\mathcal{SW}\colon \Gamma(\Omega^1) \oplus \Gamma(\mathbb{S}^+) \to \Gamma(\mathbb{S}^-) \oplus \Gamma(\Omega^+) \oplus \Gamma(\Omega^0). \tag{5.9}$$

The quantity $\mathcal{SW}(a, \varphi)$ is defined as the left-hand side of Eq. (5.8). Sometimes one removes the d^*a part of the map, at the expense of replacing $\Gamma(\Omega^1)$ in the domain with $\Gamma(\ker d^*|_{\Omega^1})$, using that d^* is injective; this will not really effect anything we have to say later. We reiterate that the Seiberg-Witten map depends on a choice of metric g on M.

By the above discussion, $\mathcal{SW}$ is an S^1-equivariant map (each element of S^1 acts on forms by the identity). The domain and codomain are both infinite-dimensional vector spaces. The map $\mathcal{SW}$ is *not* linear, but the only nonlinear parts come from $\varrho(a)\varphi$ and the component of the equation in Ω^+ coming for φ. The Seiberg-Witten map depends on a choice of metric, but we would like to have, in some sense, that the "homotopy class" of the Seiberg-Witten map is well-defined.

Unfortunately, infinite-dimensional spaces are poorly-behaved, in that they are too well-behaved. There are many manifestations of this; one which will come up later is:

Theorem 5.1.8 (Kuiper) *The general linear group of a Hilbert space (and most reasonable Banach spaces) is contractible.*

Homotopy classes of maps between Hilbert spaces are not so interesting (both domain and range are of course contractible), and this is still true for proper maps between Hilbert spaces [AR20]. However, the Seiberg-Witten equations satisfy further properties.

We shall not prove compactness for the moduli space of the Seiberg-Witten equations, but the following result is fundamental:

Proposition 5.1.9 ([Wit94]) *The space of solutions to the Seiberg-Witten equations (5.8) is compact.*

In fact, a (a priori much stronger, though in practice hard to imagine a difference between the two) property is true.

Proposition 5.1.10 ([Fur01]) *The preimage of a bounded disk, under the Seiberg-Witten map (5.9), is bounded.*

In fact, say g_t for $t \in [0, 1]$ is a family of metrics on M, with associated Seiberg-Witten functions $\mathcal{SW}_t$. If B is a bounded disk in $\Gamma(\Omega^+) \oplus \Gamma(\Omega^0) \oplus \Gamma(\mathbb{S}^-)$ the preimage $\cup_{t \in [0,1]} \mathcal{SW}_t^{-1}(B)$ is bounded.

This is a version of properness in the setting of non-locally compact spaces.

5.2 The Bauer-Furuta Construction

Let H_1, H_2 be infinite-dimensional separable Hilbert spaces over $\mathbb{R}$ and write $B_i(r)$ for the closed ball of radius r in H_i. We will call a map

$$F = L + Q \colon H_1 \to H_2$$

between Hilbert spaces H_1, H_2, *very Fredholm* if:

1. L is a linear Fredholm map,
2. Q is compact (and possibly nonlinear!). *Compact* for a nonlinear operator Q means that $Q(B_1(r))$ is a set with compact closure in H_2, for all r.
3. For each $r > 0$, $F^{-1}(B_2(r))$ is bounded.

For a compact connected space X, we also have the notion of a *family* of very Fredholm maps parameterized by X, which is a map

$$F = L + Q \colon X \times H_1 \to H_2$$

where

1. L is *fiberwise* linear and Fredholm on each fiber.
2. Q is compact as in the ordinary definition of very Fredholm.
3. For each $r > 0$, and each $x \in X$, there exists a neighborhood $x \in U_x \subset X$ and some $R > 0$ so that has $F^{-1}|_{U_x \times H_1}(B_2(r)) \subset B_1(R)$.

A family is virtually the same data as a continuously varying collection of very Fredholm maps.

In particular, a *homotopy* of very Fredholm maps is a family F_t parameterized by the interval $[0, 1]$.

We define the *index* of a very Fredholm map to be the Fredholm index of L. Note that the index is independent of the presentation of F (that is, if we can write

$F = L + Q$ and $F = L' + Q'$, then $\mathrm{ind}(L) = \mathrm{ind}(L')$), and is constant for a family of very Fredholm maps.

There is the following, potentially surprising, result of Schwarz:[1]

Theorem 5.2.1 ([Š64]) *There is a bijection:*

$$\frac{\{\text{ very Fredholm maps of index } n\}}{\{\text{very Fredholm homotopy}\}} \leftrightarrow \tilde{\pi}_n(S^0) = colim_{r\to\infty}[S^{n+r}, S^r]$$

The right-hand side is the *n-th stable stem*, or the set of stable maps, up to homotopy, $\{S^n, S^0\}$.

We will say a bit about the proof of this theorem. However, it is our viewpoint that the map $L+Q$ itself, rather than its associated cohomotopy class, is the relevant invariant from the geometry.

Choose an increasing sequence of finite-dimensional subspaces $V_k \subset H_2$, so that each V_k is transverse to $\mathrm{im}(L)$ (the condition on $\mathrm{im}(L)$ is in some sense to ensure genericity. Recall that the complement of $\mathrm{im}(L)$ is finite-dimensional). Let W_k be the preimages of V_k under L. Let π_k be the projection to V_k; we will sometimes also use projection π_k to W_k; the overloaded notation should not cause confusion. We require that the sequence π_k converges to the identity id_{H_2} in the *strong operator topology* on H_2, in that: for each $x \in H_2$, and all $\varepsilon > 0$, there exists k_ε so that

$$||\pi_k x - \mathrm{id}\, x|| < \varepsilon$$

for all $k \geq k_\varepsilon$. This requirement on the collection $\{V_k\}$ is equivalent to requiring $\overline{\bigcup_k V_k} = H_2$.

Idea of the Proof of Theorem 5.2.1 We wish to replace $L + Q$ with a map on our finite-dimensional approximation $W_k \to V_k$. Eventually, we will take closed balls $B_1 \subset H_1$ and $B_2 \subset H_2$ and consider a sequence of approximations on $B_{W_k} = B_1 \cap W_k$ and $B_{V_k} = B_2 \cap V_k$, and

$$F_k : B_{W_k} \to B_{V_k}.$$

It will turn out that the F_k send the boundary ∂B_1 far from the origin of V_k, and so we may perform a sort of Thom construction to produce a map of (finite-dimensional) spheres:

$$\tilde{F}_k : B_{W_k}/\partial B_{W_k} \to B'_{V_k}/\partial B'_{V_k}$$

where we have replaced B_{V_k} with a smaller ball B'_{V_k}.

[1] We have somewhat rearranged the statement from its original form.

Proof of Theorem 5.2.1 To start the approximation, let us fix the ball $B_2(\varepsilon)$ of radius $\varepsilon > 0$ in H_2; by condition (3), we have, without loss of generality by possibly rescaling the metric on H_1, that $F^{-1}(B_2(\varepsilon)) \subset B_1^\circ(1)$.

We restrict all our maps F, L, Q to the ball $B_1(1)$ in what follows. We also note that by the assumption that $\mathrm{im}(L)$ and V_k are transverse, $\dim W_k = n + \dim V_k$. Let $B_{W_k} = B_1(1) \cap W_k$, and fix a ball $B_2 \subset H_2$ so large that it contains the image of F (recall that F is restricted to $B_1(1)$); by construction $B_2(\varepsilon) \subset B_2$. Let $B_{V_k} = B_2 \cap V_k$.

We would like the restriction $F|_{W_k}$ to be a map to V_k. This is easily enough arranged—set

$$F_k = \pi_k F \colon W_k \to V_k.$$

□

Claim 5.2.2 The sequence $\{F_k\}_k$ converges to F in the norm topology. That is,

$$\lim_{k\to\infty} \big(\sup_{x\in B_{W_k}} ||F_k(x) - F(x)||\big) = 0.$$

Proof of Claim 5.2.2 First, we reduce this to the claim that $\pi_k Q \to Q$ in norm topology. To see this, we note:

$$F_k(x) - F(x) = \pi_k(L+Q)(x) - (L+Q)x = \pi_k Qx - Qx.$$

This explains our choice of W_k as approximating subspaces—for these approximating subspaces, the contribution to $F_k - F$ from the linear terms vanishes.

Now, we show $\pi_k Q \to Q$. Indeed, $\mathrm{im}\, Q$ is compact by (2). For each $\varepsilon > 0$, and for each $y \in Q$, pick $N_y \in \mathbb{N}$ and neighborhood $y \in U_y$ so that:

$$||\pi_k z - z|| < \varepsilon, \text{ for all } z \in U_y \text{ and } k > N_y. \tag{5.10}$$

This is possible since for each $y \in Q$, $\pi_k y \to y$ as $k \to \infty$, and then we obtain (5.10) for some neighborhood of y, using continuity of π_k.

Choose a finite subcover $\{y_1, \ldots, y_r\}$ and let

$$K = \sup_{y\in\{y_i\}_{i=1,\ldots,r}} N_y.$$

Then $||\pi_k Q(x) - x|| < \varepsilon$ for all $k > K$ and all $x \in B_1$. This suffices to establish $\pi_k Q \to Q$, finishing the proof of the claim. (Really, we have just proved the general fact that the composite of a strongly convergent sequence of operators with a compact operator is norm convergent.) □

Claim 5.2.2 is really the heart of the matter—it says our approximations are very close to the real thing.

Now, choose k so large that $||F_k - F|| < \varepsilon/2$; read that expression as $||F_k(x) - F(x)|| < \varepsilon$ for all $x \in H_1$. Although we will be interested in F_k with domain B_{W_k}, it also makes sense to consider F_k on all of H_1. We have, for $x \in \partial B_{W_k}$,

$$||F_k(x)|| \geq ||F(x)|| - ||F_k(x) - F(x)|| \geq ||F(x)|| - \frac{\varepsilon}{2}.$$

Thus, F_k induces a map of pairs:

$$F_k\colon (B_{W_k}, \partial B_{W_k}) \to (B_{V_k}, B_{V_k} - B_{V_k}(\varepsilon/2)). \tag{5.11}$$

In turn, (5.11) gives a map:

$$\tilde{F}_k\colon S^{\dim V_k + n} \cong B_{W_k}/\partial B_{W_k} \to B_{V_k}(\varepsilon/2)/\partial B_{V_k}(\varepsilon/2) \cong S^{\dim V_k}. \tag{5.12}$$

The map $\tilde{F}_k$ is exactly the class associated, in Theorem 5.2.1, to F.

We must show that $\tilde{F}_k$ is independent of the choices involved in its construction. This means:

1. first relating F_k to F_{k+1} for k large,
2. allowing changes in the radius of the balls $B_1(1)$, $B_2(\varepsilon)$,
3. changes in the family V_k,
4. different expressions for F as $L + Q$,
5. finally, also checking that the homotopy class of F_k is independent of F up to very Fredholm homotopy.

We will only check a few of these things; all of them are rather straightforward. Finally, in addition to all of that, we must show that every map f in $[S^{n+r}, S^r]$ gives a very Fredholm map. This can be done by choosing some model for the map f so that it takes $\mathbb{R}^{n+r} \to \mathbb{R}^r$, and then taking a product of f with a linear map on an infinite-dimensional Hilbert space H.

We show (1), that $\tilde{F}_k$ is independent of the choice of k in its construction. This requires some interpretation—of course, the maps $\tilde{F}_k$ and $\tilde{F}_{k+1}$ have domains of different dimensions!

Consider the family

$$F_{k+1,t} = L_{k+1} + (1-t)\pi_{k+1} Q \pi_{k+1} + t\pi_k Q \pi_k.$$

of maps defined $W_{k+1} \to V_{k+1}$. Let ε be so that the preimage (for all t) $F_{k+1,t}(B_2(\varepsilon)) \subset B_1^\circ(1)$. The same argument as for $F_{k+1,t}$ applies to show that: for all k sufficiently large: $F_{k+1,t}$ sends $\partial B_{W_{k+1}}(1)$ to $B_{V_{k+1}} - B_{V_{k+1}}(\varepsilon/2)$ for all t.

That is the same as saying that $F_{k+1,t}$ defines a homotopy of maps of pairs between:

$$F_{k+1,0} = F_{k+1} \text{ and } F_{k+1,k} = F_k \pi_k + L\pi_k',$$

$$\text{as maps } (B_{W_{k+1}}, \partial B_{W_{k+1}} \to (B_{V_{k+1}}, B_{V_{k+1}} - B_{V_{k+1}}(\varepsilon/2)).$$

where π'_k is the projection from $W_{k+1} \to W_k^\perp$. We can choose a further homotopy of F:

$$F_{k+1,t} = F_k\pi_k + ((2-t) + (t-1)\pi'_k)L\pi'_k$$

for $t \in [1, 2]$, where this π'_k will be the projection $V_{k+1} \to V_k^\perp \subset V_{k+1}$. Again, the argument as for F_k shows that $F_{k+1,t}$ defines a homotopy of pairs:

$$F_{k+1,1}, F_{k+1,2} \colon (B_{W_{k+1}}, \partial B_{W_{k+1}} \to (B_{V_{k+1}}, B_{V_{k+1}} - B_{V_{k+1}}(\varepsilon/2)).$$

Note that $\pi'_k L\pi'_k$ extends to define a homeomorphism of one-point compactifications:

$$\pi'_k L\pi'_k \colon (W_k^\perp)^+ \to (V_k^\perp)^+.$$

With this notation, $F_{k+1,2}$ is exactly the smash product

$$F_k \wedge \pi'_k L\pi'_k \colon (B_{W_{k+1}}(1), \partial B_{W_{k+1}}(1)) \wedge (W_k^\perp)^+$$
$$\to (B_{V_{k+1}}, B_{V_{k+1}} - B_{V_{k+1}}(\varepsilon/2)) \wedge (V_k^\perp)^+$$

That is to say, $F_{k+1} \colon S^{n+\dim V_{k+1}} \to S^{\dim V_{k+1}}$ is homotopic to the suspension

$$F_k \wedge \mathrm{id} \colon S^{n+\dim V_k} \wedge S^{\dim V_k^\perp} \to S^{\dim V_k} \wedge S^{\dim V_k^\perp}$$

We note one ambiguity in the proof. The domain and codomain of the maps $\tilde{F}_k$ we constructed were not *canonically* identified with the spheres $S^{\dim V_k+n}$ and $S^{\dim V_k}$. Also, of course the sphere S^n admits a non-trivial homotopy-equivalence up to homotopy; namely, the map of degree -1. So the association in the statement is not canonical. However, we can define a sphere associated to L:

$$S^{\mathrm{ind}(L)} := \Sigma^{-\mathrm{coker}(L)}(\ker L)^+.$$

This is the 'formal desuspension' of the kernel by the cokernel. It would take some time to make that precise, but the theorem does define a canonical element in the set of stable homotopy classes:

$$\{(\ker L)^+, (\mathrm{coker} L)^+\} = colim_{k\to\infty}[S^k \wedge (\ker L)^+, S^k \wedge (\mathrm{coker} L)^+]$$

To close the section, we observe that Proposition 5.1.10 implies that $\mathcal{SW}$ of (5.9) is a very Fredholm map (in fact, we will see later that it is an equivariant very Fredholm map, with a corresponding version Theorem 5.3.1 of the nonequivariant Theorem 5.2.1). That is to say, Sect. 5.1 takes us from the geometry to a certain situation in functional analysis, and the present section transforms that to a situation belonging to homotopy theory.

5.3 Equivariant Topology

In proving the $\frac{10}{8}$-Theorem, we will need to have some equivariant topology. A convenient source for the material in this section is Adams' introduction [Ada06].

Much as there is the category of topological spaces Top, for G a compact Lie group, we can form the category of G-spaces, Top_G, where the objects are G-spaces, and the maps $X \to Y$ are continuous maps that are G-equivariant. Much as there is homotopy of topological spaces, it is fine (and unambiguous) to also consider *equivariant* homotopies of G-maps; to be explicit, a (equivariant) homotopy between $f_0, f_1 : X \to Y$ is a G-equivariant map

$$f : X \times [0, 1] \to Y$$

where the G-action on $X \times [0, 1]$ is the product G-action, where G acts trivially on $[0, 1]$. For G-spaces X, Y, we will write

$$[X, Y]_G$$

for the set of homotopy classes of G-equivariant maps from X to Y.

Since the Bauer-Furuta construction started with a G-equivariant map of Hilbert spaces, where $G = \mathrm{Pin}(2)$, it is natural to expect that the Bauer-Furuta invariant should really be a $Pin(2)$-equivariant *stable homotopy class*; so let us define such an equivariant version of stable homotopy classes.

First, it will be convenient to work with pointed topological spaces always, and we will usually assume all spaces are pointed; the basepoint of a G-space is required to be fixed by the action of G. We write $\mathrm{Top}_{G,*}$ for the category of G-spaces with basepoint.

Recall that suspension defines a functor $\Sigma : \mathrm{Top} \to \mathrm{Top}$; stable (nonequivariant) homotopy classes are exactly the homotopy classes of maps 'after we have suspended enough' by $\Sigma^{n \gg 0}$.[2]

But for G-spaces, we have the mantra that *all integers should be replaced by representations*. In particular, for V an orthogonal, finite-dimensional G-representation, we have a functor

$$\Sigma^V : \mathrm{Top}_{G,*} \to \mathrm{Top}_{G,*}, \qquad X \rightsquigarrow (V^+) \wedge X =: \Sigma^V X,$$

where the space on the right is given the G-action $g(y, x) = (gy, gx)$ (and we recall that V^+ is the one-point compactification of V. We let G act trivially on the point at infinity, and also declare the point at infinity to be the basepoint of V^+).

[2] Technically speaking, we should be more careful here for spaces which are not finite CW complexes. For finite CW complexes, there exists some n for which the sequence $[X, Y] \to [\Sigma X, \Sigma Y] \to [\Sigma^2 X, \Sigma^2 Y] \to \ldots$ stabilizes.

For finite CW complexes, the sequence $[X, Y] \to [\Sigma X, \Sigma Y] \to [\Sigma^2 X, \Sigma^2 Y] \to \ldots$ eventually stabilizes, and there is an analogue (for appropriate spaces) in the equivariant category, where one now has a directed system:

$$
\begin{array}{ccccccc}
 & & & & & & \cdots \\
 & & & & & \nearrow & \\
 & & & & [\Sigma^V \oplus V X, \Sigma^{V\oplus V} Y]_G & & \\
 & & & \nearrow & & \searrow & \\
 & & [\Sigma^V X, \Sigma^V Y]_G & & & & \cdots \\
 & \nearrow & & \searrow & & \nearrow & \\
[X, Y]_G & & & & [\Sigma^{V\oplus W} X, \Sigma^{V\oplus W} Y]_G & & \\
 & \searrow & & \nearrow & & \searrow & \\
 & & [\Sigma^W X, \Sigma^W Y]_G & & & & \cdots \\
 & & & \searrow & & \nearrow & \\
 & & & & [\Sigma^{W\oplus W} X, \Sigma^{W\oplus W} Y]_G & & \\
 & & & & & \searrow & \\
 & & & & & & \cdots
\end{array}
$$

with many more arrows not drawn (and potentially all sorts of arrows going between what is drawn in the diagram, depending on G, V, W).

In any event, we define the homotopy classes of equivariant stable maps $X \to Y$ by

$$\{X, Y\}_G = colim_V [\Sigma^V X, \Sigma^V Y]_G. \tag{5.13}$$

The colimit runs over all finite-dimensional representations of G.[3] This colimit makes sense even if the sequence in the diagram did not eventually stabilize (which may happen, if, e.g. X, Y are infinite CW complexes). For notational convenience, we write S^V for V^+.

In the study of the Seiberg-Witten equations, the following point is sometimes relevant. Instead of taking the colimit (5.13) over *all* isomorphism classes of finite-dimensional representations of G, we can restrict to a smaller collection, satisfying certain properties, (a *universe* $\mathcal{U}$) of isomorphism classes of representations (which should at least contain infinitely many copies of the trivial representation $\mathbb{R}$). Different universes may give different homotopy classes of equivariant stable maps $X \to Y$.

In particular, we can define stable homotopy groups:

$$\tilde{\pi}_V(X) := \{S^V, X\}_G$$

[3] We ignore set-theoretic issues.

If we need to distinguish which universe is used, we write $\tilde{\pi}^G_{V,\mathcal{U}}(X)$ for the V-th homotopy group of X with respect to the universe $\mathcal{U}$. It is also convenient to consider sometimes formal linear combinations of finite-dimensional representations of G. Let $RO(G)$ be the *Grothendieck group* of G; that is, the free abelian group on the set isomorphism classes of finite-dimensional real representations V of G, modulo

$$[V \oplus W] = [V] + [W].$$

We also define

$$\tilde{\pi}^G_{V-W}(X) = \{S^V, \Sigma^W X\}_G,$$

where $V - W \in RO(G)$. Adams [Ada06] makes quite a point, of biblical proportions, that one should not write this expression (it is ambiguous in that $V - W$ may be written in many ways, and the expressions for different ways of presenting $V - W \in RO(G)$ are not canonically identified), but the difficulty can be overcome, and for us maybe it is best to leave our V-th equivariant stable homotopy group in this, slightly ambiguous, format.

Certainly this is all rather sensible, but it is not so clear it is useful, and might be chalked up to abstract nonsense. Let us connect it back to the Bauer-Furuta invariant. Here, let us put down our algebraic topology hat and back on our functional analysis hat.

Recall that we have defined an *index* for a Fredholm operator L as

$$\text{ind}L = \dim\ker L - \dim\text{coker}L.$$

However, if L is a G-equivariant Fredholm operator between (now possibly infinite dimensional) Hilbert spaces H_1, H_2 with G-action, then $\ker L$ and coker$L$ are no longer just vector spaces—but finite-dimensional G-representations! We define the index as:

$$\text{ind}L = \ker L - \text{coker}L \in RO(G).$$

We can state the equivariant version of the Schwarz theorem. Let H_1, H_2 be Hilbert spaces that are also orthogonal representations of G, so that each H_j contains a dense G-equivariant subspace $\oplus_{V\in I} V^\infty$, where I is a collection of finite-dimensional orthogonal G-representations.

Theorem 5.3.1 *Let $F\colon H_1 \to H_2$ be a G-equivariant very Fredholm map (in the previous notation). Then:*

$$\frac{\{G-\textit{equivariant very Fredholm maps}\}}{\{\textit{homotopy}\}} \cong \tilde{\pi}^G_{\text{ind}L}(S^0) = \{(\ker\ L)^+, (\text{coker}\ L)^+\}_G$$

The *universe* in Theorem 5.3.1 is the collection of approximations for the Hilbert spaces H_1, H_2, if indeed these collections form universes.

Keeping on our gauge theory hat, we pull the Atiyah-Singer index theorem out of that same hat:

Theorem 5.3.2 (Atiyah-Singer) *For our Dirac operator* $\not\partial$*, on* (M^4, g)*, the nonequivariant index is given by*

$$\operatorname{ind} \not\partial = (-\sigma(M)/4)$$

and

$$\operatorname{ind}(d + d^*) = -b_2^+.$$

The equivariant indices are given by:

$$\operatorname{ind} \not\partial = \mathbb{H}^{-\sigma(M)/16}$$

and

$$\operatorname{ind}(d + d^*) = \tilde{\mathbb{R}}^{-b_2^+(M)}$$

For the notation in the theorem: recall that $b_2^+(M)$ is the rank of the positive part of the intersection form of M, and $\sigma(M)$ is the signature of M. The Pin(2)-representation $\mathbb{H}$ is the quaternions considered as a Pin(2)-space (from the right), and $\tilde{\mathbb{R}}$ is the Pin(2) representation where j acts as -1 and $e^{i\theta} \in \text{Pin}(2)$ acts by 1.

Write G for Pin(2) for now. The relevant universe for us is $\mathcal{U} = \mathbb{R}^\infty \oplus \mathbb{H}^\infty \oplus \tilde{\mathbb{R}}^\infty$, though this will be suppressed from the notation. The grand finale of this section is the conclusion:

Theorem 5.3.3 (Furuta, [Fur01]) *Let* (M, g) *be a simply-connected closed spin 4-manifold. Then there is an element:*

$$BF_G(M) \in \{(\mathbb{H}^{-\sigma(M)/16)})^+, (\tilde{\mathbb{R}}^{b_2^+(M)})^+\}_G = \pi^G_{\operatorname{ind} L}(S^0). \tag{5.14}$$

The element $BF_G(M)$ *only depends on our choice of identification of* $\mathbb{H}^{-\sigma(M)/16)} - \tilde{\mathbb{R}}^{b_2^+(M)}$ *with* $\operatorname{ind} L$.

Proof By Proposition 5.1.10, associated to the Seiberg-Witten equations there is a very Fredholm Pin(2)-equivariant map, whose equivariant index is given by Theorem 5.3.2. Theorem 5.3.1 then yields an associated equivariant stable homotopy class. The second part of Proposition 5.1.10 implies that if we change the metric, in combination with Theorem 5.3.1, that the equivariant stable homotopy class is unchanged. Thus, associated to M (without reference to the metric g), we obtain a well-defined map as in (5.14). □

5.4 Proof of the 10/8-Theorem

Theorem 5.4.1 (Furuta) *Let (M, g) be a simply-connected closed spin 4-manifold. Then*

$$b_2(M) \geq \frac{10}{8}|\sigma(M)|.$$

By changing orientation, we may assume $\sigma(M) \geq 0$. Note, every such indefinite M has a decomposition of $H_2(M) = p(-E_8) \oplus qH$, where E_8 is the intersection form of the E_8 plumbing, and H is

$$H = \begin{pmatrix} 0 & 1 \\ 1 & 0 \end{pmatrix}$$

A quick calculation says that Theorem 5.4.1 is equivalent to:

$$q \geq p.$$

In fact, Furuta shows somewhat more "$10/8 + 2$"; namely, that if $p > 0$ then

$$q \geq p + 1.$$

The general plan of Furuta's proof is to show that for a manifold M violating the Theorem, there is no stable map that could be the Bauer-Furuta invariant $BF_G(M)$.

From last time, we have that for (M, g) we have a map:

$$BF_G(M) \in \{(\mathbb{H}^{-\sigma(M)/16})^+, (\tilde{\mathbb{R}}^{b_2^+(M)})^+\}_G = \{S^{p\mathbb{H}}, S^{q\tilde{\mathbb{R}}}\}_G.$$

Now, there are certainly many maps in that group on the right; let's call elements of $\{S^{p\mathbb{H}}, S^{q\tilde{\mathbb{R}}}\}_G$ *candidates*. Indeed, to see that the group on the right is nonempty, we note that there is always the *zero map* from $S^{p\mathbb{H}}$ to $S^{q\tilde{\mathbb{R}}}$ by unstably sending all of $\mathbb{H}^p$ to the basepoint.

Unfortunately, we then know that this set of homotopy classes at least has the trivial element. Let us put on our gauge theory hat again and see what other constraints our map $BF_G(M)$ must satisfy.

First, recall that for maps $[f]$ in $[X, Y]_G$, we could restrict a representative f to G-fixed points X^G, to get a new well-defined homotopy class $[f^G] \in [X^G, Y^G]$. A similar claim works perfectly well stably: we have an operation $\Phi^G\colon \mathrm{Top}_{G,*} \to \mathrm{Top}_*$, that sends stable homotopy classes $[f] \in \{X, Y\}_G$ to the stable homotopy class of the fixed point map $[f^G]$.

Indeed, let us consider $BF_G(M)^G$—this is the restriction of BF_G to a map on fixed points. Squinting at the proof of the Schwarz Theorem, we note that we have the following commutative square:

$$\begin{array}{ccc}\{G - \text{very Fredholm maps mod homotopy}\} & \longrightarrow & \pi^{G}_{\mathrm{ind}L}(S^0) \\ \text{fixed points}\big\downarrow & & \big\downarrow\text{fixed points} \\ \{\text{very Fredholm maps mod homotopy}\} & \longrightarrow & \pi_{(\mathrm{ind}L)^G}(S^0)\end{array}$$

So, to compute $BF_G(M)^G$, it suffices to consider the Seiberg-Witten map restricted to fixed points.

However, the fixed points of $\Gamma(\mathbb{S}^+) \oplus \Omega^1(M)$ are exactly the origin. The fixed points of $\Gamma(\mathbb{S}^-) \oplus \Omega^+(M) \oplus \Omega^0(M)$ are also just the origin. The Seiberg-Witten map of course takes the origin to the origin. The upshot:

$$BF_G(M)^G = \mathrm{id} \in \{(S^{p\mathbb{H}})^G, (S^{q\tilde{\mathbb{R}}})^G\} = \{S^0, S^0\} = \mathbb{Z}. \tag{5.15}$$

Call a map in $\{S^{p\mathbb{H}}, S^{q\tilde{\mathbb{R}}}\}_G$ *good* if it satisfies (5.15). We prove:

Proposition 5.4.2 *If $p > 0$, there are no good maps unless $q \geq p + 1$.*

For the proof, we will need some way to transfer data about maps between spheres to a more user-friendly receptacle, and G-equivariant K-theory is very effective for this problem.

5.4.1 G-Equivariant K-Theory

In this subsection, G can be any compact Lie group. Let X be a compact G-space, and define $K_G(X)$ to be the Grothendieck group of G-equivariant complex vector bundles over X (in this subsection, all of our vector bundles will be finite-dimensional). A *G-equivariant (complex) vector bundle* is a complex vector bundle $\pi : E \to X$, where there is an action of G on the total space E, covering the action of G on the base; i.e. for all $g \in G$ there is a commutative diagram:

$$\begin{array}{ccc} E & \xrightarrow{g} & E \\ \pi\big\downarrow & & \big\downarrow\pi \\ X & \xrightarrow[g]{} & X \end{array}$$

For a morphism $f \colon X \to Y$ of compact G-spaces, pullback of vector bundles f^* induces a map (still called f^*) $f^* \colon K_G(Y) \to K_G(X)$. The fact that, for *nonequivariant* maps, $f \simeq h \colon X \to Y$ implies $f^*(E) \cong h^*(E)$ for all vector bundles E goes through without change to the equivariant situation to say that if $f \simeq h \colon X \to Y$ (by a G-equivariant homotopy), then $f^*(E) \cong h^*(E)$ as G-bundles.

Indeed, $K_G(-)$ is an *equivariant cohomology theory*. We will not need a precise definition of this, but the notion is intuitive enough.

For X a compact G-space with a basepoint $*$, preserved by G, we define $\tilde{K}_G(X)$ to be the subgroup of $K_G(X)$ restricting to the 0 element at the basepoint $*$.

We will use only the group $\tilde{K}_G(X) = \tilde{K}^0_G(X)$. We list the key facts we will need:

1. $K_G(*)$, the K-theory of a single point, is $R(G)$, the complex representation ring of G. Equivalently, the reduced K-theory $\tilde{K}_G(S^0)$ is $R(G)$.
2. For any compact G-space X, $\tilde{K}_G(X)$ is a module over $R(G)$, where the module structure is given by tensor product of representations.
3. Let V be a complex G-representation. Then there is a natural (in X) *Bott periodicity* isomorphism:

$$\tilde{K}_G(\Sigma^V X) \xrightarrow{\mathbb{B}} \tilde{K}_G(X).$$

Moreover, there is a canonical map $\iota \colon X \to \Sigma^V X$, by sending $x \in X$ to $(0, x) \in \Sigma^V X$. The composite $\iota \circ \mathbb{B}^{-1} \colon \tilde{K}_G(X) \to \tilde{K}_G(X)$ is multiplication by the *equivariant Euler class*:

$$\lambda_{-1}(V) = \sum_{i=0}^{\dim V} (-1)^i [\wedge^i V].$$

We will also need information about $K_G(X)$ for the case $G = \mathrm{Pin}(2)$. For $G = \mathrm{Pin}(2)$, it turns out that $R(G) = \mathbb{Z}[w, z]/(2w - w^2, wz - 2w)$, where $w = \tilde{\mathbb{R}} - 1$ and $z = \mathbb{H} - 2$. (Here 1, 2 refer to the trivial 1− and two-dimensional representations, respectively.) The reader is encouraged to check the relations $w^2 = 2w$ and $wz = 2w$; it is not quite as direct to see that these are the only relations needed to describe $R(G)$.

Proof of Proposition 5.4.2 Say, on the contrary, that there exists such a (stable) map

$$f \colon S^{p\mathbb{H}} \to S^{q\tilde{\mathbb{R}}}.$$

We have a commutative diagram, where ι are the inclusions of Pin(2)-fixed points:

$$\begin{array}{ccc} S^{\mathbb{H}^p} & \xrightarrow{\;f\;} & S^{\tilde{\mathbb{R}}^q} \\ \uparrow{\scriptstyle \iota} & & \uparrow{\scriptstyle \iota} \\ S^0 & \xrightarrow[\;f^{S^1}\;]{} & S^0 \end{array}$$

inducing a diagram in K-theory as on the left below. By points (1)–(3), the left diagram is a diagram of $R(G)$-modules, with the given arrows being morphism of $R(G)$-modules:

$$
\begin{array}{ccc}
\tilde{K}_G(S^{\mathbb{H}^p}) & \xleftarrow{f^*} & \tilde{K}_G(S^{\tilde{\mathbb{R}}^q}) \\
\downarrow{\scriptstyle \iota^*} & & \downarrow{\scriptstyle \iota^*} \\
\tilde{K}_G(S^0) & \xleftarrow[(f^{S^1})^*]{} & \tilde{K}_G(S^0)
\end{array}
\qquad\qquad
\begin{array}{ccc}
R(G) & \xleftarrow{f^*} & R(G) \\
\downarrow{\scriptstyle z^p} & & \downarrow{\scriptstyle w^q} \\
R(G) & \xleftarrow[(f^{S^1})^* = 1]{} & R(G)
\end{array}
$$

Thus, $w^q = f^*(1)z^p$, which is to say $(w^q) \subset (z^p)$. This is only possible if either $p = 0, q = 0$, or, if $q \geq p + 1$ (as in the case, for example, that $w^{p+1} = wz^p$). □

So far, we have restricted attention only to *spin* manifolds, but this was not necessary. Indeed, for any closed oriented simply-connected 4-manifold M equipped with a spin^c structure $\mathfrak{s}$, there is a Bauer-Furuta invariant:

$$BF_{S^1}(M) \in \pi^{S^1}_{\mathrm{ind}L}(S^0) \cong \{S^{\mathbb{C}^{(c_1(\mathfrak{s})^2-\sigma(M))/8}}, S^{\mathbb{R}^{b_2^+(M)}}\}_{S^1} \tag{5.16}$$

To be more precise, the invariant is well-defined in $\pi^{S^1}_{\mathrm{ind}L}(S^0)$. In order to view $BF_{S^1}(M)$ in the group on the right, we must identify $\mathrm{ind}\, L$ with $\mathbb{C}^{(c_1(\mathfrak{s})^2-\sigma(M))/8} \oplus \mathbb{R}^{b_2^+(M)}$; up to deformation, there are two such choices, corresponding to orientations on $H^+(M; \mathbb{R}) \cong \mathbb{R}^{b_2^+(M)}$. A choice of orientation on $H^+(M; \mathbb{R})$ will be called a *homology orientation*.[4] The simple-connectedness assumption may be removed as well, at the cost of changing the domain.

To define $BF_{S^1}(M)$, we note that the spinor bundles $\mathbb{S}^+$ and $\mathbb{S}^-$ can still be defined only in the presence of a spin^c structure, as can the Clifford multiplication (5.1), in addition to the other operators from Sect. 5.1 necessary to define the Seiberg-Witten equations. However the spinor bundles $\mathbb{S}^{\pm}$ are now only complex bundles. This explains the fact that the Bauer-Furuta map in this setting is only an S^1-homotopy class, as opposed to a Pin(2)-homotopy class.

Finally, we also note that there is a strengthening of Furuta's theorem, by Hopkins-Lin-Shi-Xu [HLSX22].

For the reader's convenience, we also explain how to recover, following [BF04, Proposition 3.3], the ordinary Seiberg-Witten invariant from the S^1-equivariant Bauer-Furuta invariant.

Proposition 5.4.3 *Let* $(M, \mathfrak{s})$ *be a closed, oriented* spin^c *4-manifold with* $b^+ > 1$ *and* $b_1(M) = 0$*, equipped with a homology-orientation. Then the Seiberg-Witten*

---

[4] A homology orientation usually refers to an orientation of $H_1(M) \oplus H^+(M)$; recall that we have here assumed $H_1(M; \mathbb{Q}) = 0$.

invariant $SW(M) = t(BF_{S^1}(M))$, *for a certain morphism*

$$t\colon \pi_{S^1}^{b^+}(S^{p\mathbb{C}}) = \{S^{p\mathbb{C}}, S^{b^+}\}_{S^1} \to \mathbb{Z}.$$

Proof Say that we have a map as in (5.16), where we set $p = (c_1(\mathfrak{s})^2 - \sigma(M))/8$ and $q = b_2^+(M)$. Associated to the cofiber sequence of S^1-spaces

$$S^0 \to S^{p\mathbb{C}} \to S^{p\mathbb{C}}/S^0$$

we have an exact sequence of S^1-equivariant stable homotopy classes:

$$\{S^0, S^{q\mathbb{R}}\}_{S^1} \leftarrow \{S^{p\mathbb{C}}, S^{q\mathbb{R}}\}_{S^1} \leftarrow \{S^{p\mathbb{C}}/S^0, S^{q\mathbb{R}}\}_{S^1} \leftarrow \{\Sigma S^0, S^{q\mathbb{R}}\}_{S^1}.$$

By the assumption that $b^+ > 0$, the leftmost group vanishes. The assumption that $b^+ > 1$ implies that the rightmost group vanishes.[5] Thus we have an isomorphism:

$$\{S^{p\mathbb{C}}, S^{q\mathbb{R}}\}_{S^1} \overset{\cong}{\leftarrow} \{S^{p\mathbb{C}}/S^0, S^{q\mathbb{R}}\}_{S^1}.$$

The right-hand group is a certain cohomotopy group of the space $S^{p\mathbb{C}}/S^0 = \Sigma S(p\mathbb{C})$, where we write $S(p\mathbb{C})$ for the unit sphere in $\mathbb{C}^p$. This space $\Sigma S(p\mathbb{C})$ is free away from the basepoint, and so (using Theorem 5.3 of [Ada06])

$$\{\Sigma\mathbb{CP}^{p-1}, S^q\} = \pi^{q-1}(\mathbb{CP}^{p-1}).$$

To associate an integer to an element of $\pi^{q-1}(\mathbb{CP}^{p-1})$, we use *Borel homology*, which we will need only a few properties of:

1. For a compact Lie group G, $\tilde{H}^*_G(S^0) = H^*(BG)$; effectively by definition. More generally, for any pointed G-space X, $\tilde{H}^*(X)$ is a $H^*(BG)$-module.
2. $\tilde{H}^*_{S^1}(S^{q\mathbb{R}}) \cong \tilde{H}^{*-q}_{S^1}(S^0)$.
3. For a pointed space X with a G-action free away from the basepoint, $\tilde{H}^*_G(X) = \tilde{H}^*(X/G)$. In the case $G = S^1$ and $X = S^{p\mathbb{C}}/S^0$, $\tilde{H}^*_{S^1}(X)$ agrees with $\tilde{H}^*(\mathbb{CP}^{p-1}_+)$ as a $H^*(BS^1) = H^*(\mathbb{CP}^\infty) = \mathbb{Z}[U]$-module (where $\tilde{H}^*(\mathbb{CP}^{p-1}_+)$ is a module over $H^*(\mathbb{CP}^\infty)$ from the inclusion $\mathbb{CP}^{p-1} \to \mathbb{CP}^\infty$).

Then, for a map $f \in \{S^{p\mathbb{C}}/S^0, S^{q\mathbb{R}}\}_{S^1}$, we consider the induced map

$$f^*\colon H^{q-1}_{S^1}(S^{q-1}) = H^0(\mathbb{CP}^\infty) \to H^{q-1}(\mathbb{CP}^{p-1}).$$

[5] This is another explanation of the wall-crossing phenomenon. If $b^+ = 1$, there is a possibly $\mathbb{Z} = \{S^1, S^1\}$-worth of lifts from $\{S^{p\mathbb{C}}, S^{q\mathbb{R}}\}_{S^1}$ to $\{S^{p\mathbb{C}}/S^0, S^{q\mathbb{R}}\}_{S^1}$. The different Seiberg-Witten invariants (associated to a choice of chamber) are associated to different choices of lift here.

We have $U^{(q-1)/2}\colon H^{q-1}(\mathbb{CP}^{p-1}) \to H^0(\mathbb{CP}^{p-1})$ is an isomorphism for q odd and $q < 2p - 1$. We define

$$t\colon \{S^{p\mathbb{C}}, S^{q\mathbb{R}}\}_{S^1} \to \mathbb{Z}$$

by $t(f) = U^{(q-1)/2} f^*$. We note that if q is *even*, the group $H^{q-1}(\mathbb{CP}^{p-1})$ is zero, and so there is no homological information to extract! However, $\{S^{p\mathbb{C}}, S^{q\mathbb{R}}\}_{S^1}$ is not generally zero when q is even; see Proposition 5.4.4 below. □

The following is taken directly from [BF04]:

Proposition 5.4.4 ([BF04, Lemma 3.5]) *Let $d > 1$ be an integer. The Hurewicz homomorphism*

$$h^{2d-2-k}\colon \tilde{\pi}^{2d-2-k}(\mathbb{CP}^{d-1}) \to \tilde{H}^{2d-2-k}(\mathbb{CP}^{d-1})$$

has the properties:

1. *For $k = 0$, is an isomorphism,*
2. *for $k = 1$ has kernel isomorphic to $\mathbb{Z}/gcd(2, d)$.*
3. *For $k = 2$, the kernel is $\mathbb{Z}/gcd(2, d)$ and the cokernel is $\mathbb{Z}/gcd(2, d-1)$.*
4. *For $k = 3$, the kernel is $\mathbb{Z}/\ell$ with $\ell = gcd(24, d)$ if d is even and $\ell = gcd(24, d-3)/2$ otherwise.*
5. *For $k = 4$, the kernel is trivial, and for $d > 2$, the cokernel is isomorphic to $\mathbb{Z}/m$ with $\ell m = 48$ if d is even, and $\ell m = 12$ otherwise.*

Indeed, generally it seems difficult to calculate the relevant groups; but we see that at least in principle the Bauer-Furuta invariant $BF_{S^1}(M)$ is capable of detecting more than the Seiberg-Witten invariant.

Finally, an introduction to using G-equivariant homotopy theory would not be complete without mentioning the G-Whitehead theorem, which is very convenient for Floer spectra:

Theorem 5.4.5 *Let $f\colon X \to Y$ be a stable map of finite G-CW complexes. Then f is a stable G-weak equivalence if and only if it is a stable weak equivalence on all H-fixed point sets, as H runs over closed subgroups of G.*

5.5 Floer Spectra for 3-Manifolds

In this section, we consider the following problem:

Problem 5.5.1 Let $M = M_1 \cup_Y M_2$ be a closed 4-manifold decomposed along 4-manifolds M_1, M_2 with boundary $Y, -Y$ respectively. Recover $BF_G(M)$ in terms of data associated to M_1, M_2.

The ideal solution to this problem would be the construction of spectra $SWF(Y)$, along with an invariant $BF(M)$, for M an oriented spin^c manifold with boundary Y. Here, $BF(M)$ is supposed to be an S^1-equivariant stable homotopy class

$$BF(M)\colon S^{\mathrm{ind} L} \to SWF(Y). \tag{5.17}$$

To make sense of this, we note that the *index* ind L of the Seiberg-Witten equations on a manifold-with-boundary continues to make sense. For a splitting as in Problem 5.5.1, we expect there to be maps $BF(M_i)$, $i = 1, 2$, as in (5.17), *and* there to exist a "pairing" (or "duality") stable map:

$$\varepsilon\colon SWF(Y) \wedge SWF(-Y) \to S^0.$$

so that

$$\varepsilon(BF(M_1), BF(M_2)) = BF(M).$$

This expression is, in fact, ambiguous. Indeed, the domain on the left is $S^{\mathrm{ind} L_1} \wedge S^{\mathrm{ind} L_2}$, and the domain on the right is $S^{\mathrm{ind} L}$. To make sense of the expression, we must also fix a stable equivalence $S^{\mathrm{ind} L_1} \wedge S^{\mathrm{ind} L_2} \to S^{\mathrm{ind} L}$. In fact, such an identification is possible (stably); we will neglect to give the details of this, but it is treated in [KLS18b, Section 6].

Also, ideally there would be a TQFT; perhaps as a functor

$$BF\colon Cob \to Sp_{S^1}$$

where *Cob* is the category whose objects are closed oriented 3-manifolds equipped with a spin^c structure, and where the morphisms $Y_1 \to Y_2$ are spin^c 4-manifolds X with $\partial X = Y_2 \amalg -Y_1$. Further, Sp_{S^1} denotes the category of S^1-equivariant spectra.

As stated, these do not quite work out; however, much of this discussion does at least morally proceed, with some remaining unresolved problems, which we include at the end of the section. At this point, we discuss the construction of $SWF(Y)$, in the cases for which it is available.

5.5.1 *The Seiberg-Witten Floer Spectrum $SWF(Y)$ for $b_1(Y) = 0$*

In the case that $b_1(Y) = 0$, the story above has a favorable resolution and produces a "Seiberg-Witten Floer stable homotopy type" $SWF(Y, \mathfrak{s})$ (or, shorter, the "Monopole Floer stable homotopy type"). The expression that it is a "stable homotopy type" rather than a "spectrum" refers to the fact that originally it is only asserted that $SWF(Y, \mathfrak{s})$ is constructed subject to a certain set of choices (say for different choices we produce spectra $SWF(Y, \mathfrak{s})_i$ for $i = 1, 2$) and for different choices, Manolescu [Man03] proves that there exists a not necessarily well-defined equivalence $SWF(Y, \mathfrak{s})_1 \rightarrow SWF(Y, \mathfrak{s})_2$. As Heegaard Floer practioners are well-aware, working with an equivalence class of an invariant is rather inconvenient, and one would like to work with a spectrum instead of just the equivalence class of some. Later, we will specify the choices in the construction of $SWF(Y, \mathfrak{s})$ and view $SWF(Y, \mathfrak{s})$ as a *parameterized spectrum* over the set of choices in its construction. This is all rather more important in the $b_1(Y) > 0$ case, so we delay its discussion (in the case $b_1(Y) = 0$, it is largely a rephrasing of the original construction).

Building on the finite-dimensional approximation approach above, Manolescu [Man03] showed:

Theorem 5.5.2 ([Man03, Man16]) *Let* $(Y, \mathfrak{s})$ *be a* spin^c *rational homology sphere. Then there exists an* S^1*-equivariant stable homotopy type* $SWF(Y, \mathfrak{s})$*. For an oriented* $(X, \mathfrak{t})$ *with boundary* $(Y, \mathfrak{s})$ *(where* Y *is not necessarily connected) with* $b_1(Y) = 0$*, there is a morphism*

$$S^{\mathrm{ind} L} \rightarrow SWF(Y, \mathfrak{s}).$$

where L is the linearization of the Seiberg-Witten equations on $(X, \mathfrak{t})$*.*

Moreover, Manolescu showed that, given a metric, one obtains a well-defined spectrum, rather than only a stable homotopy type of a spectrum. In the event that Y in the theorem is disconnected with components Y_i, we let $SWF(Y, \mathfrak{s}) = \bigwedge_i SWF(Y_i, \mathfrak{s}_i)$.

Remark 5.5.3 The construction of the 4-manifold invariant is very similar (with some slight modifications needed for Conley index theory) to the construction of the Bauer-Furuta map. The most intricate part of the theorem is really the construction of a suitable receptacle $SWF(Y, \mathfrak{s})$.

We will not set up the geometry of the Seiberg-Witten equations in dimension three, but just give the general format of the Seiberg-Witten equations there. Indeed, let $X = Y \times \mathbb{R}$ be a 4-manifold obtained by taking a product of a 3-manifold Y with the line $\mathbb{R}$; we can consider the Seiberg-Witten equations on such X, and use the preferred decomposition of the tangent bundle (from the $\mathbb{R}$ factor) to rephrase the Seiberg-Witten equations as equations in certain bundles over Y. Indeed, as for 4-manifolds, for 3-manifolds Y, we have that there exist spin^c structures, and for any spin^c structure, we have spinor bundles, spin^c connections, and Clifford multiplication. Let $\mathbb{S}$ be the spinor bundle on the spin^c manifold $(Y, \mathfrak{s})$ (there is no analog of the decomposition of the splitting of the spinor bundle on a 4-manifold). Write $d^*\colon \Omega^2(Y) \rightarrow \Omega^1(Y)$ for the dual of the exterior derivative on Y. The Seiberg-Witten map can be rewritten using a certain "vector field" $\mathcal{X}$ on $\Gamma(\mathbb{S}) \oplus i \ker d^*$, defined as follows:

$$\begin{aligned} \mathcal{X}\colon C(Y, \mathfrak{s}) := \Gamma(\mathbb{S}) \oplus i \ker d^* &\rightarrow \Gamma(\mathbb{S}) \oplus i \ker d^* \\ (\phi, a) &\mapsto (\partial\!\!\!/ \phi + \varrho(a)\phi, i d^+ a - q(\phi)). \end{aligned} \tag{5.18}$$

A (smooth) Seiberg-Witten configuration on $X = Y \times \mathbb{R}$ is a (smooth) map $\mathbb{R} \to \Gamma(\mathbb{S}) \oplus i \ker d^*$ (tacit in this is the choice of a certain gauge on X). For $(\phi, a) = \{(\phi_t, a_t)\}_{t\in\mathbb{R}}$ such a configuration, the Seiberg-Witten map for X can be expressed as:

$$\mathcal{SW}(\phi, a) = \mathcal{X}(\phi, a) + \frac{\partial}{\partial t}(\phi, a). \tag{5.19}$$

That is to say, a solution of the Seiberg-Witten equations ($\mathcal{SW} = 0$) on X is a *downward gradient trajectory* (henceforth, just a "trajectory") of the formal vector field $\mathcal{X}$. We call $\mathcal{X}$ only a "formal" vector field, because it does not define a vector field on the Sobolev completions of $C(Y, \mathfrak{s})$ (we have been lazy about distinguishing between $C(Y, \mathfrak{s})$ and its completions, and will continue to do so, but this is one point where it is significant. Recall that vector fields do not define flows on Fréchet manifolds, but do define flows on Banach manifolds).

It turns out to be most interesting to consider those trajectories of *finite energy*, namely trajectories that limit to *stationary* solutions of the Seiberg-Witten equations. A stationary solution is simply a smooth 4-manifold configuration (ϕ, a) where ϕ, a are constant and $\mathcal{SW}(\phi, a) = 0$. For (ϕ, a) to satisfy Eq. (5.19) and simultaneously $\partial/\partial t(\phi, a) = 0$ is equivalent to $(\phi, a) \in \Gamma(\mathbb{S}) \oplus i \ker d^*$ satisfying $\mathcal{X}(\phi, a) = 0$.

That is all to say, finite-energy Seiberg-Witten solutions are exactly trajectories of $\mathcal{X}$ on $C(Y, \mathfrak{s})$ with well-defined start and end points; by definition the end points are solutions of $\mathcal{X}(\phi, a) = 0$; if we were on a finite-dimensional manifold, the end points would be critical points of the vector field.

We have not mentioned the symmetries so far. As for 4-manifolds, there is a residual action by S^1 (acting on spinors $\Gamma(\mathbb{S})$); the formal vector field $\mathcal{X}$ is invariant with respect to this action, in the sense that

$$e^{i\theta}\mathcal{X}(\phi, a) = \mathcal{X}(e^{i\theta}\phi, a).$$

If $\mathfrak{s}$ is a self-conjugate spin^c structure (equivalent, because $b_1(Y) = 0$, a spin structure), then there is also a Pin(2)-action.

As was the case with the 4-manifold invariant, at this point we can mostly forget the geometric situation, and (at least for the moment, and schematically) imagine we are dealing with the situation of any old infinite-dimensional vector space V, equipped with a vector field $\mathcal{X}$. Compactness in the 4-manifold case has its companion in the 3-manifold case, which we now state (we remind the reader that we are still in the setting $b_1(Y) = 0$).[6]

Proposition 5.5.4 ([KM07], [Man03, Proposition 1]) *Let* $||(\phi, a)||_m$ *denote the* L^2_m *norm of a Seiberg-Witten configuration* (ϕ, a). *There exists a constant* C_m *so*

[6] We cite [Man03], but this result goes back much further.

that any $(\phi, a) \in L^2(\Gamma(\mathbb{S})) \oplus L^2(i\Omega^1(Y))$ *that occurs on a finite energy trajectory satisfies* $||(\phi, a)||_m \leq C_m$.

Now, we would like to perform finite-dimensional approximation to replace $C, \mathcal{X}$ with a finite-dimensional vector space V_n (say) together with a (genuine) vector field $\mathcal{X}_n$ on V_n.

That is, we are, roughly, performing finite-dimensional approximation (as in Sect. 5.2) on the map

$$\mathcal{X}\colon \Gamma(\mathbb{S}) \oplus i \ker d^* \to \Gamma(\mathbb{S}) \oplus i \ker d^*.$$

Following Sect. 5.2 closely, we could try to choose finite-dimensional subspaces of both the domain and the range. However, there is an added wrinkle in that in our present situation the domain and the range *agree*, and we also want our approximations in both the domain and the range to be the same (in order to get a vector field on a finite-dimensional space). The problem of making the domain and range approximations compatible does not appear in the 4-manifold problem; it makes the 3-manifold problem more complicated, and adds an obstruction so that in fact (as we will see below) it may not be possible to perform finite-dimensional approximation at all.

For this approximation it helps to use a linear/quadratic decomposition as before. We have that $\mathcal{X} = L + Q$, where L is the linear map $L(\phi, a) = (\not{\partial}\phi, *da)$ and $Q(\phi, a) = (a\phi, -q(\phi))$. We recall that in Sect. 5.2, we used that the linear map, roughly, 'dominated' the nonlinear term as approximations became large. We expect the same here. As with the four-dimensional situation, L comes from an elliptic operator. The analysis of elliptic operators gives that L is self-adjoint, Fredholm, and has infinitely many positive, and negative, eigenvalues, without accumulation point.

Finally, in choosing our approximations, it is reasonable to try to choose approximations so that $\mathcal{X}_n$ is as close to $\mathcal{X}$ as possible. Since $\mathcal{X}$ is mostly built from L, this can be stated as choosing our approximations so that $\pi_n L \sim L$, where π_n is the projection to V_n. This would be most easily arranged if the approximations V_n were just invariant subspaces of the operator L; i.e., sums of eigenspaces of L.

Note that any smooth vector field on a finite-dimensional manifold Z defines an honest *flow* $\mathbb{R} \times Z \to Z$ on Z.

Indeed, set $V^n_{-n} = \bigoplus_{|\lambda|<n} V(\lambda)$, where $V(\lambda)$ is the λ-eigenspace of L. More generally, we write V^{μ}_{λ} for the sum of the eigenspaces of L with eigenvalue e with $\lambda < e < \mu$.

The close relationship between V_n and the linear part of $\mathcal{X}_n$ allows one to prove the following proposition:

Proposition 5.5.5 ([Man03]) *Fix a sequence of approximations* $(V_n, \mathcal{X}_n)$ *as above. There exists a constant* C_m *so that any* $(\phi, a) \in \Gamma(\mathbb{S}) \oplus L^2(i\Omega^1(Y))$ *that occurs on a finite energy approximate trajectory, for n large, satisfies* $||(\phi, a)||_m \leq C_m$.

Moreover, as $n \to \infty$, approximate trajectories converge uniformly to genuine trajectories.

Moreover, we note that on the face of things, it is not necessary to perform finite-dimensional approximation only for *linear subspaces* V_n. Fix a general finite-dimensional smooth submanifold V_n of $C(Y, \mathfrak{s})$. Here "smooth" carries the double meaning that the manifold V_n is itself smooth, but also that V_n lies in the smooth part of $C(Y, \mathfrak{s})$, not just in some completion $L^2_m(C(Y, \mathfrak{s})$. Let π_n denote the (L^2) projection $TC(Y, \mathfrak{s})|_{V_n} \to TV_n$. Then $\pi_n \mathcal{X}|_{V_n}$ defines a genuine vector field on the finite-dimensional manifold V_n. One can prove a generalization of Proposition 5.5.5 to a broader class of sequences of approximations $\{V_n\}_n$. This is desirable in that there are problems in gauge theory where the configuration space is not contractible, as in Sect. 5.5.3.

5.5.2 Spaces From Flows

It is not immediately clear that Proposition 5.5.5 is useful; however, there is an invariant called the *Conley index* [Con78] that can transform the output of Proposition 5.5.5 into space (and eventually spectrum) information.

Let A be a locally compact space, together with a flow φ on A. An *isolated invariant set* S of φ is defined as follows. First, a compact subset N of A is called an *isolating neighborhood* if

$$\mathrm{Inv}(N) := \{x \in N : \varphi_t(x) \in N \text{ for all } t\} \subset \mathrm{Int}\, N.$$

Here, $\mathrm{Int}\, N$ is the interior of N. A set S is called an isolated invariant set (of φ) if $S = \mathrm{Inv}(N)$ for some isolating neighborhood N.

To connect back to the previous section, Propositions 5.5.4 and 5.5.5 together tell us that, for sufficiently large n, $(V_n, \mathcal{X}_n)$ is a flow with a particular isolated invariant set given by

$$S_n = \mathrm{Inv} B(V_n, C_m + 1),$$

where $B(V_n, C_m + 1)$ is the ball of radius $C_m + 1$ in the L^2_m norm on V_n. (Since V_n is finite-dimensional, the restriction of all the L^2_k norms are equivalent, however the particular constant C_m used here of course depends on the norm itself.) It turns out that S_n is the invariant set of $B(V_n, R)$ under the flow $\mathcal{X}_n$, for all R large enough.[7]

Given an isolated invariant set S of a flow φ, an *index pair* of S is a pair of compact sets (N_1, N_2) where $N_2 \subset N_1$. N_1 is required to be an *isolating*

[7] Technically, an approximate trajectory on V_n may go to infinity in finite time. To avoid this, Manolescu "truncates" the flow at infinity, but we will suppress this point.

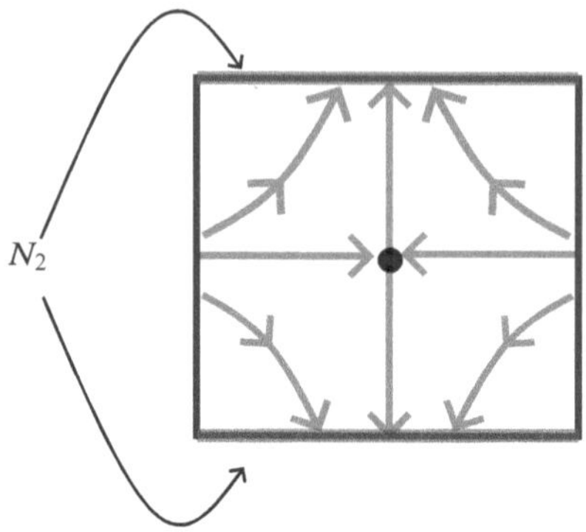

Fig. 5.1 A Morse critical point of index 1, with neighborhood N_1. The Conley index is N_1/N_2, which is homotopy equivalent to S^1

neighborhood of S, and N_2 will be a so-called "exit set" of N_1. Generally, the role of N_1, N_2, as in the items below, is summarized in Fig. 5.1.

We require, more explicitly:

1. $S = \text{Inv}(N_1 - N_2)$, and $N_1 - N_2$ is a neighborhood of S.
2. For all $x \in N_2$ and $t > 0$, if $\varphi([0,t],x) \subset N_1$ (read: the flow starting at x remains in N_1 until time t) then $\varphi([0,t],x) \subset N_2$.
3. For all $x \in N_1$ and $t > 0$, if $\varphi(t,x) \notin N_1$, then there exists t' with $0 \leq t' \leq t$ so that $\varphi(t',x) \in N_2$.

If (N_1, N_2) are as above, then N_2 is an exit set of N_1. Practically speaking, a special case of the above definition is easier to visualize, though not strictly necessary for our discussion:

Definition 5.5.6 For a compact set N in a flow (X, φ), introduce the notation:

$$n^+(N) := \{x \in \partial N \mid \exists \varepsilon > 0 \text{ so that } \varphi((-\varepsilon, 0), x) \cap N = \emptyset\},$$

$$n^-(N) := \{x \in \partial N \mid \exists \varepsilon > 0 \text{ so that } \varphi((0, \varepsilon), x) \cap N = \emptyset\}.$$

Then N is called an *isolating block* if $\partial N = n^+(N) \cup n^-(N)$.

An isolating block N defines an index pair $(N, n^-(N))$. (The advantage of this definition is that there is an automatic choice of the exit set N_2). In our setting, we may always take isolating blocks, by a result of Conley-Easton [CE71].

Given an index pair (N_1, N_2), we obtain the *Conley index* as the pointed space $(N_1/N_2, [N_2])$. One should compare with Fig. 5.1—in that case the Conley index is a particular sphere, whose dimension is the Morse index. In this sense, the Conley index is a 'spacification' of the Morse index.

Of course, for an isolated invariant set S, there may be many index pairs. Conley [Con78] shows that for any two index pairs (N_1, N_2) and (N_1', N_2') for S, there is a preferred homotopy class of homotopy equivalence (induced by φ):

$$\Phi_{(N_1,N_2),(N_1',N_2')} \colon (N_1/N_2, [N_2]) \to (N_1'/N_2', [N_2']).$$

Moreover, the 'connecting maps' Φ are compatible with composition in the expected ways:

$$\Phi_{(N_1',N_2'),(N_1'',N_2'')} \circ \Phi_{(N_1,N_2),(N_1',N_2')} \simeq \Phi_{(N_1,N_2),(N_1'',N_2'')}$$
$$\Phi_{(N_1,N_2),(N_1,N_2)} \simeq \text{id}.$$

The $\simeq$ here are interpreted as the two sides being homotopic; it is not asserted that the homotopy itself is well-defined. This is all to say that the data of the *Conley indices* $(N_1/N_2, [N_2])$, ranging over all index pairs (N_1, N_2), form a *connected simple system*, as defined by Conley (see also [Sal85, Definition 2.6]). Thus, up to homotopy, we may unambiguously talk about the Conley index $I(S) = (N_1/N_2, [N_2])$ of an isolated invariant set S. It is important that we are *not* talking about only the homotopy type of $I(S)$, since any two representatives of $I(S)$ have a *preferred* equivalence between them, not merely that there exists some equivalence.

The Conley index also enjoys some very nice dynamical properties:

1. Let φ_t be a family of flows on X, together with a family of isolated invariant sets S_t. Here, we interpret the claim that S_t lie in a family to mean there is a set $N \subset X$ that is an isolating neighborhood, for all t, for $S_t \subset X$ under φ_t. Then $I(S_0, \varphi_0) \simeq I(S_1, \varphi_1)$. In fact, there is a preferred homotopy equivalence $I(S_0, \varphi_0) \to I(S_1, \varphi_1)$, well-defined up to homotopy (even stronger, it is a morphism of connected simple systems).
2. Let $\varphi_1 \times \varphi_2$ on $X_1 \times X_2$ be a product flow. Then if S_i are isolated invariant sets of φ_i, the set $S_1 \times S_2$ is an isolated invariant set of $\varphi_1 \times \varphi_2$ with $I(S_1 \times S_2, \varphi_1 \times \varphi_2) = I(S_1, \varphi_1) \wedge I(S_2, \varphi_2)$.

The first item (1) is delicate, see [Sal85, Section 6]. The second item the reader should prove!

In any event, at this point, we have the schematic:

$$\begin{aligned}(\text{Seiberg-Witten equations on } (Y, \mathfrak{s})) \rightsquigarrow (C(Y), \mathcal{X}) \overset{Appx}{\rightsquigarrow} (V_n, \mathcal{X}_n) \rightsquigarrow \\ \rightsquigarrow \text{Isolated Invariant set } S_n \subset V_n \rightsquigarrow I(S_n).\end{aligned} \tag{5.20}$$

In the last step, we note that it is not the isolated invariant set S_n alone, but rather the data of the flow $(V_n, \mathcal{X}_n)$ and the isolated invariant set S_n that determines $I(S_n)$.

As with the case of the Bauer-Furuta invariant, this can all be made equivariant by simply saying a few magic words. First, a *G-equivariant* flow on a G-space X will be a flow φ on X so that $g(\varphi(t, x)) = \varphi(t, gx)$. Floer [Flo87] and Pruszko [Pru99] showed that Conley index theory goes through without change in the G-equivariant case, and the Conley index is naturally a G-space $I(S)$, with a well-defined G-equivariant (unstable) homotopy type. For us, if $(Y, \mathfrak{s})$ is a spin^c 3-manifold, then $I(S_n)$ is an S^1-space, and if $\mathfrak{s}$ is a spin structure, $I(S_n)$ is naturally a Pin(2)-space.

Indeed, Schematic (5.20) essentially completes the construction of $SWF(Y, \mathfrak{s})$. That is, to each approximation $(V_n, \mathcal{X}_n)$ we have a space $I(S_n)$; to cut down on notation let us just call $I(S_n) = I(\mathcal{X}_n)$. We must ask how the Conley indices $I(\mathcal{X}_n)$ and $I(\mathcal{X}_{n+1})$ are related. The proof here is not essentially harder than the corresponding claim in the construction of the Bauer-Furuta invariant. Indeed, we show that the flow $\mathcal{X}_{n+1}$ on V_{n+1} can be deformed to a product flow

$$\mathcal{X}_n \oplus L_{n+1} : V_{n+1} \to V_{n+1},$$

where L_{n+1} is the linearization of the Seiberg-Witten equations on the orthogonal complement $V_n^{\perp}$ to V_n in V_{n+1}. The expression $\mathcal{X}_n$ on V_{n+1} is to be interpreted as projection $V_{n+1} \to V_n$, followed by $\mathcal{X}_n$ defined as usual, followed by inclusion $V_n \to V_{n+1}$. By items (1)–(2), $I(V_{n+1}, \mathcal{X}_{n+1})$ is then identified with $I(V_n, \mathcal{X}_n) \wedge I(V_n^{\perp}, L_{n+1})$.

However, as in the discussion near Fig. 5.1, the Conley index of a linear flow is exactly a sphere, of dimension the Morse index. That is all to say:

$$I(V_{n+1}, \mathcal{X}_{n+1}) \simeq S^{\dim V^{-n}_{-n-1}} \wedge I(V_n, \mathcal{X}_n).$$

The equivalence can be made well-defined up to (equivariant) homotopy.

Finally, Manolescu [Man03] chooses a normalization to define $SWF(Y, \mathfrak{s})$ from the collection of $I(V_n, \mathcal{X}_n)$ (all related by suspension), to remove the ambiguity of the number of suspensions.

To give the normalization, we need a number $n(Y, \mathfrak{s}, g)$. Here g is the choice of a metric on Y. The number n may be interpreted using the spectral geometry of Y, or by considering a compact spin^c 4-manifold X with boundary Y.

Most directly,

$$n(Y, \mathfrak{s}, g) = \frac{1}{2}(\eta_{dir} - k(\not\partial) - \frac{\eta_{sign}}{4}),$$

where η_{dir} is the Atiyah-Patiodi-Singer η invariant of the Dirac operator, and η_{sign} is the same, for the signature operator, and where $k(\not\partial)$ is the (complex) rank of the kernel of the Dirac operator. Alternatively, let $(X, \mathfrak{t})$ be a bounding 4-manifold for $(Y, \mathfrak{s})$, and set

$$n(Y, \mathfrak{s}, g) = \operatorname{ind}_{\mathbb{C}}(D_A^+) - \frac{c_1(\mathfrak{s})^2 - \sigma(X)}{8}.$$

(The definitions agree by [APS75].)

Finally, define:

$$SWF(Y, \mathfrak{s}) = \Sigma^{-\dim V^0_{-n} - n(Y,\mathfrak{s},g)\mathbb{C}} I(V_n, \mathcal{X}_n).$$

As stated, this is only an equivariant *stable homotopy type*. However, for most purposes it is not hard to arrange for the invariant to be a *spectrum*, as below.

We go back to our algebraic topology viewpoint from Sect. 5.3. We follow the notation of Adams [Ada95], which is not so common nowadays, but is suitable for us. What we call below a spectrum is usually called a *pre-spectrum*, or a *sequential pre-spectrum*.

Definition 5.5.7 We call a sequence of spaces $\{X_n\}_{n\in\mathbb{N}}$, equipped with maps

$$f_n\colon \Sigma X_n \to X_{n+1}$$

a *spectrum*.

We will also consider equivariant spectra. As in Sect. 5.3, we will work with a universe.

Definition 5.5.8 Fix G a compact Lie group. A *G-universe* is an infinite-dimensional inner product space $\mathcal{U}$ that is also an orthogonal G-representation. We require that $\mathcal{U}$ is isomorphic to the direct sum of countably many copies of a given set of finite-dimensional representations of G, and also that $\mathbb{R}^\infty$ is a summand of $\mathcal{U}$.

Often, one works with a *complete* universe, where all finite-dimensional representations of G occur in $\mathcal{U}$. However, the naturally-occurring universe for Seiberg-Witten theory is, in the S^1-setting, $\mathbb{C}^\infty \oplus \mathbb{R}^\infty$ (as S^1-representations). In the Pin(2)-setting, the natural universe is $\mathbb{H}^\infty \oplus \tilde{\mathbb{R}}^\infty \oplus \mathbb{R}^\infty$. We will usually suppress these from the notation going forward.

Definition 5.5.9 Let G be a compact Lie group. A G-spectrum on a universe $\mathcal{U}$ is a collection of G-spaces X_V indexed on finite-dimensional G-equivariant subspaces V of $\mathcal{U}$, together with G-maps whenever $V \subset W$,

$$f_{V,W}\colon \Sigma^{W-V} X_V \to X_W.$$

Here $W - V$ is the orthogonal complement of V in W, as a G-representation. We require commutativity for the structure maps of a triple $Z \to V \to W$ in $\mathcal{U}$, and also $f_{V,V} = \mathrm{id}$.

One way of storing the information of the sequence $I(V_n, \mathcal{X}_n)$ from Sect. 5.5.2 is then to consider these data as a collection of maps:

$$f_{n,n+1}\colon I(V_{n+1} - V_n, L_{n+1}) \wedge I(V_n, \mathcal{X}_n) \to I(V_{n+1}, \mathcal{X}_{n+1}), \tag{5.21}$$

where the universe $\mathcal{U}$ is the direct sum of the nonpositive eigenspaces of the linearization of the Seiberg-Witten equations on $(Y, \mathfrak{s})$ (this $\mathcal{U}$ is isomorphic to the universe considered above over S^1, but with a more invariant description).

The collection of maps (5.21) does not quite suffice to define a spectrum in the sense of Definition 5.5.9. However, we will somewhat abuse notation and also consider as a spectrum any collection of spaces X_V defined on a cofinal collection of $V \subset \mathcal{U}$, and where the structure maps $f_{V_i,V_{i+1}}$ are (compatibly) defined for a cofinal sequence $V_i \subset V_{i+1} \subset \ldots$ of $\mathcal{U}$.

Thus stated, for any metric g on Y, we have defined a Seiberg-Witten Floer spectrum (aka monopole Floer spectrum) $SWF(Y, \mathfrak{s})$:

Definition 5.5.10 We let the *Seiberg-Witten Floer spectrum* $SWF(Y, \mathfrak{s}, g)$ denote the S^1-equivariant prespectrum formed from the collection of Conley indices $I(V_n, \mathcal{X}_n)$ with structure maps

$$f_{n,n+1} : I(V_{n+1} - V_n, L_{n+1}) \wedge I(V_n, \mathcal{X}_n) \to I(V_{n+1}, \mathcal{X}_{n+1})$$

There is analogous claim over Pin(2). Instead of making the spectrum depend on g explicitly, one can also parameterize the $SWF(Y, \mathfrak{s})$ by *Floer framings* as in [SS21]; for a choice of a Floer framing $\mathfrak{P}$ we have a spectrum $SWF(Y, \mathfrak{s}, \mathfrak{P})$. This is particularly relevant in the case of Y with $b_1(Y) > 0$, where normalizing using the metric is not possible. For now, if $b_1(Y) = 0$, suffice it to say that a metric g determines a certain Floer framing $\mathfrak{P}_g$, and we can consider the (well-defined up to homotopy) S^1-spectrum $SWF(Y, \mathfrak{s}, g) = SWF(Y, \mathfrak{s}, \mathfrak{P}_g)$.

5.5.3 *SWF(Y) for $b_1(Y) > 0$*

In order to set up the TQFT, we must also describe Floer spectra for 3-manifolds Y with $b_1(Y) > 0$. Here, we will try to be brief and proceed to the output of the construction.

The functional-analytic setup is similar to the previous discussion. There is a configuration space

$$C(Y, \mathfrak{s}) \cong \Gamma(\mathbb{S}) \oplus i \ker d^*,$$

though this time the isomorphism is not canonical (it depends on the choice of a *base connection*). However, $C(Y, \mathfrak{s})$ inherits an action by harmonic maps $\mathrm{Harm}(Y, S^1)$ from $Y \to S^1$ (aka "harmonic gauge transformations"), in correspondence with $H^1(Y; \mathbb{Z})$. Quotienting by this action gives $\tilde{C}(Y, \mathfrak{s})$, a bundle of Hilbert spaces over the Picard torus $\mathrm{Pic}(Y, \mathfrak{s}) \cong H^1(Y; \mathbb{R})/2H^1(Y; \mathbb{Z})$, the collection of flat spin^c connections on $\mathfrak{s}$, modulo equivalence. The space $\tilde{C}(Y, \mathfrak{s})$ is the natural home of the Seiberg-Witten equations of $(Y, \mathfrak{s})$.

However, we frequently encounter the following situation. Let $(X, \mathfrak{t})$ be a compact, oriented 4-manifold with $\partial(X, \mathfrak{t}) = (Y, \mathfrak{s})$, and so that $H^1(Y; \mathbb{Z}) \to H^1(X, \mathbb{Z})$ has nontrivial kernel. Then there are harmonic gauge transformations of Y which do not extend to harmonic gauge transformations on X. That is to say, the Seiberg-Witten equations on X naturally reduce to equations on Y where we have not quotiented out by the action of (certain) harmonic gauge transformations on Y.

In particular, for $b_1(Y) > 0$, it makes sense to consider the Harm(Y, S^1)-equivariant equations on $C(Y, \mathfrak{s})$. As before, we have a formal vector field:

$$\mathcal{X} : C(Y, \mathfrak{s}) \to C(Y, \mathfrak{s}).$$

However, Proposition 5.5.4 does not hold in this setting; instead we have a weaker claim, that finite-energy trajectories must remain a bounded distance from the harmonic forms $0 \oplus \mathcal{H}^1(Y) \subset \Gamma(\mathbb{S}) \oplus i \ker d^*$ (see [KLS18a, Theorem 3.2] for a precise statement).

To overcome this, Kronheimer-Manolescu [KM02], for $b_1 = 1$ and later [KLS18a] for all Y define finite-dimensional approximation in an increasing family of subsets of $C(Y, \mathfrak{s})$ (these subsets are not vector subspaces).

Rather than giving the construction, we instead describe the home in which it naturally resides.

Let $\mathfrak{S}_{S^1}$ be the category defined as follows.[8] Its objects will be diagrams $F : \mathbb{N} \to Sp_{S^1}$ of S^1-spectra (where a choice of universe is suppressed from the notation). Here $\mathbb{N}$ is the category with objects $\mathbb{N}$ and a single morphism $i \to j$ exactly when $i \leq j$. We will write an object of $\mathfrak{S}_{S^1}$ as:

$$Z : Z_1 \xrightarrow{j_1} Z_2 \xrightarrow{j_2} \dots \tag{5.22}$$

An expression as in (5.22) is called an ind-object, or an inductive system.

Morphisms in $\mathfrak{S}_{S^1}$ between objects Z and Z' are given by

$$\mathrm{Hom}_{\mathfrak{C}_{S^1}}(Z, Z') = \lim_{\infty \leftarrow m} \lim_{n \to \infty} \mathrm{Hom}_{Sp_{S^1}}(Z_m, Z'_n).$$

Note that with these definitions, the isomorphism classes of the Z_i are not invariants of the isomorphism class of Z; Z should be regarded as a kind of 'formal colimit' of the pieces Z_i.

There is also a weaker version of $\mathfrak{S}_{S^1}$, called $\mathfrak{S}'_{S^1}$, where we require the objects Z_n are finite S^1-CW spectra, and where the morphism spectra above are replaced with homotopy classes:

$$\mathrm{Hom}_{\mathfrak{S}'_{S^1}}(Z, Z') = \lim_{\infty \leftarrow m} \lim_{n \to \infty} \pi_0(\mathrm{Hom}_{Sp_{S^1}}(Z_m, Z'_n)).$$

Similarly, $\mathfrak{S}^*_{S^1}$ is defined in a similar way to $\mathfrak{S}_{S^1}$, but dually; the objects are inverse systems:

$$\bar{Z} : \bar{Z}_1 \xleftarrow{\bar{j}_1} \bar{Z}_2 \xleftarrow{\bar{j}_2} \dots$$

[8] In [KLS18a], $\mathfrak{S}$ refers to a category formed from considering only homotopy classes of morphisms; the category $\mathfrak{S}'_{S^1}$ below corresponds to [KLS18a].

The morphisms are now defined by:

$$\mathrm{Hom}_{\mathfrak{S}^*_{S^1}}(\bar{Z}, \bar{Z}') = \lim_{\infty \leftarrow n} \lim_{m \to \infty} \mathrm{Hom}_{Sp_{S^1}}(\bar{Z}_m, \bar{Z}'_n),$$

and similarly for the 'homotopy category' $\mathfrak{S}^*_{S^1}$. The categories $\mathfrak{S}_{S^1}$ and $\mathfrak{S}^*_{S^1}$ are respectively categories of *ind* and *pro* (equivariant) spectra.

One of the principal functions of pro- and ind- categories is to expand the setting in which there is some version of Spanier-Whitehead duality. In the *nonequivariant* case, this is made precise in [CI04, Corollary 8.6].[9] The following definition is due to Khandhawit-Lin-Sasahira [KLS18a]:

Definition 5.5.11 We will say that $Z \in \mathrm{ob}\,\mathfrak{S}_{S^1}$ is Spanier-Whitehead dual to $\bar{Z} \in \mathrm{ob}\,\mathfrak{S}^*_{S^1}$ if there are elements

$$\varepsilon \in \lim_{\infty \leftarrow m} \lim_{n \to \infty} \pi_0 \mathrm{Hom}_{Sp_{S^1}}(\bar{Z}_n \wedge Z_m, S^0),$$

$$\eta \in \lim_{\infty \leftarrow n} \lim_{m \to \infty} \pi_0 \mathrm{Hom}_{Sp_{S^1}}(S^0, Z_m \wedge \bar{Z}_n),$$

which satisfy certain conditions:

1. For any $m > 0$ there exists n large enough relative to m, and m' large enough relative to n so that the composition:

$$Z_m \cong S^0 \wedge Z_m \xrightarrow{\eta_{m',n} \wedge \mathrm{id}} Z_{m'} \wedge \bar{Z}_n \wedge Z_m \xrightarrow{\mathrm{id} \wedge \varepsilon_{m,n}} Z_{m'} \wedge S^0 \cong Z_{m'}$$

is homotopic to the connecting morphism $Z_m \to Z_{m'}$ in the inductive system for Z.
2. For any $n > 0$, there exists m large enough relative to n, and n' large enough relative to m so that the composition:

$$\bar{Z}_{n'} \cong \bar{Z}_{n'} \wedge S^0 \xrightarrow{\mathrm{id} \wedge \eta_{m,n}} \bar{Z}_{n'} \wedge Z_m \wedge \bar{Z}_n \xrightarrow{\varepsilon_{m,n'} \wedge \mathrm{id}} S^0 \wedge \bar{Z}_n \cong \bar{Z}_n$$

is equal to the connecting morphism $\bar{Z}_{n'} \to \bar{Z}_n$ of the projective system $\bar{Z}$.

Remark 5.5.12 If $\mathfrak{S}_{S^1}$ were better understood, one would hope to construct, from an object Z of $\mathfrak{S}_{S^1}$ a *function pro-spectrum* $F(Z, S^0)$. More explicitly, we conjecture that there are functors:

$$F(-, S^0) \colon \mathfrak{S}_{S^1} \to \mathfrak{S}^*_{S^1}, \qquad F(-, S^0) \colon \mathfrak{S}^*_{S^1} \to \mathfrak{S}_{S_1},$$

and X and $F(X, S^0)$ are Spanier-Whitehead duals as in Definition 5.5.11.

[9] As far as I am aware, an analog of Corollary 8.6 of [CI04] in equivariant pro- and ind- spectra has not been established.

For any finite S^1-spectrum X, we can regard X as an ind (or pro) spectrum using the constant directed system $\underline{X} = X \xrightarrow{\text{id}} X \xrightarrow{\text{id}} X \xrightarrow{\text{id}} \dots$.

The reason for introducing all of this is the following Theorem of [KLS18a]:

Theorem 5.5.13 ([KLS18a, Theorem 1.1]) *Let $(Y, \mathfrak{s})$ be a closed* spin^c 3-*manifold. Given a Riemannian metric g on Y and a base connection A_0, there are well-defined (up to homotopy) objects*

$$\underline{SWF}^A(Y, \mathfrak{s}, A_0, g) \in \mathrm{Ob}\,\mathfrak{S}_{S^1} \textit{ and } \underline{SWF}^R(Y, \mathfrak{s}, A_0, g) \in \mathrm{Ob}\,\mathfrak{S}^*_{S^1}$$

We will usually suppress the base connection and the metric from the notation; changes in the base connection and metric change the spectrum by suspension (also, as before, we made remove the 'metric dependence' to say that the construction instead depends on a choice of "Floer framing"). We will also often omit the underline when clear from context.[10]

We note that the considerations involving the Picard torus for 3-manifolds in the present section also have a four-dimensional aspect. Indeed, closed, spin^c 4-manifolds $(X, \mathfrak{t})$ with $b_1(X) > 0$ define a 'family' of Seiberg-Witten maps; roughly, one for each element of the Picard torus $\mathrm{Pic}(X, \mathfrak{t}) \cong H^1(X; \mathbb{R})/2H^1(X; \mathbb{Z})$. That is, for each point $p \in \mathrm{Pic}(X, \mathfrak{t})$, there is a linear Seiberg-Witten map L_p, instead of just the single map L of Sect. 5.2. Indeed, for closed X, the Bauer-Furuta invariant takes the form of a stable homotopy class

$$BF_{S^1}(X, \mathfrak{t}) \in [T(X, \mathfrak{t}), S^0]_{S^1}.$$

Here $T(X, \mathfrak{t})$ replaces the term $S^{\mathrm{ind} L}$ that came in the $b_1(X) = 0$ case. The space $T(X, \mathfrak{t})$ is the *Thom spectrum* of the virtual index bundle $\mathrm{ind}L$ over $\mathrm{Pic}(X, \mathfrak{t})$. In the case $\dim \mathrm{ind}L > 0$, this space is simply the Thom construction applied to the bundle $\mathrm{ind}L \to \mathrm{Pic}(X, \mathfrak{t})$. As one might expect, there is also a version for Pin(2)-equivariance.

Khandhawit-Lin-Sasahira [KLS18b] define cobordism maps for 4-manifolds with boundary. We will need the following notation for a cobordism $X \colon Y_{in} \to Y_{out}$. We write

1. V_X^+ for a maximal positive subspace of $\mathrm{Im}(H^2(X, \partial X; \mathbb{R}) \to H^2(X; \mathbb{R}))$ (the space of such choices is contractible),
2. V_{in} for the cokernel of $\iota^* \colon H^1(X; \mathbb{R}) \to H^1(Y_{in}; \mathbb{R})$,
3. V_{out} for the cokernel of $\iota^* \colon H^1(X; \mathbb{R}) \to H^1(Y_{out}; \mathbb{R})$.

[10] In [KLS18a], the authors reserve the notation $\underline{SWF}(Y, \mathfrak{s})$ for the Floer spectrum when $c_1(\mathfrak{s})$ is torsion, where the homotopy type does not depend on a choice of metric, and write $\underline{swf}$ for the invariants in Theorem 5.5.13.

Theorem 5.5.14 ([KLS18b, Theorem 1.2]) *Let* $(X, \mathfrak{t})$ *be a* spin^c *cobordism* $(Y_{in}, \mathfrak{s}_{in}) \to (Y_{out}, \mathfrak{s}_{out})$. *There are well-defined morphisms in* $\mathfrak{S}_{S^1}$, *and* $\mathfrak{S}^*_{S^1}$, *respectively:*

$$BF^A(X, \mathfrak{t})\colon \Sigma^{-(V_X^+ \oplus V_{in})} T(X, \mathfrak{t}) \wedge SWF^A(Y_{in}, \mathfrak{s}_{in}) \to SWF^A(Y_{out}, \mathfrak{s}_{out})$$

and

$$BF^R(X, \mathfrak{t})\colon \Sigma^{-(V_X^+ \oplus V_{out})} T(X, \mathfrak{t}) SWF^R(Y_{in}, \mathfrak{s}_{in}) \to SWF^R(Y_{out}, \mathfrak{s}_{out})$$

The V-terms record the index of the relevant Seiberg-Witten equations. As in Remark 1.3 of [KLS18b], this invariant is defined only over the *relative* Picard torus

$$\mathrm{Pic}^0(X, Y, \mathfrak{t}) \cong \ker(H^1(X; \mathbb{R}) \to H^1(Y; \mathbb{R}))/\ker(H^1(X; \mathbb{Z}) \to H^1(Y; \mathbb{Z})),$$

and $T(X, \mathfrak{t})$ is a Thom construction of a virtual index bundle (see Theorem 1.2 of [KLS18b] for details).

Khandhawit-Lin-Sasahira [KLS18b] prove the following *pairing theorem* to recover the Bauer-Furuta invariant of a manifold in terms of its pieces, building on work in a special case by [Man07]:

Theorem 5.5.15 ([KLS18b, Theorem 1.5]) *Let*

$$(X_0, \mathfrak{t}_0)\colon (Y_0, \mathfrak{s}_0) \to (Y_2, \mathfrak{s}_2), \qquad (X_1, \mathfrak{t}_1)\colon (Y_1, \mathfrak{s}_1) \to (-Y_2, \mathfrak{s}_2)$$

be connected spin^c *cobordisms and*

$$(X, \mathfrak{t})\colon (Y_0, \mathfrak{s}_0) \amalg (Y_1, \mathfrak{s}_1) \to \emptyset$$

be the glued cobordism. If:

1. Y_2 *is connected.*
2. $b_1(Y_0) = b_1(Y_1) = 0$,
3. $\mathrm{im}(H^1(X_0; \mathbb{R}) \to H^1(Y_2; \mathbb{R})) \subset \mathrm{im}(H^1(X_1; \mathbb{R}) \to H^1(Y_2; \mathbb{R}))$,

then, under the natural identification between domains and targets, one has

$$BF(X, \mathfrak{t})|_{\mathrm{Pic}^0(X, Y_2)} = \tilde{\varepsilon}(BF^A(X_0, \mathfrak{t}_0), BF^R(X_1, \mathfrak{t}_1)).$$

The term $\mathrm{Pic}^0(X, Y_2)$ *is the relative Picard torus*

$$\ker(H^1(X; \mathbb{R}) \to H^1(Y_2; \mathbb{R}))/\ker(H^1(X; \mathbb{Z}) \to H^1(Y_2; \mathbb{Z})).$$

The restriction on the homological behavior should be possible to remove, and indeed it would be very desirable to do so. For example, Theorem 5.5.15 cannot be applied to Fintushel-Stern knot surgery [FS98]. Toward removing this restriction, there is a further generalization of SWF for Y with $b_1(Y) > 0$. For this, we also have a notion of *parameterized spectra*, as introduced and studied by May-Sigurdsson [MS06]. We will dispense with a careful definition, but refer the reader to [SS21]. Roughly speaking, a parameterized spectrum over a base B (which is usually a finite CW complex) will be a family of spectra, one for each point of B (the development of the homotopy theory of such objects is quite intricate). Recall that the configuration space $\tilde{C}(Y, \mathfrak{s})$ is naturally a bundle over $\mathrm{Pic}(Y, \mathfrak{s})$. We have:

Theorem 5.5.16 ([SS21]) *Let* $(Y, \mathfrak{s})$ *be a closed* spin^c *3-manifold where the form* $\cup^3\colon H^1(Y;\mathbb{Z})^{\otimes 3} \to \mathbb{Z}$*, given by* $\cup^3(a\otimes b\otimes c) = a\cup b\cup c$*, vanishes identically. Say* $c_1(\mathfrak{s})$ *is torsion. Let* $\mathfrak{P}$ *be a Floer framing for* $(Y, \mathfrak{s})$*. Then there is a well-defined (up to parameterized homotopy)* S^1*-equivariant parameterized spectrum*

$$SWF(Y, \mathfrak{s}, \mathfrak{P})$$

over $\mathrm{Pic}(Y, \mathfrak{s})$*, called the* folded *Floer spectrum of* $(Y, \mathfrak{s}, \mathfrak{P})$.

If $\mathfrak{s}$ is self-conjugate there is also a Pin(2)-analog of Theorem 5.5.16. The appearance of the triple cup product is perhaps strange in the above theorem. Indeed, as mentioned after Proposition 5.5.4, there is an obstruction already to choosing candidate finite-dimensional approximations (the *polarization class*, in the sense of [CJS95]). Kronheimer-Manolescu [KM02] calculate the polarization class for the Seiberg-Witten equations, and it is vanishing exactly when that condition on $\cup^3$ is met. The assumption that $c_1(\mathfrak{s})$ is torsion is related; indeed, if $c_1(\mathfrak{s})$ is nontorsion, it is not expected that there is a Floer *spectrum*, but a slightly different type of invariant. (We note however that Theorem 5.5.16 has a generalization to the nontorsion case.) As usual, the assertion of the existence of a 4-manifold invariant is a relatively easy consequence of the construction of the 3-manifold construction.

Theorem 5.5.17 ([SS21]) *Let* $(X, \mathfrak{t})$ *be a* spin^c *cobordism* $(Y_{in}, \mathfrak{s}_{in}) \to (Y_{out}, \mathfrak{s}_{out})$*, where* $c_1(\mathfrak{s}_{in})$ *and* $c_1(\mathfrak{s}_{out})$ *are both torsion, where the triple cup product vanishes on both ends. Say that* $H^1(X;\mathbb{R}) \to H^1(Y_{in};\mathbb{R})$ *and* $H^1(X;\mathbb{R}) \to H^1(Y_{out};\mathbb{R})$ *are isomorphisms. Fix Floer framings* $\mathfrak{P}_{in}, \mathfrak{P}_{out}$*. Then, there is a well-defined morphism of* S^1*-equivariant parameterized spectra, over* $\mathrm{Pic}(X;\mathfrak{t})$*:*

$$\begin{aligned} BF(X, \mathfrak{t})\colon \Sigma^{-(V_X^+ \oplus V_{in}) + \mathrm{ind}(\mathfrak{P})} T(X, \mathfrak{t}) \wedge SWF(Y_{in}, \mathfrak{s}_{in}, \mathfrak{P}_{in}) \\ \to SWF(Y_{out}, \mathfrak{s}_{out}, \mathfrak{P}_{out}) \end{aligned}$$

Here, the term $\mathrm{ind}(\mathfrak{P})$ *is a virtual vector space that depends on the Floer framings on the ends (the map for any pair of Floer framings is determined by that of any other pair).*

Theorem 5.5.17 is not the most general statement available; for cobordisms without any assumption on $H^1(X)$, $H^1(Y)$, the Bauer-Furuta map should go to a combination of the unfolded and folded Floer spectra.

5.6 Lattice Homology and Spectra

5.6.1 Lattice Homology

We finally get to calculate the invariants $SWF(Y, \mathfrak{s})$ for some rational homology three-spheres. We define the *Lattice Spectrum* associated to a plumbing graph, and also certain subspectra thereof. Our discussion of the definitions is mostly lifted directly from [DSS23].

Let us start off with a negative-definite plumbing graph Γ, and we obtain a plumbed 4-manifold W_Γ by plumbing spheres according to the graph Γ (the negative-definiteness is not strictly necessary for the following constructions). Let $Y_\Gamma = \partial W_\Gamma$.

Recall that an element $k \in \mathrm{Hom}(L_\Gamma, \mathbb{Z})$ is *characteristic* if $k(x) \equiv x^2 \bmod 2$ for all $x \in L_\Gamma$. Denote the set of characteristic elements by Char_Γ. The set Char_Γ has an action of L_Γ which sends $k \mapsto k + 2x^*$, where x^* is the Poincaré dual $x^* = (x, -)_\Gamma$. The action of L_Γ partitions the elements of Char_Γ into equivalence classes; we denote a generic equivalence class by $[k]$. It is a standard fact that the spin^c-structures on W_Γ are in bijection with Char_Γ, and that the spin^c-structures on Y_Γ are in bijection with the set of equivalence classes $[k]$.

Extend $(-, -)_\Gamma$ to a $\mathbb{Q}$-valued pairing on $L_\Gamma \otimes \mathbb{Q}$. Since $(-, -)_\Gamma$ has non-zero determinant, any element $k \in \mathrm{Hom}(L_\Gamma, \mathbb{Z})$ may be written as $(x, -)_\Gamma$ for some rational vector $x \in L_\Gamma \otimes \mathbb{Q}$. This correspondence allows us to define an intersection pairing on $\mathrm{Hom}(L_\Gamma, \mathbb{Z})$, which by abuse of notation we also denote by $(-, -)_\Gamma$.

Finally, note that each equivalence class $[k]$ is an affine lattice over L_Γ. This gives a cubical subdivision of $[k] \otimes \mathbb{R} \cong \mathbb{R}^{|\Gamma|}$ coming from the obvious cubical subdivision of $L_\Gamma \otimes \mathbb{R} \cong \mathbb{R}^{|\Gamma|}$:

Definition 5.6.1 A *d-dimensional cube* of $[k]$ is the convex hull (in $[k] \otimes \mathbb{R}$) of a set of 2^d elements of the form

$$\left\{ k + \sum_{i=1}^{d} c_i (2v_i^*) \mid (c_1, \ldots, c_d) \in \{0, 1\}^d \right\}$$

for some $k \in [k]$ and set of distinct vertices $v_1, \ldots, v_d$ in Γ. A zero-dimensional cube will be a vertex of $[k]$.

For example, a square in $[k]$ is the convex hull of $\{k, k + 2v^*, k + 2w^*, k + 2v^* + 2w^*\}$ for some characteristic element $k \in [k]$ and pair of distinct vertices v and w. We denote a typical cube of dimension d by $\square_d$. Strictly speaking, each cube (as a geometric object) is a subset of $[k] \otimes \mathbb{R}$, but we refer to these as being in $[k]$ for convenience. We often refer to zero-dimensional cubes as *lattice points* and one-dimensional cubes as *lattice edges*.

Following [Ném08, Definition 3.1.8], we now recall the definition of lattice homology.

Definition 5.6.2 Let $[k]$ be an equivalence class in Char_Γ. Define the *weight function* w on $[k]$ as follows. For any zero-dimensional cube k in $[k]$, define

$$w(k) = \frac{1}{4}\left(k^2 + |\Gamma|\right).$$

For any d-dimensional cube $\Box_d$ in $[k]$, define $w(\Box_d)$ to be the minimum over the weights of all lattice points appearing as vertices of $\Box_d$:

$$w(\Box_d) = \min_k \{w(k) \mid k \text{ a vertex of } \Box_d\}.$$

Thus, w is a function on the set of all cubes (of any dimension) in $[k]$.

For any fixed equivalence class $[k]$, the weight function w takes values in a coset of $2\mathbb{Z}$ in $\mathbb{Q}$. To see this, note that for any pair of adjacent lattice points, we have

$$w(k + 2v^*) - w(k) = (k, v^*)_\Gamma + (v^*, v^*)_\Gamma = k(v) + v^2 \equiv 0 \bmod 2$$

since k is characteristic.

Definition 5.6.3 Define the *lattice complex* $\mathbf{C}(\Gamma, [k])$ as follows. Let $\mathbf{C}_d(\Gamma, [k])$ be the free (infinitely-generated) $\mathbb{Z}[U]$-module formally spanned by the set of d-dimensional cubes in $[k]$:

$$\mathbf{C}_d(\Gamma, [k]) = \mathrm{span}_{\mathbb{Z}[U]}\{\Box_d \mid \Box_d \text{ a } d\text{-dimensional cube in } [k]\}.$$

Set

$$\mathbf{C}(\Gamma, [k]) = \bigoplus_{d \geq 0} \mathbf{C}_d(\Gamma, [k]).$$

This has two gradings. The first is just d, which we call the *dimensional grading*. The second is the *Maslov grading*, which is defined on cubes by setting $\mathrm{gr}(\Box_d) = w(\Box_d) + d$. We extend this to all of $\mathbf{C}(\Gamma, [k])$ by setting $\mathrm{gr}(U) = -2$. The differential is given by inserting powers of U into the usual cellular/cubical differential to make $\mathrm{gr}(\partial) = -1$. That is, let

$$\partial(\Box_d) = \sum_{\Box_{d-1} \in \text{ cellular boundary of } \Box_d} U^{(w(\Box_{d-1}) - w(\Box_d))/2} \varepsilon(\Box_{d-1}) \Box_{d-1}$$

and extend $\mathbb{Z}[U]$-linearly. Here, the signs $\varepsilon(\Box_{d-1}) = \pm 1$ are chosen in the usual way by thinking of the cubes of $[k]$ as coming from a cubical decomposition of $\mathbb{R}^{|\Gamma|}$ and choosing orientations. Note that ∂ drops both the dimensional and Maslov gradings by one. The homology of $(\mathbf{C}(\Gamma,[k]),\partial)$ is called the *lattice homology* and is denoted by

$$\mathbb{H}(\Gamma,[k])=\bigoplus_{d\geq 0}\mathbb{H}_d(\Gamma,[k]),$$

where $\mathbb{H}_d(\Gamma,[k])$ is the homology in dimensional grading d. We write $\mathbb{H}(\Gamma)$ to denote the sum of $\mathbb{H}(\Gamma,[k])$ over all equivalence classes $[k]$, and similarly write $\mathbb{H}_d(\Gamma)$ to denote the analogous sum of the groups $\mathbb{H}_d(\Gamma,[k])$.

In [Ném08, Proposition 3.4.2], it is shown that the lattice homology (in fact, the chain homotopy equivalence class of the lattice chain complex) is an invariant of $(Y_\Gamma,\mathfrak{s})$, rather than the plumbing graph Γ.

5.6.2 Relationship with 3-Manifolds

Indeed, for the relationship to 3-manifolds that is established in Sect. 5.6.3, it suffices to only consider $\mathbb{H}_0$—it is a fundamental result of Némethi that if Y_Γ is an almost-rational ('AR') plumbed manifold then $\mathbb{H}_d(\Gamma,[k])=0$ for all $d\geq 1$ [Ném08, Theorem 4.3.3].

The individual cells of $\mathbb{H}_0(\Gamma,[k])$ correspond to elements of Char_Γ, and the edges of the lattice complex relate the cells of k and $k+2v^*$.

Associated to any characteristic vector k there is a Bauer-Furuta map

$$BF(W_\Gamma,k)\colon S^{\mathrm{ind}(k)}\to SWF(Y_\Gamma,[k]) \tag{5.23}$$

and it turns out that these maps, for varying k, are related to each other using the spectrum map $U\colon S^0\to S^{\mathbb{C}}$ (we abuse notation and also write U for the induced map $U\colon \tilde{H}^{S^1}_*(S^0)\to\tilde{H}^{S^1}_*(S^{\mathbb{C}})=\tilde{H}^{S^1}_{*-2}(S^0)$). These relations will be given in Sect. 5.6.4, and geometrically are a consequence, in both homology and spectra, of the blow-up formula for Heegaard Floer homology (respectively monopole Floer spectra). These relationships, working in Heegaard Floer homology instead of monopole Floer spectra, are what allow Ozsváth-Szabó and Némethi [OS03], [Ném08] to calculate the Heegaard Floer homology of AR plumbings. For more general plumbings, it is necessary to understand 'the relationships between these relationships'. I.e. (5.23) gives a map on homology $H_*(BF(W_\Gamma,k))$, which is equal to $U^{(w(k)-w(k+2v^*))/2}H_*(BF(W_\Gamma,k+2v^*))$ (in the event $w(k)-w(k+2v^*)\geq 0$). Indeed, more is true, namely that $BF(W_\Gamma,k)$ and $U^{(w(k)-w(k+2v^*))/2}BF(W_\Gamma,k+2v^*)$ are homotopic maps of spectra! However, the homotopy is not defined in an obvious way, and it is not clear how the homotopies for different v are related to each other.

The Heegaard Floer analog of the maps (5.23), considered on homology, defines a map

$$\mathbb{T}\colon \mathbb{H}_0(\Gamma, [k]) \to HF^-(Y_\Gamma, [k]),$$

and it is a fundamental result of Ozsváth and Szabó [OS03, Theorem 1.2] and Némethi [Ném05, Theorem 8.3] that $\mathbb{T}$ is an isomorphism if Γ is (for example) an AR plumbing. The proof of this rests on certain formal properties of the Heegaard Floer package, chief among which is the presence of a surgery exact sequence. A similar surgery exact sequence holds for Seiberg-Witten-Floer spectra [SS23]. Following the same arguments as in [OS03, Ném05] will show that a generalization of $\mathbb{T}$ is an isomorphism in our setting.

More generally, the groups $\mathbb{H}_d(\Gamma, [k])$ are also necessary. Indeed, for any plumbing tree, Zemke [Zem21b] has recently shown that the lattice homology $\mathbb{H}(\Gamma, [k])$ is isomorphic to $HF^-(Y_\Gamma, [k])$. This more general statement requires a much greater array of machinery, and is a consequence of understanding the behavior of the link surgery formula of Manolescu-Ozsváth [MO10] under connected sum [Zem21a]. Not all of this machinery has an analog in our Bauer-Furuta setting, but it is natural to conjecture that many of the same ideas from Zemke's proof are relevant in the spectral setting.

5.6.3 Lattice Spectrum: The General Construction

Let Γ be a negative-definite plumbing and $[k]$ be an equivalence class in Char_Γ. Recall that the weight function w of Definition 5.6.2 is valued in a coset of $2\mathbb{Z}$ in $\mathbb{Q}$; below we will often encounter terms $S^{w(\square_d)\mathbb{C}}$, where the term in the superscript may be a fraction. We recall that we work in a category of S^1-spectra where we allow desuspension by 'fractional' representations. This is made sense of as follows. We have that for spaces X, Y, there are no stable maps $S^{n\mathbb{C}}X \to S^{m\mathbb{C}}Y$ unless $n - m \in \mathbb{Z}$, and if $n - m \in \mathbb{Z}$, such stable maps are in bijection with maps $S^{(n-m)\mathbb{C}}X \to Y$, which does not involve fractional representations.

For each d-dimensional cube in $[k]$, let

$$\mathcal{F}(\square_d) = S^{(w(\square_d)/2)\mathbb{C}} \wedge \square_d^+. \tag{5.24}$$

We make $\mathcal{F}(\square_d)$ into a pointed S^1-spectrum by giving $\square_d^+$ the trivial S^1-action (recall that $+$ denotes the disjoint basepoint). Note that $\mathcal{F}(\square_d)$ is just the product $S^{(w(\square_d)/2)\mathbb{C}} \times \square_d$, except that $\{\infty\} \times \square_d$ is collapsed to a single basepoint to stay in the category of pointed S^1-spaces. If $d = 0$, then $\mathcal{F}(\square_d)$ is just the sphere $S^{(w(\square_d)/2)\mathbb{C}}$.

Now suppose $\square_{d-1}$ is a face of $\square_d$, so that we have an inclusion $f\colon \square^+_{d-1} \to \square^+_d$. This gives a way to glue $\mathcal{F}(\square_{d-1})$ and $\mathcal{F}(\square_d)$ together: for any $x \in S^{(w(\square_d)/2)\mathbb{C}}$ and $t \in \square^+_{d-1}$, identify the points

$$(i(x), t) \sim (x, f(t)) \tag{5.25}$$

in $S^{(w(\square_{d-1})/2)\mathbb{C}} \wedge \square^+_{d-1}$ and $S^{(w(\square_d)/2)\mathbb{C}} \wedge \square^+_d$, respectively. Here, i is an inclusion map of S^1-spheres; we are using the fact that if $\square_{d-1}$ is a face of $\square_d$, then $w(\square_d) \leq w(\square_{d-1})$. We then let

$$X = \left(\bigsqcup_{\square_d \text{ in } [k] \text{ with } w(\square_d) \geq h} \mathcal{F}(\square_d) \right) / \sim, \tag{5.26}$$

where the quotient $\sim$ glues together $\mathcal{F}(\square_{d-1})$ and $\mathcal{F}(\square_d)$ according to (5.25) whenever $\square_{d-1}$ is a face of $\square_d$. See Fig. 5.2.

We should be quite careful about the choice of map

$$i\colon S^{(w(\square_d)/2)\mathbb{C}} \to S^{(w(\square_{d-1})/2)\mathbb{C}};$$

the space of such maps is the limit of the space of complex linear subspaces (of codimension $w(\square_{d-1})/2 - w(\square_d)/2$) of $\mathbb{C}^t$, as $t \to \infty$; i.e. the Grassmannian $\mathrm{Gr}(w(\square_{d-1})/2 - w(\square_d)/2, \infty)$, although we only care about these maps up to

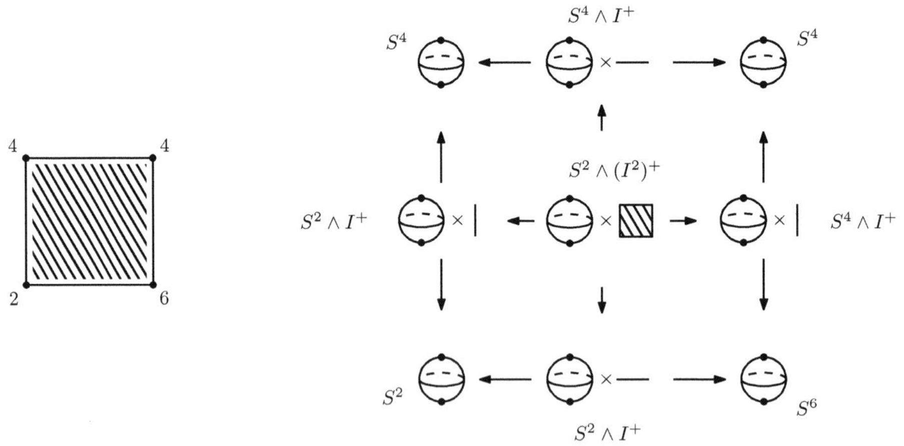

Fig. 5.2 Figure 3 of [DSS23]. A schematic depiction of the construction (5.26) restricted to a single square in $[k]$. Left: a square in $[k]$ with example values of w on its vertices. Right: taking $h = 0$, the pointed S^1-space constructed according to (5.26). Each arrow represents gluing a portion of the boundary of the domain to a subset (or the entirety) of the codomain. For example, the upper edge $S^2 \times I \subset \partial(S^2 \times I^2)$ of the central square is glued homeomorphically to the subset $S^2 \times I \subset S^4 \times I$ along the top row of the diagram

homotopy. Alternatively, we could build a model of a spectrum with the same homology as lattice homology using *any* maps $S^{(w(\square_d)/2)\mathbb{C}} \to S^{(w(\square_{d-1})/2)\mathbb{C}}$ inducing the correct map on (co)Borel homology—it is not at all obvious that different choices (even in the linear case) give equivalent spectra, and Dai-Sasahira and the author conjecture: for suitable weight functions on $\mathbb{R}^{|\Gamma|}$, the resulting spectra for different i are not equivalent.

For our i we simply take the inclusion by putting 0 in each of the missing coordinate entries. We note that this choice has some defects—e.g. it is not immediately obvious how to make this selection Pin(2)-equivariant. It seems quite plausible that for general plumbings, a more careful choice of i will be needed. For the plumbings in this survey though, this choice of i yields spectra that coincide with the Floer spectra.

It is convenient to write S_h for the subset of $\mathbb{R}^{|\Gamma|}$ on which the weight function is at most h. For h sufficiently negative, the inclusion of S_h into S_{h-2} is a homotopy equivalence, and in fact the latter deformation retracts onto the former [Ném08, Corollary 3.2.5]; this says that in some sense there is only a finite amount of information contained in the above gluing process.

Definition 5.6.4 The (S^1-)*lattice spectrum* is defined to be

$$\mathcal{H}(\Gamma, [k]) = X \in \mathrm{Ob}(\mathfrak{C}_{S^1}).$$

Its (stable) homotopy type will be called the lattice homotopy type of Γ.

Several theorems from lattice homology have analogous statements in the context of lattice homotopy. In particular, it is possible to upgrade [Ném08, Proposition 3.4.2] to show that $\mathcal{H}(\Gamma, [k])$ is an invariant of $(Y_\Gamma, \mathfrak{s})$, rather than Γ.

5.6.4 Path Lattice Homotopy

As in Sect. 5.6.2, it is clear that we may restrict the index set in Eq. (5.26) to any cubical subcomplex of $[k] \otimes \mathbb{R}$. In particular, let γ be a path in $[k]$. Let X be defined as in Eq. (5.26), but with index set restricted to the zero- and one-dimensional cubes present in γ. We thus obtain a well-defined homotopy type:

Definition 5.6.5 If γ is any path in $[k]$, the (S^1-)*path spectrum* (or *path homotopy type*) is defined to be

$$\mathcal{H}(\gamma, [k]) = X \in \mathrm{Ob}(\mathfrak{C}_{S^1}).$$

In fact, for an AR plumbed manifold, there is a path $\gamma \subset [k] \otimes \mathbb{R}$ so that $\mathbb{H}_0(\gamma, [k]) \to \mathbb{H}(\Gamma, [k])$ is an isomorphism; see [DSS23, Theorem 4.9] for references. We call such a γ a path *carrying* the lattice homology, and define the zeroth lattice spectrum of an AR plumbed manifold to be the path spectrum associated to any path carrying the lattice homology.

Definition 5.6.6 Let Γ be an AR plumbing and $[k]$ be an equivalence class in Char_Γ. Define the *(zeroth)* $(S^1$*-)lattice spectrum* to be

$$\mathcal{H}_0(\Gamma, [k]) = \mathcal{H}(\gamma, [k])$$

for γ any path carrying the lattice homology.

It is not obvious that $\mathcal{H}_0(\Gamma, [k])$ is independent of the choice of γ, although this turns out to be true. For a path γ as above, we have maps as in $F_{W_\Gamma,k}$ as in (5.23) for each of the vertices $k \in \gamma$. Moreover, the blow-up formula (applied to the spheres in the plumbing) holds for Bauer-Furuta invariants to give the following. Recall that $U: S^0 \to S^{\mathbb{C}}$ is the map of spectra induced by the inclusion $\{0\} \to \mathbb{C}$. Denote

$$n = \frac{1}{2}(w(k+2v^*) - w(k)) = \frac{1}{2}(k(v) + v^2).$$

Then:

$$\begin{cases} BF(W_\Gamma, k) \simeq BF(W_\Gamma, k+2v^*) \circ U^n \colon S^{\mathrm{ind}(k)} \to SWF(Y_\Gamma, [k]) \text{ if } n \geq 0 \\ BF(W_\Gamma, k) \circ U^{-n} \simeq BF(W_\Gamma, k+2v^*) \colon S^{\mathrm{ind}(k+2v^*)} \to SWF(Y_\Gamma, [k]) \text{ if } n < 0. \end{cases} \tag{5.27}$$

The homotopies in these relations are not explicit, but their existence is given by the blow-up formula. Also, note that, unlike in the definition of the domains for lattice homology, the domains for the maps $BF(W_\Gamma, k)$ and $BF(W_\Gamma, k+2v^*) \circ U^n$ (similarly for second line of (5.27)) are not canonically identified. However, *up to homotopy* we may identify the domains, since there is a unique orientation-preserving homotopy class between complex representation spheres of the same dimension.

By making *a choice* of homotopy in (5.27) for each edge of γ, we obtain, by construction $\mathcal{H}(\gamma, [k]) := \mathcal{H}_0(\gamma, [k])$, a map

$$\mathcal{T} \colon \mathcal{H}(\gamma, [k]) \to SWF(Y_\Gamma, [k]).$$

In [DSS23] it is shown that $\mathcal{T}$ induces an isomorphism on homology, and also that it induces a homotopy equivalence on fixed-point sets. By the equivariant Whitehead Theorem 5.4.5, it follows that $\mathcal{T}$ is a homotopy equivalence, and so $\mathcal{H}(\Gamma, [k])$ and $SWF(Y_\Gamma, [k])$ agree.

With substantially more work, it is possible to find the Pin(2)-equivariant spectrum $SWF(Y_\Gamma, [k])$, giving the following theorem:

Theorem 5.6.7 ([DSS23, Theorem 1.2]) *Let Y_Γ be an almost-rational plumbed homology sphere and $\mathfrak{s}$ be a self-conjugate* spin^c *structure on Y_Γ. Then we have a* Pin(2)*-equivariant homotopy equivalence*

$$\mathcal{H}(\Gamma, [k]) = SWF(Y_\Gamma, \mathfrak{s}).$$

Acknowledgments I would like to thank İnanç Baykur, Maciej Borodzik, Kristen Hendricks, Marco Marengon, Tom Mrowka, András Stipsicz and Ian Zemke for helpful conversations, as well as Irving Dai and Hirofumi Sasahira for both many fruitful discussions about the subject matter, and also for helpful comments on an earlier version of these notes. These lectures took place at the Rényi Institute during the Singularities and Low Dimensional Topology—2023 Spring Semester program. Many thanks to the program organizers, in particular Marco and András, and to the participants in these lectures for much helpful feedback. The author is supported by NSF grant DMS-2203828.

References

[Ada95] J. F. Adams. *Stable homotopy and generalised homology*. Chicago Lectures in Mathematics. University of Chicago Press, Chicago, IL, 1995. Reprint of the 1974 original.

[Ada06] John Frank Adams. Prerequisites (on equivariant stable homotopy) for Carlssons's lecture. In *Algebraic Topology Aarhus 1982: Proceedings of a conference held in Aarhus, Denmark, August 1–7, 1982*, pages 483–532. Springer, 2006.

[APS75] M. F. Atiyah, V. K. Patodi, and I. M. Singer. Spectral asymmetry and Riemannian geometry. I. *Math. Proc. Cambridge Philos. Soc.*, 77:43–69, 1975.

[AR20] Alberto Abbondandolo and Thomas O. Rot. On the homotopy classification of proper Fredholm maps into a Hilbert space. *J. Reine Angew. Math.*, 759:161–200, 2020.

[BF04] Stefan Bauer and Mikio Furuta. A stable cohomotopy refinement of Seiberg-Witten invariants: I. *Inventiones mathematicae*, 155(1):1–19, 2004.

[BK20] David Baraglia and Hokuto Konno. A gluing formula for families Seiberg-Witten invariants. *Geom. Topol.*, 24(3):1381–1456, 2020.

[CE71] C. Conley and R. Easton. Isolated invariant sets and isolating blocks. *Trans. Amer. Math. Soc.*, 158:35–61, 1971.

[CI04] J. Daniel Christensen and Daniel C. Isaksen. Duality and pro-spectra. *Algebr. Geom. Topol.*, 4:781–812, 2004.

[CJS95] R. L. Cohen, J. D. S. Jones, and G. B. Segal. Floer's infinite-dimensional Morse theory and homotopy theory. In *The Floer memorial volume*, volume 133 of *Progr. Math.*, pages 297–325. Birkhäuser, Basel, 1995.

[Con78] Charles Conley. *Isolated invariant sets and the Morse index*, volume 38. American Mathematical Society, Providence, R.I., 1978.

[DSS23] Irving Dai, Hirofumi Sasahira, and Matthew Stoffregen. Lattice homology and Seiberg-Witten Floer spectra. *preprint*, 2023.

[Flo87] Andreas Floer. A refinement of the Conley index and an application to the stability of hyperbolic invariant sets. *Ergodic Theory Dynam. Systems*, 7(1):93–103, 1987.

[FS98] Ronald Fintushel and Ronald J. Stern. Knots, links, and 4-manifolds. *Invent. Math.*, 134(2):363–400, 1998.

[Fur01] M. Furuta. Monopole equation and the $\frac{11}{8}$-conjecture. *Math. Res. Lett.*, 8(3):279–291, 2001.

[HLSX22] Michael J. Hopkins, Jianfeng Lin, XiaoLin Danny Shi, and Zhouli Xu. Intersection forms of spin 4-manifolds and the Pin(2)-equivariant Mahowald invariant. *Comm. Amer. Math. Soc.*, 2:22–132, 2022.

[KLS18a] Tirasan Khandhawit, Jianfeng Lin, and Hirofumi Sasahira. Unfolded Seiberg–Witten Floer spectra, i: Definition and invariance. *Geometry & Topology*, 22(4):2027–2114, 2018.

[KLS18b] Tirasan Khandhawit, Jianfeng Lin, and Hirofumi Sasahira. Unfolded Seiberg-Witten Floer spectra, ii: Relative invariants and the gluing theorem. *arXiv preprint arXiv:1809.09151*, 2018.

[KM02] Peter B Kronheimer and Ciprian Manolescu. Periodic Floer pro-spectra from the Seiberg-Witten equations. *arXiv preprint math/0203243*, 2002.

[KM07] Peter Kronheimer and Tomasz Mrowka. *Monopoles and three-manifolds*, volume 10. Cambridge University Press, 2007.

[KM20] P. B. Kronheimer and T. S. Mrowka. The Dehn twist on a sum of two $K3$ surfaces. *Math. Res. Lett.*, 27(6):1767–1783, 2020.

[Lin16] Francesco Lin. Lectures on monopole Floer homology. In *Proceedings of the Gökova Geometry-Topology Conference 2015*, pages 39–80. Gökova Geometry/Topology Conference (GGT), Gökova, 2016.

[LM21] Jianfeng Lin and Anubhav Mukherjee. Family Bauer–Furuta invariant, exotic surfaces and Smale conjecture. *arXiv preprint arXiv:2110.09686*, 2021.

[Man03] Ciprian Manolescu. Seiberg-Witten-Floer stable homotopy type of three-manifolds with $b_1 = 0$. *Geom. Topol.*, 7:889–932 (electronic), 2003.

[Man07] Ciprian Manolescu. A gluing theorem for the relative Bauer-Furuta invariants. *J. Differential Geom.*, 76(1):117–153, 2007.

[Man14] Ciprian Manolescu. On the intersection forms of spin four-manifolds with boundary. *Math. Ann.*, 359(3–4):695–728, 2014.

[Man16] Ciprian Manolescu. Pin(2)-equivariant Seiberg-Witten Floer homology and the triangulation conjecture. *J. Amer. Math. Soc.*, 29(1):147–176, 2016.

[MO10] Ciprian Manolescu and Peter Ozsváth. Heegaard Floer homology and integer surgeries on links. *arXiv preprint arXiv:1011.1317*, 2010.

[MS06] J. P. May and J. Sigurdsson. *Parametrized homotopy theory*, volume 132 of *Mathematical Surveys and Monographs*. American Mathematical Society, Providence, RI, 2006.

[Ném05] András Némethi. On the Ozsváth-Szabó invariant of negative definite plumbed 3-manifolds. *Geom. Topol.*, 9:991–1042, 2005.

[Ném08] András Némethi. Lattice cohomology of normal surface singularities. *Publ. Res. Inst. Math. Sci.*, 44(2):507–543, 2008.

[OS03] Peter Ozsváth and Zoltán Szabó. On the Floer homology of plumbed three-manifolds. *Geom. Topol.*, 7:185–224, 2003.

[Pru99] Artur M. Pruszko. The Conley index for flows preserving generalized symmetries. In *Conley index theory (Warsaw, 1997)*, volume 47 of *Banach Center Publ.*, pages 193–217. Polish Acad. Sci. Inst. Math., Warsaw, 1999.

[Sal85] Dietmar Salamon. Connected simple systems and the Conley index of isolated invariant sets. *Trans. Amer. Math. Soc.*, 291(1):1–41, 1985.

[SS21] Hirofumi Sasahira and Matthew Stoffregen. Seiberg-Witten Floer spectra for b_1 > 0. *arXiv preprint arXiv:2103.16536*, 2021.

[SS23] Hirofumi Sasahira and Matthew Stoffregen. Surgery exact triangles in Monopole Floer spectra. *preprint*, 2023.

[Š64] A. S. Švarc. On the homotopic topology of Banach spaces. *Dokl. Akad. Nauk SSSR*, pages 61–63, 1964.

[Wit94] Edward Witten. Monopoles and four-manifolds. *Mathematical Research Letters*, 1:769–796, 1994.

[Zem21a] Ian Zemke. Bordered manifolds with torus boundary and the link surgery formula. *arXiv preprint arXiv:2109.11520*, 2021.

[Zem21b] Ian Zemke. The equivalence of lattice and Heegaard Floer homology. *arXiv preprint arXiv:2111.14962*, 2021.

Chapter 6
Spectra in Khovanov and Knot Floer Theories

Sucharit Sarkar, Marco Marengon, and András Stipsicz

Contents

6.1 Introduction ... 187
6.2 Framed Flow Categories ... 188
  6.2.1 A Recap of Morse Theory ... 189
  6.2.2 From Morse Functions to Framed Flow Categories ... 190
  6.2.3 The Hypercube Flow Category ... 191
  6.2.4 The Cohen-Jones-Segal Construction ... 195
6.3 A Khovanov Stable Homotopy Type ... 200
  6.3.1 The Khovanov Complex ... 200
  6.3.2 The Lipshitz-Sarkar Framed Flow Category ... 203
6.4 A Knot Floer Stable Homotopy Type ... 208
  6.4.1 Grid Diagrams ... 209
  6.4.2 Grid Homology ... 209
  6.4.3 Maslov Index 1 Domains ... 212
  6.4.4 Maslov Index 2 Domains ... 213
  6.4.5 Maslov Index 3 Domains ... 214
  6.4.6 Maslov Index k Domains ... 215
  6.4.7 Bubbles ... 218
References ... 223

6.1 Introduction

Invariants of knots and links witnessed an unparalleled development in the past couple of decades. The subject of knot theory was born at the end of the nineteenth century, while a systematic mathematical study began with work of Reidemeister

S. Sarkar
UCLA, Los Angeles, CA, USA
e-mail: sucharit@math.ucla.edu

M. Marengon · A. Stipsicz (✉)
Alfréd Rényi Institute of Mathematics, Budapest, Hungary
e-mail: marengon@renyi.hu; stipsicz@renyi.hu

© The Author(s), under exclusive license to Springer Nature Switzerland AG 2024

J. Fernández de Bobadilla et al. (eds.), *Singularities and Low Dimensional Topology*, Bolyai Society Mathematical Studies 30,
https://doi.org/10.1007/978-3-031-56611-0_6

in the 1920s [Rei27]. Alexander's work (and in particular the introduction of the Alexander polynomial invariant [Ale28]) in the late 1920s was a remarkable step in understanding this subject. The next major development was brought by work of Vaughan Jones in the 80s: his introduction of the Jones polynomial [Jon85], and subsequent generalizations, the HOMFLY-PT polynomial and the definition of quantum invariants, reconnected the subject with theoretical physics.

After the turn of the century a new paradigm, *categorification*, provided deep insight into the (still mysterious) world of knots. Khovanov's definition of his homologies [Kho00] refined the Jones polynomial, while work of Ozsváth and Szabó [OS04], and independently of Rasmussen [Ras03] produced a family of homology theories having the Alexander polynomial as their (graded) Euler characteristic. It was a routine exercise to find knots which are not distinguished by their Jones (or Alexander) polynomials, but have distinct Khovanov (or knot Floer) homologies. Indeed, both Khovanov and Heegaard Floer homologies are *unknot detectors* (i.e. the unknot is characterized by its homology in either theory). For comparison, this statement is false for the Alexander polynomial, and an open problem for the Jones polynomial.

Recently a further level of abstraction, *spacification*, brought new tools in knot theory. In this context one wants to construct topological spaces (or, more generally, spectra) with the property that these objects are knot invariants, and their singular homology (in the appropriate sense) is the Khovanov homology (or knot Floer homology) of the underlying knot. Having a spectrum provides the opportunity of applying further structures (such as Steenrod operations) and possibly other functors (like K-theory) to find further knot invariants. The program has been completed for Khovanov homology [LS14], and the first step has been taken for knot Floer homology [MS21].

These notes are based on lectures of Sucharit Sarkar given at a focused week of the special semester *Singularities and low-dimensional topology* at the Erdős Center, Budapest in the Spring semester of 2023. We thank Michael Willis for his comments on an earlier version of these notes.

6.2 Framed Flow Categories

The basic idea of *spacification* via flow categories is to start from a chain complex C together with a given basis and construct a 'space' (more precisely a CW spectrum) X with the property that its reduced cellular chain complex $\widetilde{C}_*^{\mathrm{cell}}(X)$ is isomorphic to C. The following remark explains why we really need to choose a basis of C for our construction.

Remark 6.2.1 If we have a chain complex C, which we assume to be finitely generated for simplicity, and we disregard its basis and grading, then we can always make a basis change so that

$$C \cong \mathbb{Z}^k \oplus \bigoplus_i (\mathbb{Z} \xrightarrow{k_i} \mathbb{Z}),$$

where the boundary map on $\mathbb{Z}^k$ is zero and the symbol $\mathbb{Z} \xrightarrow{k_i} \mathbb{Z}$ denotes a two-step complex where the boundary map is the multiplication by k_i. From here we can construct a CW complex $M(C)$ with $\widetilde{C}_*^{\text{cell}}(X) = C$ simply by taking an appropriate wedge of spheres (for the free part) and other simple spaces (for the 2-step complexes) as described in [Hat02, Example 2.40]. The space $M(C)$ is often called a *Moore space*, and its construction can be refined to take gradings in account as well.

The drawback of Moore spaces is that they are not (and they cannot be made) functorial. More precisely, by a result of Carlsson [Car81] the functors

$$H_*, H_n \colon \text{Top}_* \to \text{Ab}$$

do not admit functorial sections $M, M_n \colon \text{Ab} \to \text{Top}_*$.

This is in sharp contrast with the case of homotopy groups, where the functor

$$\pi_n \colon \text{Top}_* \to \text{Ab},$$

defined for $n > 1$, does admit a functorial section, given by the Eilenberg-Maclane spaces.

6.2.1 A Recap of Morse Theory

Let $f \colon M \to \mathbb{R}$ be a proper Morse-Smale function with finitely many critical points on a Riemannian manifold M (Fig. 6.1). Each index-k critical point gives rise to a *downward disc* $\mathbb{D}^k$, which together with all other downward discs presents M as a (based) CW complex.

Remark 6.2.2 There is a subtlety in the presentation of M as a CW complex. If one tries to define the attaching map between cells just by flowing, then the resulting map would be discontinuous. See [Mil65] for the details on the attaching maps.

The construction of the Morse chain complex for f mirrors the construction of $\widetilde{C}_*^{\text{cell}}$ of the associated CW complex. The generators of the Morse complex are the

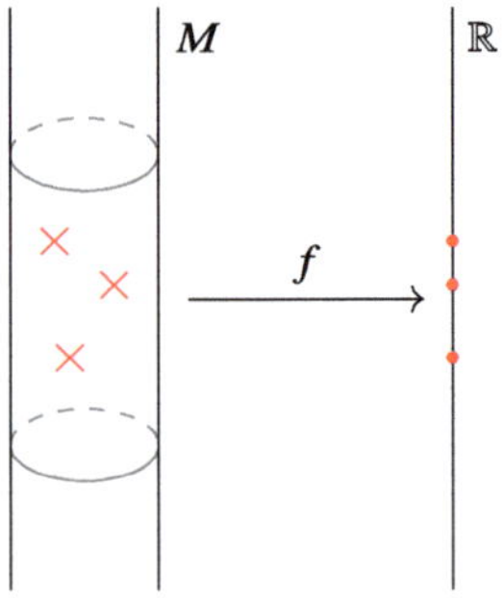

Fig. 6.1 Schematic of a proper Morse function

critical points of the Morse function f. We let $|x|$ denote the index of the critical point x. For each $x, y \in \text{Crit}\, f$ we let $\mathcal{M}(x, y)$ denote the unparameterised moduli space of flowlines from x to y, of dimension $|x| - |y| - 1$.

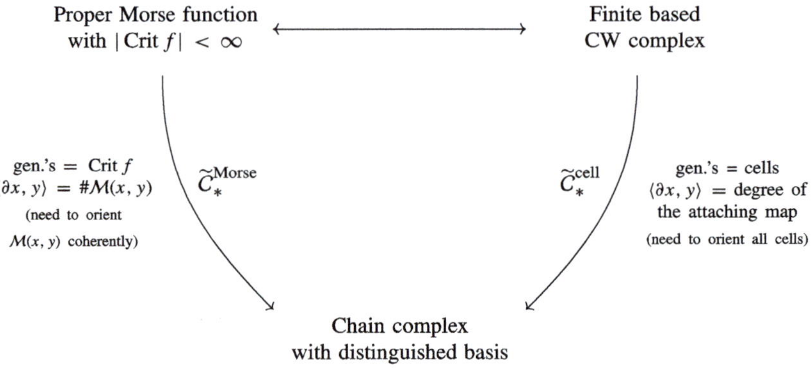

In the next subsection we will construct framed flow categories, which are 'equivalent' to Morse functions with gradient flowlines and to based CW complexes.

6.2.2 *From Morse Functions to Framed Flow Categories*

This section is included mostly for motivational purposes, we describe ideas and constructions which motivate the definitions and results in Sect. 6.3; consequently, no proofs in this section will be given.

Definition 6.2.3 Let f be a Morse-Smale function. We define the *flow category* C_f as the category whose objects are $\text{Ob}\, C_f = \text{Crit}\, f$, and whose non-identity morphisms are given by the unparameterised moduli spaces $\mathcal{M}(x, y)$ of (possibly broken) flowlines from x to y.

For a precise definition of *flow category* and of *Morse flow category*, we refer the reader to [LS14, Definitions 3.12 and 3.13].

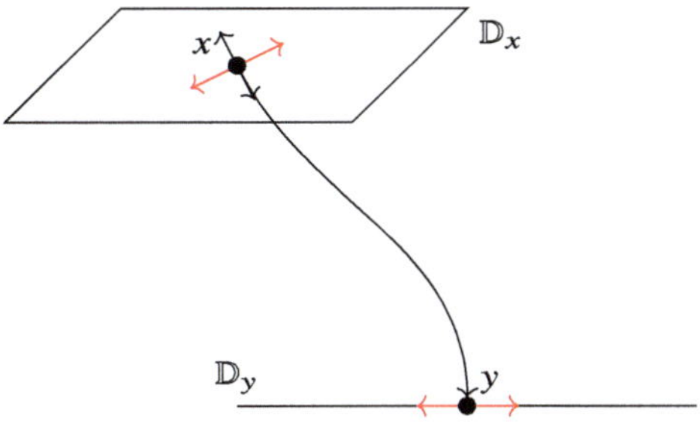

Fig. 6.2 An illustration of the equality $T\mathcal{M}(x, y) \oplus \mathbb{R} \oplus T\mathbb{D}_y = T\mathbb{D}_x$

In order to promote C_f to a *framed* flow category we need to endow each $\mathcal{M}(x, y)$ with a framing of the stable normal bundle. We do it as follows:

- for every downward disc $\mathbb{D}^k$ we choose a trivialisation of $T\mathbb{D}^k$;
- for each pair of critical points x, y with $|x| > |y|$ and connected by a flowline, there is an identification

$$T\mathcal{M}(x, y) \oplus \mathbb{R} \oplus T\mathbb{D}_y = T\mathbb{D}_x,$$

  which gives a stable framing of $T\mathcal{M}(x, y)$ (see Fig. 6.2);
- a stable framing of $T\mathcal{M}(x, y)$ induces a stable framing of the normal bundle after picking coherent embeddings of these moduli spaces in certain Euclidean spaces.

In the case when $|x| = |y| + 1$, the choice of a framing of the stable normal bundle is the same as a (coherent) system of orientations.

6.2.3 The Hypercube Flow Category

This subsection is devoted to a crucial example of flow category, which will be the fundamental building block of the Khovanov stable homotopy type.

The starting point is an innocent-looking function $f: \mathbb{R} \to \mathbb{R}$ with only two critical points, at 0 and 1, which are respectively a minimum and a maximum. For example, we can choose $f(x) = x^2 \cdot \left(\frac{3}{2} - x\right)$.

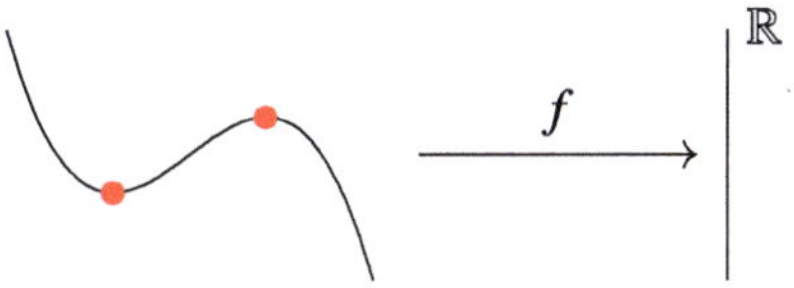

Define the function $f_n\colon \mathbb{R}^n \to \mathbb{R}$ by setting

$$f_n(x_1, \ldots, x_n) = \sum_{i=1}^{n} f(x_i). \tag{6.1}$$

Then $\mathrm{Crit}\, f_n = \{0, 1\}^n$. The index of a critical point $v \in \{0, 1\}^n$ coincide with its L^1 norm:

$$|v| = ||v||_{L^1} = \sum_{i=1}^{n} v_i .$$

Moreover, the set $\mathrm{Crit}\, f_n$ inherits a product partial order by setting $0 < 1$. For two distinct critical points $u, v \in \{0, 1\}^n$, we have that

$$\mathcal{M}(u, v) = \varnothing \qquad \text{if } u \not> v.$$

In the next examples we will try to identify $\mathcal{M}(u, v)$ when $u > v$.

Example 6.2.4 The function $f = f_1\colon \mathbb{R} \to \mathbb{R}$ has only two critical points, and the unparameterised moduli space $\mathcal{M}(1, 0)$ consists of a single point.

Example 6.2.5 The function $f_2\colon \mathbb{R}^2 \to \mathbb{R}$ has four critical points, namely 00, 01, 10, and 11. We have two kinds of moduli spaces, depending on the index difference of u and v.

Using Example 6.2.4, one can check that if the index difference of two critical points of f_2 is exactly 1, then the moduli spaces $\mathcal{M}(u, v)$ consists of a single point. We name them

$$\mathcal{M}(11, 10) = \{a\}, \qquad \mathcal{M}(10, 00) = \{b\},$$
$$\mathcal{M}(11, 01) = \{c\}, \qquad \mathcal{M}(01, 00) = \{d\}.$$

The moduli space $\mathcal{M}(11, 00)$ is 1-dimensional, and its boundary consists of the two broken flowlines labelled ab and cd. The moduli space is in fact an interval with boundary $\{ab, cd\}$. One way to visualise $\mathcal{M}(11, 00)$ is to draw the segment from 10 to 01 (in $\mathbb{R}^2$ this segment lies on the line $x + y = 1$), and noticing that such a segment intersects each flowline from 11 to 00 exactly once. The two endpoints 10 and 01 of the segment correspond to the broken flowlines. See also Fig. 6.3.

Exercise 6.2.6 Identify the flow category associated with f_3. Note that $\mathcal{M}(111, 000)$ is a hexagon: this can be visualised as the section of the unit cube given by the plane $x + y + z = \frac{3}{2}$, which intersects each (possibly broken) flowline from 111 to 000 in exactly one point. See Fig. 6.4.

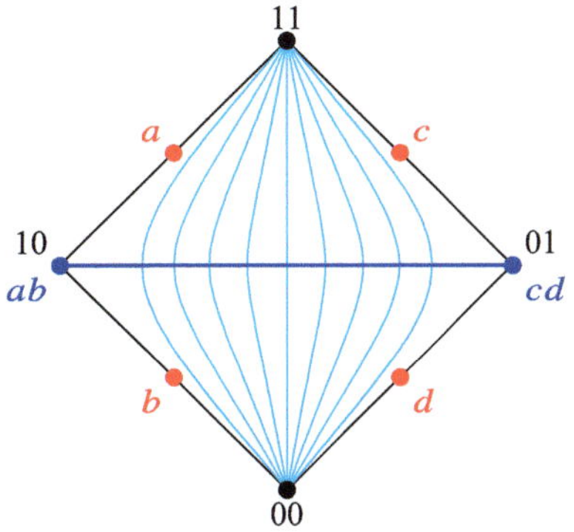

Fig. 6.3 Illustration of the flowlines from 11 to 00 and of the resulting moduli space

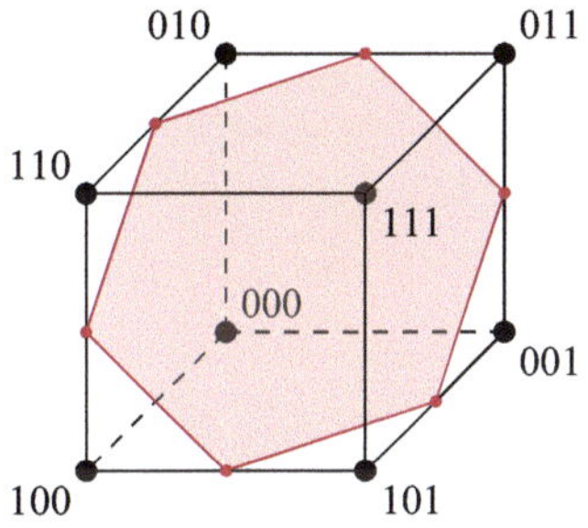

Fig. 6.4 The moduli space $\mathcal{M}(111, 000)$

More generally, the moduli space $\mathcal{M}(1 \cdots 1, 0 \cdots 0)$ in the hypercube flow category is a *permutohedron.*

Definition 6.2.7 The *n-permutohedron* Π^{n-1} is the convex hull of

$$\left\{(\sigma(1), \ldots, \sigma(n)) \in \mathbb{R}^n \mid \sigma \in S_n\right\}.$$

Note that for every permutation $\sigma \in S_n$, we have $\sum_{i=1}^{n} \sigma(i) = \frac{n(n+1)}{2}$. Thus, Π^{n-1} is contained in an $\mathbb{R}^{n-1}$-slice of $\mathbb{R}^n$, and it is in fact an $(n-1)$-dimensional manifold with corners.

The hypercube flow category can also be upgraded to a framed flow category. In order to do so, we need to endow the moduli space $\mathcal{M}(u, v)$, which is a permutohedron, with a *coherent framing of the stable normal bundle.* We hint at its construction by one explicit example.

Consider the square flow category, i.e. the one associated with f_2. A stable framing of the normal bundle of the 0-dimensional moduli space corresponds to a choice of orientation of each of them (either $+$ or $-$). We need to choose orientations such that exactly one or three of the points a, b, c, d are framed positively; for instance, we may choose orientations as in Fig. 6.5.

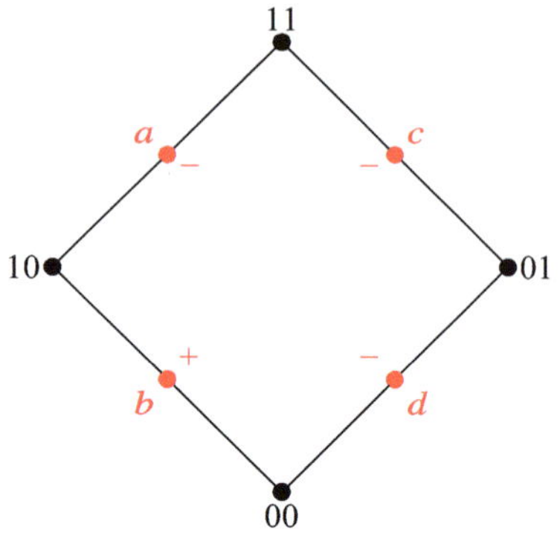

Fig. 6.5 A choice of orientation of the 0-dimensional moduli spaces

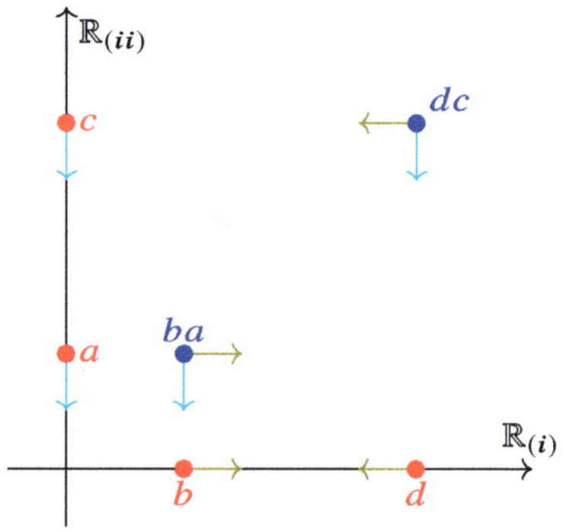

Fig. 6.6 The product orientation on the broken flowlines in the square flow category

We envision the moduli spaces $\mathcal{M}(01, 00)$ and $\mathcal{M}(10, 00)$ as sitting in one copy of $\mathbb{R}$, which we denote by $\mathbb{R}_{(i)}$, and the moduli spaces $\mathcal{M}(11, 10)$ and $\mathcal{M}(11, 01)$ as sitting in another copy of $\mathbb{R}$, which we denote by $\mathbb{R}_{(ii)}$. Thus, we get a product orientation for the broken flowlines ba and dc, embedded in $\mathbb{R}_{(i)} \times \mathbb{R}_{(ii)}$, as illustrated in Fig. 6.6.

We now wish to orient the stable normal bundle of the moduli space $\mathcal{M}(11, 00)$, which is an interval with (unoriented) boundary $\{ba, dc\}$. Such a moduli space should come with an embedding

$$\mathcal{M}(11, 00) \hookrightarrow \mathbb{R}_{(i)} \times \mathbb{R}_{(ii)} \times \mathbb{R}_{\geq 0}$$

whose restriction to the boundary gives the embedding

$$\{ba, dc\} \hookrightarrow \mathbb{R}_{(i)} \times \mathbb{R}_{(ii)}$$

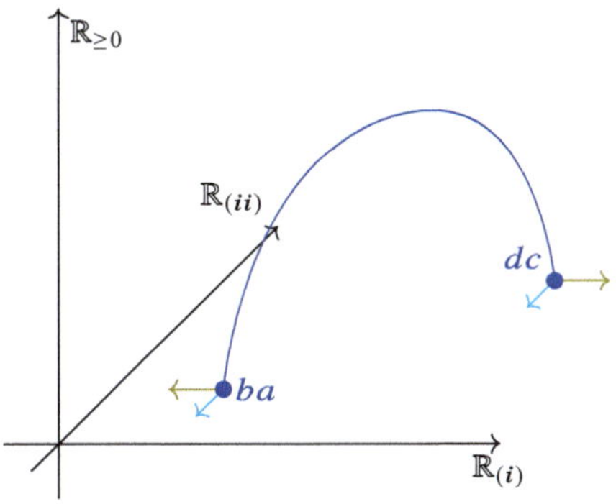

Fig. 6.7 An embedding of the moduli space $\mathcal{M}(11, 00)$ into $\mathbb{R}_{(i)} \times \mathbb{R}_{(ii)} \times \mathbb{R}_{\geq 0}$

constructed above. The coherence condition implies that we can find a framed interval embedded in $\mathbb{R}_{(i)} \times \mathbb{R}_{(ii)} \times \mathbb{R}_{\geq 0}$ with boundary the two framed points (see Fig. 6.7). The possible stable framings of such an embedded interval are a torsor over

$$\pi_1(SO) := \varinjlim \pi_1(SO(n)) \cong \mathbb{Z}/2\mathbb{Z}.$$

More generally, using obstruction theory we can frame the stable normal bundle coherently for all moduli spaces in the hypercube flow category.

6.2.4 The Cohen-Jones-Segal Construction

Cohen-Jones-Segal [CJS95] provide a construction to produce a stable based CW complex out of a framed flow category, which we will outline in one particular case in this section. Their construction shares many similarities with the Pontryagin-Thom construction.

Definition 6.2.8 Two topological spaces X and Y are called *stably homotopy equivalent* (denoted $X \sim_S Y$) if $\Sigma^n X$ and $\Sigma^n Y$ are homotopy equivalent for $n \gg 0$.

Exercise 6.2.9 $\mathbb{T}^2 \sim_S S^1 \vee S^1 \vee S^2$.
Hint: The attaching map of the 2-cell of $\mathbb{T}^2$ is a commutator in π_1 of the 1-skeleton.

Exercise 6.2.10 $\mathbb{C}P^2 \not\sim_S S^2 \vee S^4$.
Hint: Steenrod squares are preserved under stable homotopy equivalence.

We remark that Exercise 6.2.9 shows that the cup product is not preserved under stable homotopy equivalence.

We will now turn to the construction of a specific example of a stable based CW complex, namely the one associated with the square framed flow category.

Each vertex v of the square will give rise a cell of dimension $|v|+2$. The resulting stable homotopy type is that of a point.

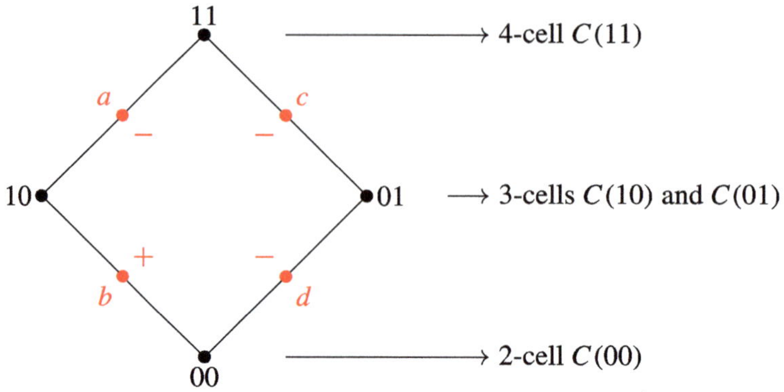

Given $R \gg \varepsilon > 0$, we parameterise the cells as follows:

$$C(11) := [0, R] \times [-R, R]^2 \times [0, R];$$
$$C(10) = C(01) := [0, R] \times [-R, R] \times [-\varepsilon, \varepsilon];$$
$$C(00) := [-\varepsilon, \varepsilon]^2.$$

We construct the based CW complex by attaching the cells in order of increasing dimension. We start with the 2-cell $C(00)$, which is attached to the basepoint $*$ via the unique possible attaching map $\varphi_{00} \colon \partial C(00) \to *$, see Fig. 6.8. We denote the resulting CW complex by X_0, i.e.

$$X_0 := C(00) \cup_{\varphi_{00}} \{*\}.$$

We now move to the 3-cells. See Fig. 6.9 for reference. We view the $[-R, R]$ factor as the $\mathbb{R}_{(i)}$ direction, and we recall that we have an embedding of b and d into $\mathbb{R}_{(i)}$, as framed points.

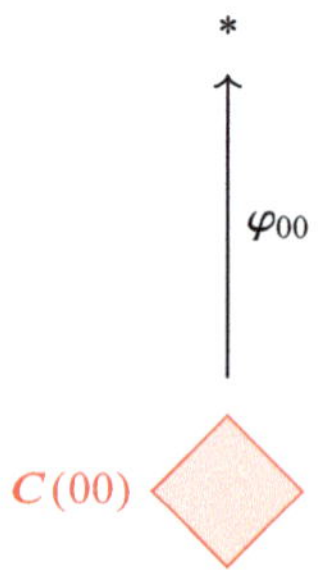

Fig. 6.8 An illustration of the CW complex X_0 built as part of the process of constructing the stable homotopy type associated with the square framed flow category

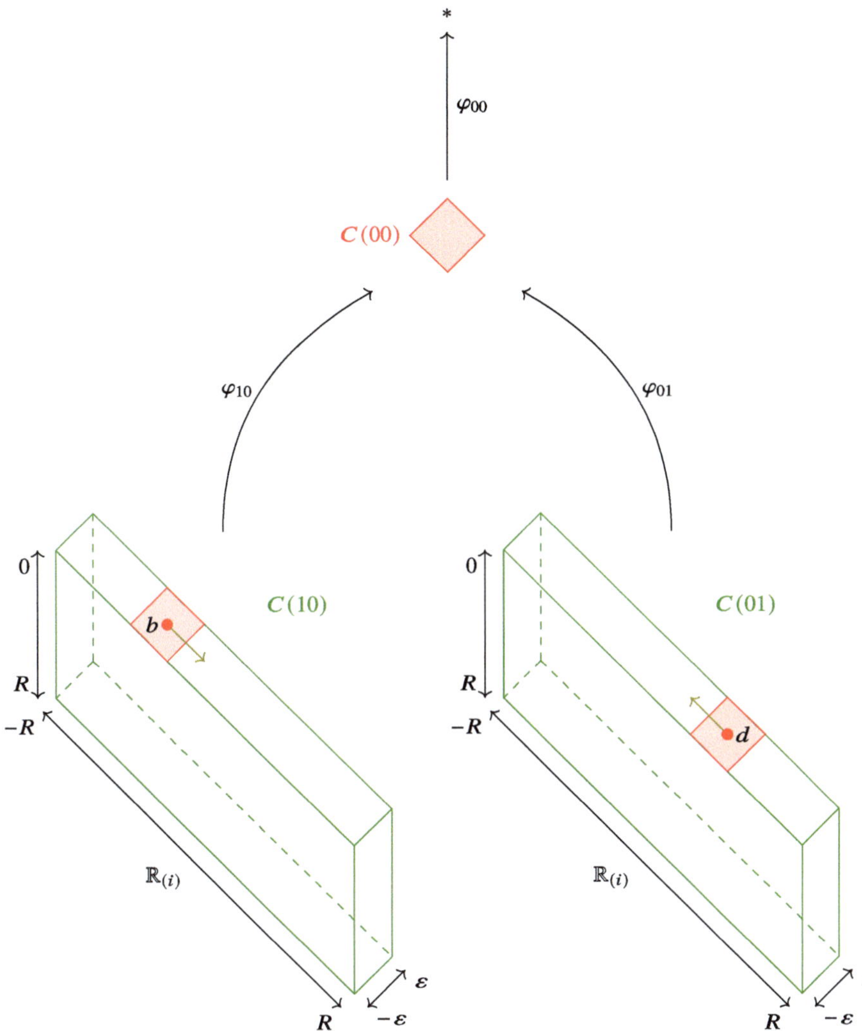

Fig. 6.9 An illustration of the CW complex X_0 built as part of the process of constructing the stable homotopy type associated with the square framed flow category

On the boundary of the 3-cell $C(10)$ (resp. $C(01)$), we draw a copy of the $[-\varepsilon, -\varepsilon]^2$ on $\{0\} \times [-R, R] \times [-\varepsilon, \varepsilon]$, centred at $(0, b, 0)$ (resp. $(0, d, 0)$).

The maps

$$\varphi_{10} : \partial C(10) \to C(00) \cup_{\varphi_{00}} \{*\}$$

$$\varphi_{01} : \partial C(01) \to C(00) \cup_{\varphi_{00}} \{*\}$$

are defined as follows:

- on the embedded copies of $[-\varepsilon, \varepsilon]^2$ the maps restrict to degree ± 1 maps to $C(00)$, with the sign of the degree depending on the framing of b or d;
- on the complement of the embedded copy of $[-\varepsilon, \varepsilon]^2$, the maps send everything to the basepoint $\{*\}$.

The maps so defined descend to continuous maps to $C(00) \cup_{\varphi_{00}} \{*\}$. Using the attaching maps, we construct the CW complex

$$X_1 := (C(10) \cup C(01) \cup X_0)/\sim.$$

We now turn to the last remaining cell, namely $C(11)$. Recall that

$$C(11) = [0, R] \times [-R, R]^2 \times [0, R].$$

We need to define an attaching map

$$\varphi_{11} \colon \partial C(11) \to X_1.$$

We view $\partial C(11)$ as the union of three pieces:

- $\{0\} \times [-R, R]^2 \times [0, R]$;
- $[0, R] \times [-R, R]^2 \times \{0\}$;
- the part of $\partial C(11)$ with 1st and 4th coordinate both non-zero.

Note that the first and the second piece intersect at $\{0\} \times [-R, R]^2 \times \{0\}$.

We consider a copy of $C(01)$ and a copy of $C(10)$ sitting in

$$[0, R] \times [-R, R]^2 \times \{0\} \subset \partial C(11)$$

as the slices

$$[0, R] \times [-R, R] \times [a-\varepsilon, a+\varepsilon] \times \{0\} \quad \text{and} \quad [0, R] \times [-R, R] \times [c-\varepsilon, c+\varepsilon] \times \{0\}$$

respectively, as shown in Fig. 6.10.

We define the map φ_{11} separately on the three pieces of $\partial C(11)$. We start from the blue piece $[0, R] \times [-R, R]^2 \times \{0\}$: with the above picture in mind, it is natural to map the embedded cells $C(01)$ and $C(10)$ inside $\partial C(11)$ to the respective abstract ones via degree ± 1 maps, depending on the framings of a and c, and to extend this to a map from $[0, R] \times [-R, R]^2 \times \{0\}$ to X_1 by sending everything in the complement to $*$.

Then, a naïve approach to extend φ_{11} to all of $\partial C(11)$ would be to map the purple and the black pieces to the basepoint $*$. However, the map so defined would not be continuous, because the whole purple piece of $\partial C(11)$ would be mapped to $*$, except the two red squares in $\{0\} \times [-R, R]^2 \times \{0\}$, which lie in $\partial(\{0\} \times [-R, R]^2 \times [0, R])$. Thus, we modify the map in the purple portion

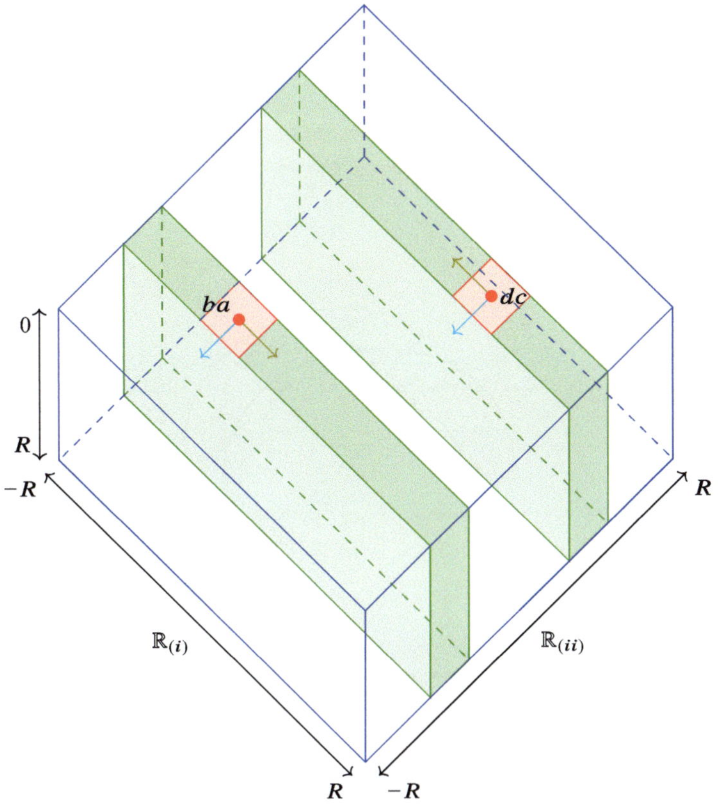

Fig. 6.10 The "blue" piece $[0, R] \times [-R, R]^2 \times \{0\}$ of $\partial C(11)$

of $\partial C(11)$ by using the framed interval from ab to cd. We envision the framed interval constructed above as sitting inside $\{0\} \times [-R, R]^2 \times [0, R]$, where the last coordinate corresponds to $\mathbb{R}_{\geq 0}$. A regular neighbourhood of it can be parameterised by $[-\varepsilon, \varepsilon]^2 \times I$. We can then define the map φ_{11} on $\{0\} \times [-R, R]^2 \times [0, R]$ by setting it on the neighbourhood of the framed interval as

$$[-\varepsilon, \varepsilon]^2 \times I \xrightarrow{pr_1} [-\varepsilon, \varepsilon]^2 \xrightarrow{\sim} C(00) \to X_0 \subset X_1,$$

and by sending the complement to $*$. See Fig. 6.11.

Thus, we have so far defined φ_{11} on $\{0\} \times [-R, R]^2 \times [0, R]$ and $[0, R] \times [-R, R]^2 \times \{0\}$. We extend it to $\partial C(11)$ by sending the rest of $\partial C(11)$ to $*$.

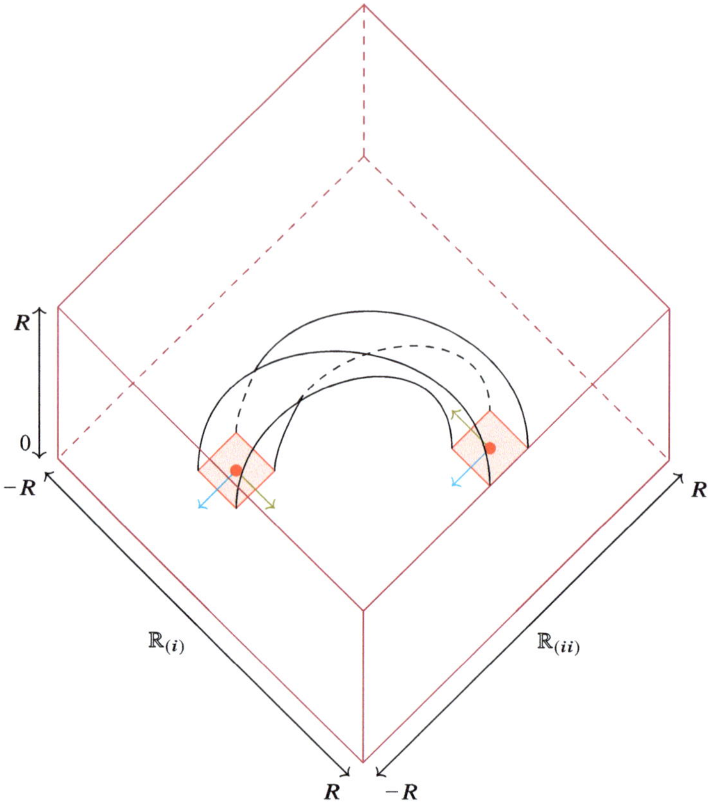

Fig. 6.11 The "purple" piece $\{0\} \times [-R, R]^2 \times [0, R]$ of $\partial C(11)$

6.3 A Khovanov Stable Homotopy Type

Our next goal is to construct a Khovanov homotopy type starting from the Khovanov complex. We quickly recall the construction of the Khovanov complex. For a more detailed exposition we refer the reader to [Kho00, BN02].

6.3.1 *The Khovanov Complex*

Given a link diagram and a crossing, we define the 0-smoothing and the 1-smoothing of the crossing as in Fig. 6.12. Let D be a link diagram with n crossings, labeled in some order from 1 to n. See Fig. 6.13 for an example. Each $v \in \{0, 1\}^n$ determines a *complete resolution* D_v by smoothing crossing i according to the i-th entry of v. Topologically, each complete resolution consists of an unlink. We can organise the

Fig. 6.12 The 0-smoothing and the 1-smoothing of a crossing

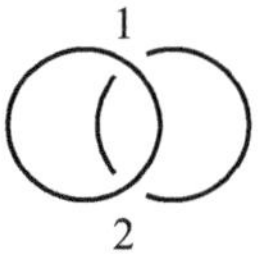

Fig. 6.13 A non-minimal diagram of the 2-component unlink, together with a labelling of the crossings

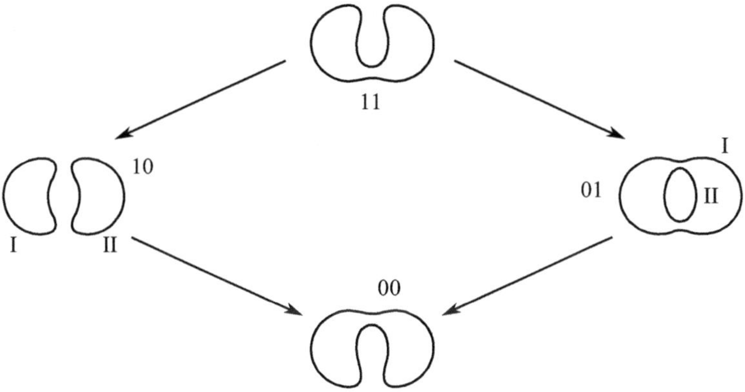

Fig. 6.14 The cube of resolutions of the diagram for the unlink in Fig. 6.13. We have also fixed a total ordering on the circles at each resolution: the resolutions, indicated by the labels I and II for the circles at the resolutions 10 and 01 (The other resolutions have only one circle, so there is only one total order there)

complete resolutions on a hypercube of dimension n, so that each v corresponds to a vertex of the hypercube. See Fig. 6.14.

Let the *weight* or *index* $|v|$ of v be the number of coordinates of v with value 1. Two vertices u and v are connected by an edge e of the hypercube if and only if u and v differ only at one entry, and in such a case the edge is oriented from the vertex with *higher* weight to the vertex with *lower* weight. Note that this convention is opposite to the standard one in the literature (e.g., [Kho00, BN02]); equivalently, in these notes we consider Khovanov homology as opposed to the more popular Khovanov cohomology.

From the cube of resolution Khovanov constructed a complex as follows. Let $V = \mathbb{Z}\langle 1, x\rangle$ be the $\mathbb{Z}$-module spanned by two elements 1 and x. To each $v \in \{0, 1\}^n$ associate the vector space

$$V_v = \bigotimes_{\text{components of } D_v} V,$$

where each factor V is considered as associated to a link component of D_v. The module underlying the Khovanov complex is defined as

$$\mathrm{CKh}(D) = \bigoplus_{v \in \{0,1\}} V_v.$$

Recall that in the construction of framed flow category *spacifying* a given chain complex C one needs to specify a basis of C as well, see Remark 6.2.1. Luckily for us, the $\mathbb{Z}$-module $\mathrm{CKh}(D)$ comes with a preferred basis. For each $v \in \{0, 1\}$, the $\mathbb{Z}$-module V_v has a basis consisting of all possible tensor products of the generators 1 and x for V; we call them the *standard generators* of V_v, and we denote them with a v subscript, e.g. if D_v has four components, after totally ordering them, the element $1x11_v$ is the Khovanov generator that labels the second component by x and other components by 1. As v varies in $\{0, 1\}$, all standard generators give a basis for $\mathrm{CKh}(D)$.

In order to define the differential on $\mathrm{CKh}(D)$, Khovanov defined two maps m and Δ, associated to the merge and split cobordisms respectively:

$$\begin{aligned} m\colon V \otimes V &\to V & \Delta\colon V &\to V \otimes V \\ 1 \otimes 1 &\mapsto 1 & 1 &\mapsto 1 \otimes x + x \otimes 1 \\ 1 \otimes x &\mapsto x & x &\mapsto x \otimes x \\ x \otimes 1 &\mapsto x & & \\ x \otimes x &\mapsto 0 & & \end{aligned}$$

If there is an edge from u to v, then there is an elementary cobordism between D_u and D_v, which is a split saddle or a merge saddle. We define a map

$$\partial_{u,v}\colon V_u \to V_v$$

as the map m or Δ on the V factors associated with the components of D_u and D_v on which the saddle cobordism happens, and as the identity map on all other V factors.

Fix a framing of all the 0-dimensional moduli spaces in the hypercube flow category associated to the function f_n defined in Eq. (6.1) at page 192, such that on every square face, exactly one or three of the points are framed positively. The differential is then defined by summing over all edges:

$$\partial := \sum_{\substack{|u-v|=1 \\ |u|=|v|+1}} (-1)^{s(u,v)} \partial_{u,v},$$

where $s(u, v) = 0$ if $\mathcal{M}(u, v)$ is framed positively and $s(u, v) = 1$ if $\mathcal{M}(u, v)$ is framed negatively.

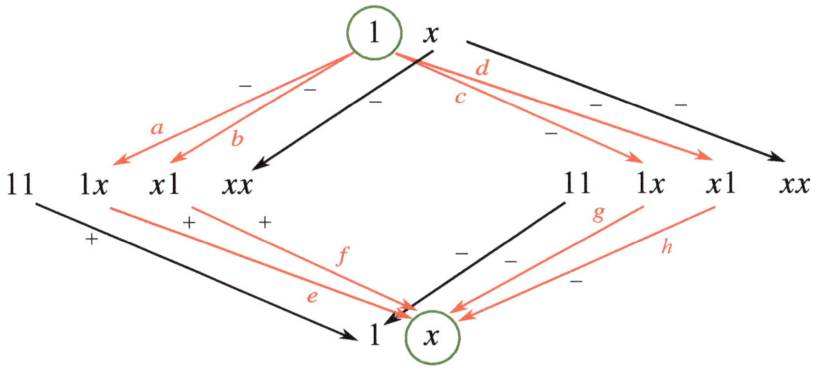

Fig. 6.15 The Khovanov complex associated with the diagram for the 2-component unlink in Fig. 6.13. To simplify the notation, we suppressed the tensor product symbol $\otimes$ and the subscripts specifying the vertex of the hypercube. The relative location of the standard generators indicates the vertex of the hypercube to which it belongs. Compare with Fig. 6.14. For example, the four standard generators located at the left are in $V_{(10)}$; among them, the generator $1x$ is the one that labels the first circle (marked I) by 1 and the second circle (marked II) by x

As an illustration, Fig. 6.15 shows the Khovanov complex for the diagram of the 2-component unlink in Fig. 6.13. The figure shows the 12 standard generators of the Khovanov complex, organised to mirror the structure of the cube of resolutions (Fig. 6.14). The arrows encode the differential: if there is an arrow from y to z, then z appears in the expansion of ∂y with coefficient ± 1 depending on the sign label on the arrow. (For the moment disregard the fact that some arrows are coloured in red and some of the generators are circled in green.) The homology of $(\mathrm{CKh}(D), \partial)$ is the Khovanov homology of the link. However, we do not take homology, but instead construct a framed flow category from the complex.

6.3.2 *The Lipshitz-Sarkar Framed Flow Category*

We will use the example from Figs. 6.13 and 6.14 to describe the framed flow category $\mathcal{X}(D)$ from [LS14].

We start from the set of objects, $\mathrm{Ob}\,\mathcal{X}(D)$, which consists of all standard Khovanov generators. Thus, for our recurring example of the diagram of the 2-component unlink in Fig. 6.13, the **objects** are the 12 generators that are shown in Fig. 6.15.

Notice that an object of $\mathcal{X}(D)$ gives rise to an object in the hypercube flow category (see Sect. 6.2.3), by taking the vertex of the hypercube where the Khovanov generator lives. For example, the object $x_{(11)}$ from Fig. 6.15 is associated to the object (11) of the square flow category.

The rest of this section will be devoted to the constructions of the **moduli spaces** of $\mathcal{X}(D)$. A crucial property is that the moduli space between two given objects of $\mathcal{X}(D)$ will always turn out to be a *trivial* cover of the moduli space connecting the associated objects in the hypercube flow category; in the case of 1-dimensional moduli spaces this property will be enforced by what we call below a choice of a ladybug matching, while for all other dimensions the triviality of the cover will be automatic.

The **0-dimensional moduli spaces** are framed points, given by the Khovanov differential. Following the convention in Fig. 6.15, the elements of 0-dimensional moduli spaces are the arrows shown in the figure, together with a framing specified by the sign label: for example,

$$\mathcal{M}(x_{(11)}, xx_{(01)}) = \{p_-\}, \qquad \mathcal{M}(x_{(11)}, x1_{(01)}) = \varnothing,$$

where p_- is a negatively framed point. More precisely, given two objects / standard generators $\mathbf{y}$ and $\mathbf{z}$ with $|\mathbf{y}| = |\mathbf{z}| + 1$, we represent $\partial\mathbf{y} \in CKh(D)$ as a linear combination of standard generators, and we take the coefficient $\langle\partial\mathbf{y}, \mathbf{z}\rangle$ of $\mathbf{z}$ in such an expansion. Note that by the definition of the Khovanov differential ∂, the only values $\langle\partial\mathbf{y}, \mathbf{z}\rangle$ can take are $+1$, -1, and 0. Then, we set

$$\mathcal{M}(\mathbf{y}, \mathbf{z}) = \begin{cases} \{\text{a positively framed point}\} & \text{if } \langle\partial\mathbf{y}, \mathbf{z}\rangle = +1 \\ \{\text{a negatively framed point}\} & \text{if } \langle\partial\mathbf{y}, \mathbf{z}\rangle = -1 \\ \varnothing & \text{if } \langle\partial\mathbf{y}, \mathbf{z}\rangle = 0 \end{cases}$$

We will construct the higher moduli spaces inductively as disjoint unions of permutohedra, using the fact that $\partial\mathcal{M}$ is the union of products of lower-dimensional moduli spaces already constructed.

Most of the higher moduli spaces are constructed without making choices, but there is one interesting case, which appears during the construction of the **1-dimensional moduli spaces**, where a choice is needed. This is the case of $\mathcal{M}(1_{(11)}, x_{(00)})$ illustrated in Fig. 6.15: $1_{(11)}$ and $x_{(00)}$ are the two standard generators circled in green; the broken flowlines between them, which constitute the boundary of the moduli space, are obtained by following the red arrows. Thus, we can see that

$$\partial\mathcal{M}(1_{(11)}, x_{(00)}) = \{ea, fb, gc, hd\},$$

where a, b, c, d, e, f, g, and h are the red arrows in Fig. 6.15. Of the four broken flowlines, two (namely ea and fb) factor through the vertex 10, and the other two (namely gc and hd) factor through the vertex 01. Note that the vertex through which the broken flowline factors determines its framing.

Thus, if we want it to be a union of permutohedra, $\mathcal{M}(1_{(11)}, x_{(00)})$ should consist of two intervals, which can be chosen in two sensible ways: either an interval connecting ea to gc and one connecting fb to hd (the *right* choice), or an interval connecting ea to hd and one connecting fb to gc (the *left* choice). See Fig. 6.16. In a bombastic moment of self-confidence, we claim that

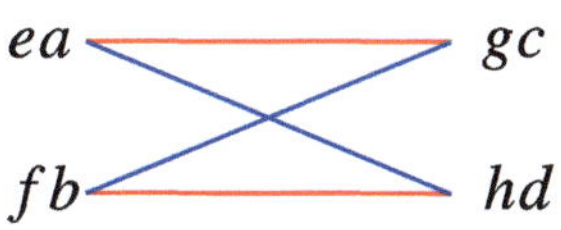

Fig. 6.16 The two possibilities for $\mathcal{M}(1_{(11)}, x_{(00)})$ from the example in Fig. 6.15. The *right* choice is shown in red, and the *left* choice is shown in blue. The two possible matchings are called *ladybug* matchings, and a choice between the two of them must be done each time a ladybug configuration as in Fig. 6.17 appears

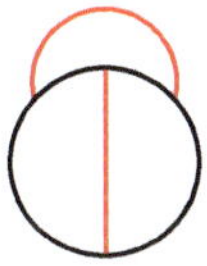

Fig. 6.17 A ladybug matching. The figure represents an all-1 resolution of a link diagram (in fact, of the diagram in Fig. 6.13). The red arcs, one for each crossing of the link diagram, trace the band surgeries that turn each 1-smoothing into a 0-smoothing

Fig. 6.18 The figure on the left shows a link diagram for the 2-component unlink. The figure on the right represents its all-1 resolution together with the arcs that turn each 1-smoothing into a 0-smoothing

we always make the right choice! The 1-dimensional moduli spaces so defined can be upgraded to *framed* moduli spaces, using a framing on the cube flow category.

Remark 6.3.1 In the language of [LS14], we just made a choice of a *ladybug matching*. The name comes from the fact that the diagram D can be alternatively specified by its all-1 resolution, together with a collection of embedded arcs that record the band surgeries needed to turn a 1-smoothing into a 0-smoothing. Figure 6.17 shows the all-1 resolution and the surgery arcs for the diagram in Fig. 6.13. The shape resembling a ladybug motivates the name *ladybug configuration*.

Each time a ladybug configuration appears we must choose a ladybug matching. The fact that we always make the right choice will be important for the next step, namely the construction of the 2-dimensional moduli spaces.

Surprisingly, we will see that the choice of the ladybug matchings on the 1-dimensional moduli spaces is *the only choice we need* to make, under the assumption that each moduli space $\mathcal{M}(\mathbf{y}, \mathbf{z})$ is a disjoint union of permutohedra.

We now turn to the construction of the **2-dimensional moduli spaces**. We start from the example in Fig. 6.18, whose associated cube of resolutions is shown in

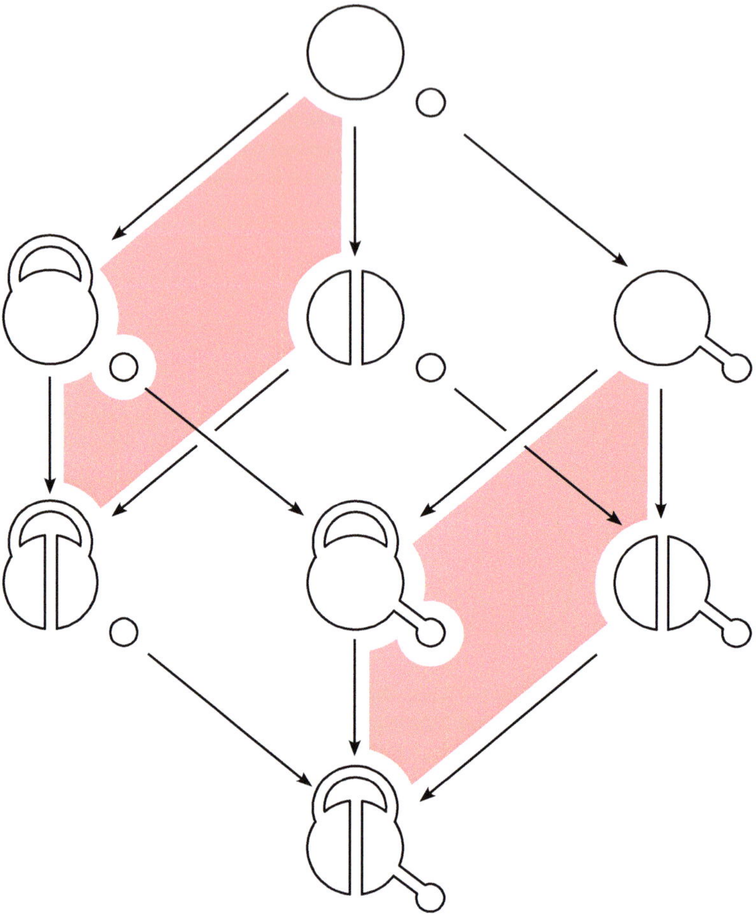

Fig. 6.19 The cube of resolutions for the link diagram in Fig. 6.18. The faces shaded in red correspond to ladybug configurations

Fig. 6.19. One can check that the 3-dimensional cube of resolutions has 2 ladybug faces, which are shaded in light red.

We consider the moduli space

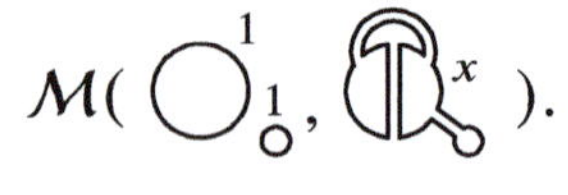

Its boundary (already defined, since it is the union of products of lower-dimensional moduli spaces) contains 12 fully broken flowlines:

- 4 fully broken flowlines for each ladybug face;

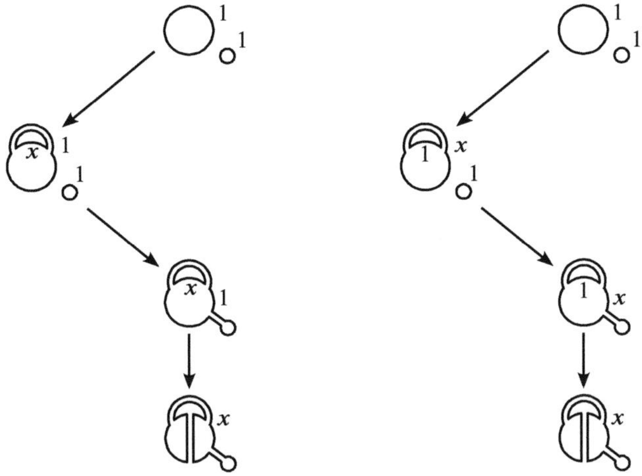

Fig. 6.20 The figure shows two fully broken flowlines in the boundary of a 2-dimensional moduli space arising from the cube of resolutions associated to the link diagram in Fig. 6.18

- 2 fully broken flowlines factoring through

 and ,

namely the ones illustrated in Fig. 6.20;
- 2 fully broken flowlines factoring through

 and ,

which the reader is encouraged to find by themselves.

The boundary of such a moduli space is therefore a 1-dimensional manifold with 12 corners (one for each fully broken trajectory). It turns out that it must be either a dodecagon (i.e., 12-gon) or the disjoint union of two hexagons (i.e., 6-gon). Lipshitz and Sarkar [LS14] showed that if we choose all the ladybug matchings to be right (or all of them to be left), then we get two hexagons. Thus, we can define the moduli space

$\mathcal{M}($ 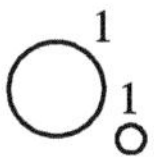,  $).$

as the disjoint union of two filled hexagons, which are permutohedra (see Exercise 6.2.6). Once again, these inherit framings from a framing of the cube flow category.

Finally, we turn to the **moduli spaces of dimension** ≥ 3. Their definition is actually easier than in the cases of lower dimension.

Assume that $\mathbf{y}$ and $\mathbf{z}$ are standard generators of the Khovanov chain complex, such that $|\mathbf{y}| > |\mathbf{z}| + 3$. By induction, we assume that the boundary of the moduli space $\mathcal{M}(\mathbf{y}, \mathbf{z})$ is already defined. It turns out that $\partial\mathcal{M}(\mathbf{y}, \mathbf{z})$ is a regular cover of the moduli space in the cube flow category, i.e. a permutohedron. If $n := |\mathbf{y}| - |\mathbf{z}| > 3$, then the permutohedron Π^{n-1} has simply connected boundary, and therefore every regular cover of it is trivial. Thus, there is no obstruction to defining the moduli space $\mathcal{M}(\mathbf{y}, \mathbf{z})$ as a disjoint union of permutohedra of dimension $n - 1$, which also inherit framings from a framing of the cube flow category.

Recall that the definition of the Khovanov chain complex over $\mathbb{Z}$ involves the choice of a sign assignment; similar further choices must be made for the cube flow category and consequently for the flow category $\mathcal{X}(D)$. As in case of Khovanov homology, those choices do not affect the stable homotopy type of the CW complex constructed from $\mathcal{X}(D)$.

6.4 A Knot Floer Stable Homotopy Type

Once we have the 'spacification' of Khovanov homology, and witness its success, it is natural to ask a similar construction for invariants of knots defined in the Heegaard Floer context. As the construction requires a good understanding of 'moduli spaces', it is instructive to start the process with the version of knot Floer homology where the most control on the moduli spaces is available: grid homology. For an introduction to grid homology see [MOS09, MOST07, OSS15]. For general (pointed or doubly pointed) Heegaard diagrams, the determination of the moduli spaces required for describing the flow category is analytically extremely challenging, and indeed in general these spaces depend on further choices (such as an almost complex structure, or perturbations), which are hard to incorporate into the theory. These difficulties can be bypassed by the combinatorial description provided by grid homology.

Despite the simplicity of the description of the chain complex involved in the definition of grid homology, the presence of 'bubbles' provides serious difficulties for setting up the stable homotopy type. These bubbles do arise already on the homological level, but they cancel out, hence in grid homology they are 'invisible' to the boundary map, resulting in a (relatively) simple theory. In the construction of the framed flow category, however, the bubbles cannot be ignored anymore, bringing in an extra level of complexity in the theory. In this section (following [MS21]) the construction of a spectrum with stable homology giving grid homology is outlined. The appropriate topological invariance of the spectrum, that is, the proof of the fact that the result is a knot-link invariant is, however, still open. This invariance would provide the real power of the theory, since it would open the way to construct further knot invariants, which might go beyond the mere homology theory given by grid homology.

6.4.1 Grid Diagrams

Recall that every knot (and even link) in the standard three-sphere S^3 can be presented by a grid diagram, a combinatorial object illustrated in Fig. 6.21, and described more formally in the next paragraph.

Consider the grid of n^2 small squares in an $n \times n$ square in the plane for some $n \in \mathbb{N}$. Suppose that each column and each row has exactly one marking X and exactly one marking O in one of the small squares in the column or row at hand. By connecting the X in a column to the O in the same column, and the O in the row to the X in the same row (oriented by this order), and with the convention that a vertical segment always passes over a horizontal one, the grid diagram provides the diagram of a (PL embedded) link in S^3. It is not hard to see that every knot and link can be presented in this way, and a method similar to the proof of Reidemeister's fundamental result in knot theory provides two types of moves with the property that grids defining isotopic knots/links can be transformed into each other by a sequence of these 'grid moves'.

Indeed, by identifying the top and bottom edges of the big square, and doing the same with the left-most and right-most edges, we get a grid diagram on the standard torus of S^3. The two grid moves of commutation and stabilization then provide a similar theory for knots in this context.

6.4.2 Grid Homology

The chain complex of grid homology is generated by grid states: a *grid state* $\mathbf{x}$ is simply a bijection between the horizontal and vertical circles in the toroidal grid. A

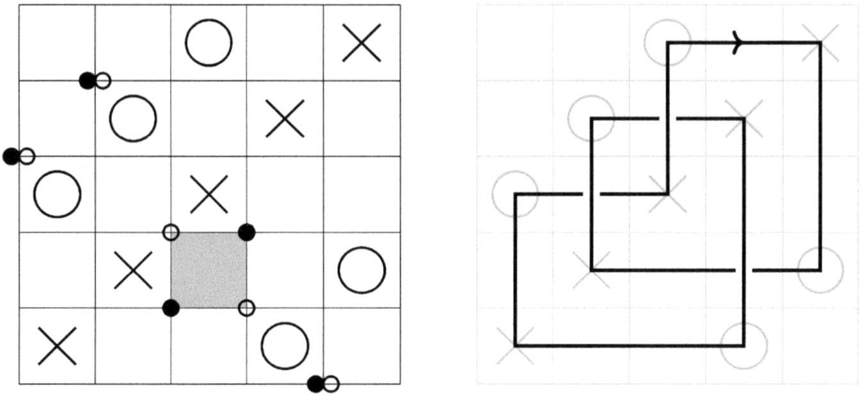

Fig. 6.21 The left hand side shows a grid diagram for the left-handed trefoil knot illustrated on the right. The full and the hollow dots determine two grid states $\mathbf{x}$ and $\mathbf{y}$ respectively. A small square (out of the 25 ones for this grid) is shaded. The shaded small square is also a rectangle from $\mathbf{x}$ to $\mathbf{y}$

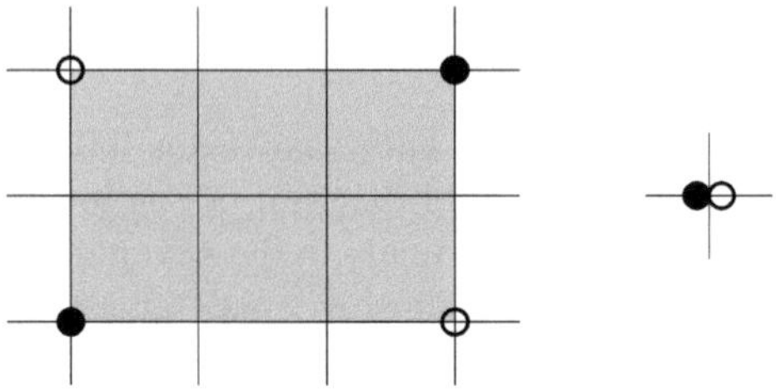

Fig. 6.22 A rectangle connecting the generators **x** (full circle) and **y** (hollow circle). Note that the two generators have all further coordinates equal, symbolized by the crossing on the right

geometric presentation of such an object can be given by choosing an intersection point of the vertical and horizontal circles in the grid in such a way that each vertical and each horizontal circle contains exactly one chosen point. The differential in the chain complex is given by counting "empty rectangles", cf. Fig. 6.22. In more detail, to a pair of two distinct grid states $\mathbf{x}, \mathbf{y}$ we associate 0 if $\mathbf{x}$ and $\mathbf{y}$ differ in more than 2 coordinates. If the difference is in exactly two circles, then the circles of these coordinates (there are two vertical and two horizontal of those) partition the torus into four rectangles, and by choosing an appropriate orientation convention, two of them will point from $\mathbf{x}$ to $\mathbf{y}$ and two from $\mathbf{y}$ to $\mathbf{x}$. Now we count the rectangles from $\mathbf{x}$ to $\mathbf{y}$ which are *empty* (i.e., which do not contain any further coordinate of $\mathbf{x}$) and do not contain X-marking. The actual set-up of the homology theory is slightly more involved (requires a system of weights determined by the O-markings, and consequently is defined over a more elaborate ring); we will not go into the details of the theory here, and advise the interested reader to consult [OSS15] for a more complete account. We just mention here that a crucial step in building the theory is to show that the boundary map ∂ (counting empty rectangles with some weights) has square zero, and therefore gives rise to a chain complex, and hence to homologies. The geometric input in the proof of $\partial^2 = 0$ is an incarnation of Gromov's compactness result, showing that the compositions of two rectangles (counted by ∂^2) come in pairs. (Over the field of two elements this fact is sufficient, over the integers a sign assignment should be also fixed so these pairs contribute zero to the final count.) In turn, this pairing follows from a simple planar geometric observation decomposing the union of two rectangles in two different ways—this is depicted in Figs. 6.23 and 6.24. The situation is somewhat more complicated in further versions of the theory (where we allow the empty rectangle to contain X-markings), as the concatenation of two rectangles might add up to a strip around the torus; this contribution will be cancelled by another such strip, as shown in Fig. 6.33. The significance of this last small problem becomes more crucial in the spacification process, as it will be hinted later on.

The spacification of grid homology will once again be done by constructing a framed flow category C associated with a grid diagram. The objects of the category are the grid states, i.e. the n-tuples of intersection points of the vertical and horizontal circles such that each vertical and each horizontal circle contains exactly one chosen point. Given two grid states $\mathbf{x}$ and $\mathbf{y}$, the moduli space $\mathcal{M}(\mathbf{x}, \mathbf{y})$ will be a smooth manifold with corners. By contrast with the Khovanov homotopy

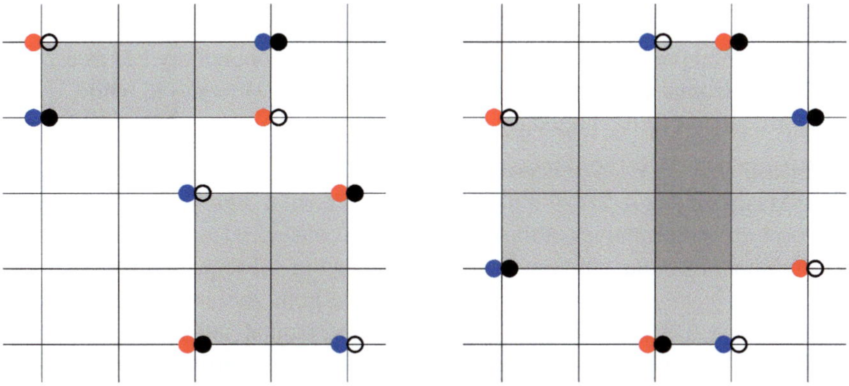

Fig. 6.23 Each of the two figures show a Maslov index 2 domain from the (black) full circle to the hollow circle. Geometrically, the domain consists of two rectangles, which can be disjoint (as in the left) or overlapping (as on the right). In either case, the Maslov index 2 domain can be expressed as the concatenation of two rectangles (i.e., doing one rectangle after the other) in two different ways, depending on which rectangle comes first. One such concatenation factors through the grid state given by red circles, and the other one factors through the grid state given by blue circles. The resulting moduli space is a segment, as in Fig. 6.25, with the endpoints (broken flowlines) given by the two concatenations just described

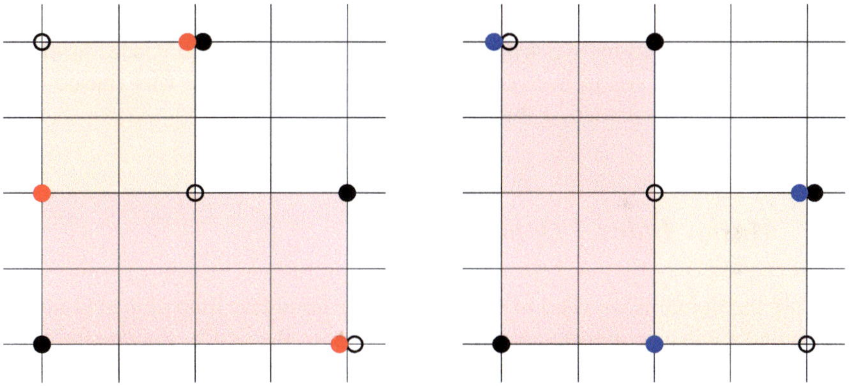

Fig. 6.24 The L-shaped region decomposes into two rectangles in two different ways. Such a region is a domain from $\mathbf{x}$ (full circles) to $\mathbf{y}$ (hollow circles); the two decompositions factor through $\mathbf{w}$ (red circles) or $\mathbf{z}$ (blue circles)

type, the construction is more involved and the moduli spaces will not always be disjoint unions of permutohedra. In the most general case, the moduli space $\mathcal{M}(\mathbf{x}, \mathbf{y})$ will be obtained by gluing together the moduli spaces $\mathcal{M}(D)$ associated with all the *domains* from $\mathbf{x}$ to $\mathbf{y}$ with Maslov index $|\mathbf{x}| - |\mathbf{y}|$. Recall from [OSS15] that a domain from $\mathbf{x}$ to $\mathbf{y}$ in a given grid diagram is a formal linear combination

$$D = \sum n_{i,j} S_{i,j}$$

of (the closures) of the small squares $\{S_{i,j}\}_{i,j=1,\dots,n}$ with the following properties:

- the horizontal part of the boundary (i.e. the part of the boundary ∂D as a 2-chain on the horizontal circles) is a 1-chain $\partial_h D$ with $\partial(\partial_h D) = \mathbf{y} - \mathbf{x}$, while
- the vertical part of the boundary $\partial_v D$ satisfies $\partial(\partial_v D) = \mathbf{x} - \mathbf{y}$.

The domain $D = \sum n_{i,j} S_{i,j}$ is *positive* if all $n_{i,j} \geq 0$.

The Maslov index $\mu(D)$ of a domain D is the formal dimension of the space of holomorphic representatives of the domain (when considered as a 2-chain); it admits a rather simple formula, which will be sufficient in our subsequent discussion. For a domain $D = \sum n_{i,j} S_{i,j}$ from $\mathbf{x}$ to $\mathbf{y}$ we define the point measure $p(D)$ as follows: for each $x_i \in \mathbf{x}$ (i.e. an intersection point of the horizontal and vertical lines which belongs to $\mathbf{x}$) we take the average of the four numbers $n_{i,j}$ corresponding to the four small squares meeting at x_i. Then $p_{\mathbf{x}}(D)$ is the sum of these local contributions for all $x_i \in \mathbf{x}$, and $p(D) = p_{\mathbf{x}}(D) + p_{\mathbf{y}}(D)$. The Maslov index of D is then

$$\mu(D) = p(D).$$

Remark 6.4.1 The general formula for the Maslov index in Heegaard Floer theory also involves a term originating from the geometry of D (called the Euler measure of D); as this term is additive and vanishes for the small squares, in a grid diagram all domains have vanishing Euler measure.

In the next subsections we will give examples of the moduli spaces associated with some domains, and in Sect. 6.4.6 we describe an inductive way (based on the Maslov index) for constructing these spaces.

6.4.3 Maslov Index 1 Domains

These are the domains we need to encounter in the boundary map of grid homology. When $\mathbf{x}$ and $\mathbf{y}$ can be connected by a rectangle (that is, they differ in two circles), the moduli space $\mathcal{M}(\mathbf{x}, \mathbf{y})$ contains two, one or zero points depending on the number of empty rectangles connecting $\mathbf{x}$ and $\mathbf{y}$. An example of an empty rectangle is shown by Fig. 6.22.

This case corresponds to the connecting domains with Maslov index $\mu(D) = 1$. The space associated to such a domain is a single point, hence the moduli space $\mathcal{M}(\mathbf{x}, \mathbf{y})$ (the morphisms between the two objects $\mathbf{x}$ and $\mathbf{y}$ in the flow category) consists of two, one or zero points.

6.4.4 *Maslov Index 2 Domains*

In the next step we consider domains with Maslov index $\mu(D) = 2$ connecting the grid states $\mathbf{x}$ and $\mathbf{y}$. These domains are relatively easy to understand: they are concatenations of two rectangles. A simple analysis shows that this can happen in two ways. In the first case the two rectangles 'move' different coordinate pairs (so the two rectangles move four coordinates in total). Geometrically in this case the domain D is the union of two rectangles (which might not be disjoint, but they do not share sides), see Fig. 6.23.

In the second case one of the moving coordinates of the first rectangle is also a moving coordinate of the second one. In this case, in geometric terms the domain D is an L-shaped region in the grid torus, see Fig. 6.24.

The domain D can be decomposed into the concatenation of two empty rectangles in two different ways, once as a rectangle D_1 from $\mathbf{x}$ to a grid state $\mathbf{w}$ composed with a rectangle D_2 from $\mathbf{w}$ to $\mathbf{y}$, and also as the composition of D'_1 from $\mathbf{x}$ to $\mathbf{z}$ with D'_2 from $\mathbf{z}$ to $\mathbf{y}$. When the two rectangles move four coordinates, the geometric presentation is quite obvious: D_1 (as a subset of the grid torus) agrees with D'_2 and D_2 with D'_1. When the rectangles share moving coordinates, the decomposition is given by the two ways an L-shaped domain decomposes as the union of two rectangles. This is indicated in Fig. 6.24.

Indeed, these decompositions correspond to the broken flowline picture, hence we associate to such a domain an interval as moduli space, with the two endpoints corresponding to the two products of one-point moduli spaces, as it is shown in Fig. 6.25. The moduli space in this case associated to the domain D is a manifold with boundary (an interval, really), and $\mathcal{M}(\mathbf{x}, \mathbf{y})$ will be (as always) the union of all the spaces associated to Maslov index 2 domains from $\mathbf{x}$ to $\mathbf{y}$.

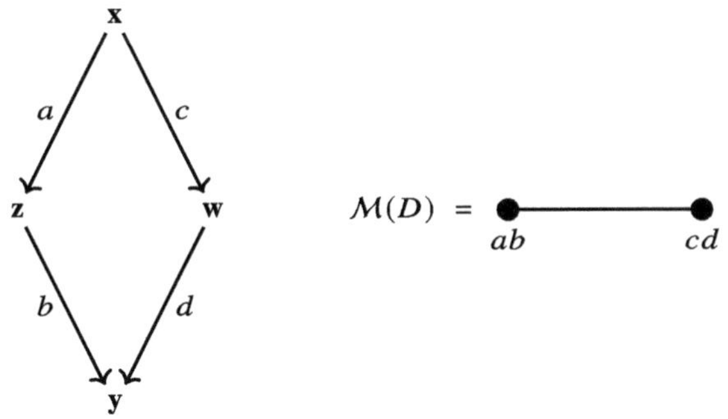

Fig. 6.25 The figure on the left shows the broken flowlines for a domain D of Maslov index 2 (such as those in Figs. 6.23 and 6.24). The figure on the right show the moduli space $\mathcal{M}(D)$

6.4.5 Maslov Index 3 Domains

The same scheme applies if the domain D has Maslov index $\mu(D) = 3$, and D moves 6 coordinates, so it can be decomposed into three 'independent' rectangles, each moving different pairs of coordinates, as shown in Fig. 6.26. The by now customary argument suggests that we need to associate to D a permutohedron, i.e. a hexagon in this dimension.

There are many more possibilities for a domain D to have Maslov index equal to 3, and to still have a hexagon as its moduli space $\mathcal{M}(D)$; see Fig. 6.27 for some examples.

The situation, however, can be much more complicated, and further configurations lead to other natural choices: the domain shown by Fig. 6.28, for example,

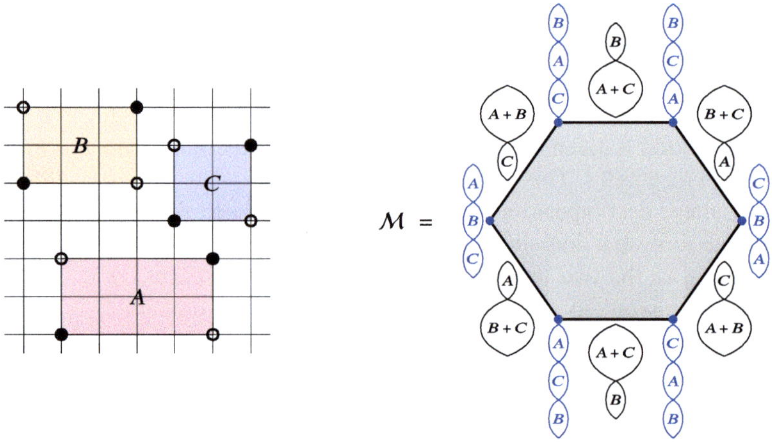

Fig. 6.26 A domain with Maslov index 3 (on the left), and its moduli space (on the right). The boundary of the moduli space consists of all possible (partially or fully) broken flowlines: in particular, the vertices of the hexagon correspond to doubly broken flowlines, whereas the edges consist of once-broken flowlines

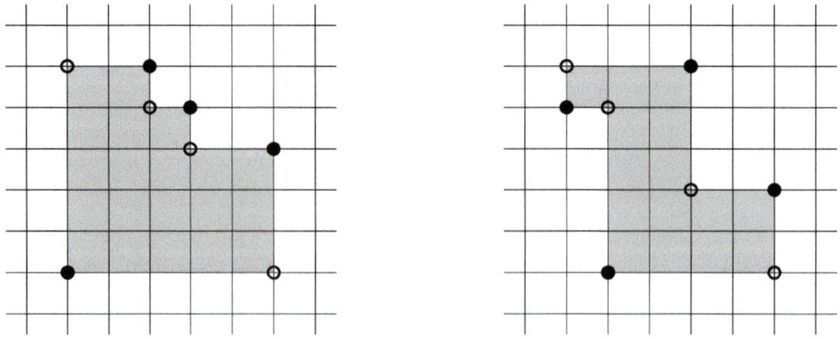

Fig. 6.27 Other domains with Maslov index 3

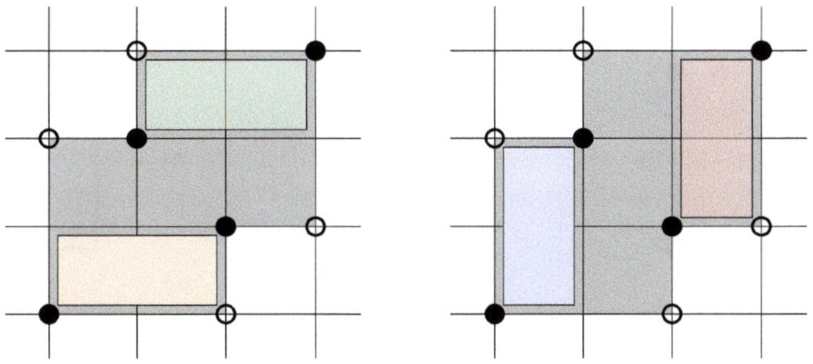

Fig. 6.28 Both sides show a more involved domain with Maslov index 3. The 4 rectangles shown within the given domain are the 4 possible starts of the 8 decompositions of the domain into rectangles. Each such decomposition corresponds to a fully broken flowlines

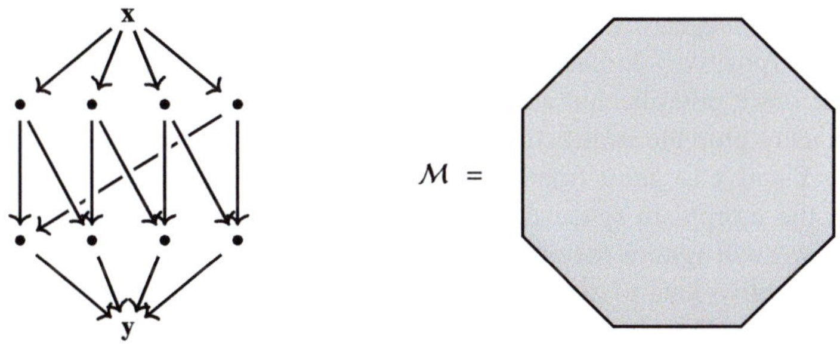

Fig. 6.29 The structure of the eight fully broken flowlines and the moduli space associated with the domain of Maslov index 3 from Fig. 6.28. The reader is invited to identify the eight fully broken flowlines (corresponding to the eight vertices of the octagon) and the partially broken flowlines (corresponding to the sides of the octagon)

provide 4 possible starts from **x** when decomposed into rectangles, and altogether there are 8 broken trajectories, hence the moduli space is an octagon (see Fig. 6.29).

A further example is provided by a rectangle which is not empty (contains a further coordinate of **x** and therefore of **y**). This example has Maslov index 3, and there are four broken trajectories in the decomposition, so the natural choice for the moduli spaces is a rectangle, depicted in Fig. 6.30.

6.4.6 *Maslov Index k Domains*

The above ad hoc arguments is extended to any Maslov index in a more systematic way by induction, using obstruction theory. There are several complications along the way, though.

First of all, remember that in the flow category we require the moduli spaces to be framed manifolds (indeed, manifolds with corners). Because of the presence

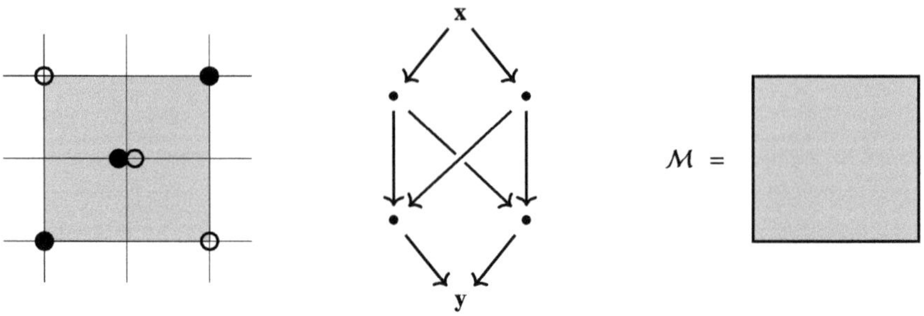

Fig. 6.30 A domain with Maslov index 3 (on the left) from **x** (full circles) to **y** (hollow circles), the fully broken flowlines (in the middle), and the moduli space (on the right). The reader is invited to work out the boundary of the moduli space

of 'bubbles' in Heegaard Floer homology, however, the natural choices for spaces associated to (positive) domains are not always manifolds with corners—we need to consider more complicated stratified spaces. On the positive side, the bubbles will guide us to glue the stratified spaces associated to the domains connecting two grid states **x** and **y** to get a framed manifold with corner, which will then be our choice for the morphism space $\mathcal{M}(\mathbf{x}, \mathbf{y})$ in the flow category. To explain the main ideas, first we will ignore the complications brought by the bubbles, and describe the naïve inductive idea to define the manifold with corner associated to a domain D of Maslov index $k+1$. We will return to the discussion about the role of bubbles in Sect. 6.4.7.

Our inductive hypothesis is therefore that we have already constructed the framed moduli spaces for all (positive) domains of Maslov index $\leq k$; consider now a domain D with Maslov index $\mu(D) = k+1$. By the inductive hypothesis, the framed boundary $\partial\mathcal{M}(D)$ of the moduli space $\mathcal{M}(D)$ is an already defined $(k-1)$-dimensional framed manifold with corners, and hence provides an element $o(D)$ of the framed cobordism group $\Omega^{\mathrm{fr}}_{k-1}$. The latter is the group of $(k-1)$-dimensional framed cornered manifolds taken up to framed cornered cobordism, with group operation given by the disjoint union.

Remark 6.4.2 In reality the boundary defines an element in a more complicated group, which is denoted by $\widetilde{\Omega}^{\mathrm{fr}}_{k-1}$ in [MS21]: while its definition differs from that of $\Omega^{\mathrm{fr}}_{k-1}$, it turns out to be isomorphic to it. Once again, we will ignore this subtlety here.

Thus, for each (positive) domain D with Maslov index $\mu(D) = k+1$, we get an element $o(D) \in \Omega^{\mathrm{fr}}_{k-1}$ (the notation indicates that it is an obstruction class). Putting these classes together, we get a k-cochain on a chain complex generated by the domains, with values in $\Omega^{\mathrm{fr}}_{k-1}$.

Before proceeding any further, let us discuss the chain complex generated by the positive domains—this chain complex will be called the *obstruction chain complex* and denoted by CD_*.

Definition 6.4.3 The chain complex $CD_* = CD_*(\mathbb{G})$ is freely generated over $\mathbb{Z}$ by the positive domains and graded by their Maslov indices:

$$CD_k(\mathbb{G}) = \mathbb{Z}\langle(\mathbf{x}, \mathbf{y}, D) \mid D \in \mathcal{D}^+(\mathbf{x}, \mathbf{y}), \mu(D) = k\rangle.$$

Moreover, CD_* is equipped with the endomorphism ∂, given by

$$\partial(D) = \sum_{\substack{(R,E)\in\mathcal{R}(\mathbf{x},\mathbf{w})\times\mathcal{D}^+(\mathbf{w},\mathbf{y})\\ R*E=D}} s(R)E \quad + \quad (-1)^k \sum_{\substack{(E,R)\in\mathcal{D}^+(\mathbf{x},\mathbf{w})\times\mathcal{R}(\mathbf{w},\mathbf{y})\\ E*R=D}} s(R)E.$$

Here $\mathcal{R}(\mathbf{x}, \mathbf{y})$ denotes the rectangles from $\mathbf{x}$ to $\mathbf{y}$, and s denoted a sign assignment.

It is not hard to see that

Lemma 6.4.4 *The pair (CD_*, ∂) is a chain complex, that is, $\partial^2 = 0$.*

On the other hand, it requires a rather tedious calculation to show that the homology of this complex is rather simple:

Lemma 6.4.5 *The homology of (CD_*, ∂) is isomorphic to $\mathbb{Z}$, supported in grading 0.*

Because of the presence of possible bubbles (a.k.a. *boundary degenerations*), however, later we are forced to consider a more complicated chain complex, an enhancement CDP_* of CD_* where also vectors of partitions and further data describing the positions of the bubbles (and their relative distances) are added to the generators. These extra data will be essential in locating possible bubbles, hence play a crucial role in gluing arguments. (We will omit the details of this extension here.)

The 'earlier defined' cochains (based on the definition of the obstruction $o(D)$) naturally live in a subcomplex of the dual CDP^* of CDP_*, and an important technical step in the construction is the calculation showing that this subcomplex is acyclic.

By associating the obstruction class $o(D)$ to a Maslov index $k+1$ positive domain, we get a $(k+1)$-cochain, which turns out to be a cocycle, that is

$$\delta(o) = 0,$$

where δ is the dual coboundary map. This is proved by showing that $o(\partial D) = 0$ for all positive domains D with $\mu(D) = k+2$. As the cochain complex is acyclic, o is a coboundary, that is, there is a map f on index k domains associating

$$E \mapsto f(E) \in \Omega^{\mathrm{fr}}_{k-1}$$

to the domain E, satisfying

$$o = \delta(f).$$

Now we 'change' the framed manifold $\mathcal{M}(E)$ (that was already constructed) to $\mathcal{M}(E) \amalg -f(E)$. With respect to these new moduli spaces, the obstruction class vanishes. In other words, for each positive domain D with $\mu(D) = k + 1$, its boundary $\partial\mathcal{M}(D)$ is framed null-cobordant, and therefore, can be filled as a framed manifold, which we define to be the moduli space $\mathcal{M}(D)$. Then the induction proceeds.

In this naïve approach we would define the moduli spaces $\mathcal{M}(\mathbf{x}, \mathbf{y})$ in the flow category by taking the disjoint union for all domains of the given Maslov index connecting the two grid states $\mathbf{x}$ and $\mathbf{y}$.

The potential (and actual) presence of bubbles, however, prevent us to proceed as outlined above. We need to allow the spaces $\mathcal{M}(D)$ to be more general than framed manifolds with corners; and these stratified spaces will be glued together in an intricate way to form $\mathcal{M}(\mathbf{x}, \mathbf{y})$, which will ultimately be a framed manifold with corners. We will discuss some of the complications caused by the bubbles in the next subsection through some examples; for the full theory the interested reader is advised to turn to [MS21].

6.4.7 *Bubbles*

Recall that for two grid states $\mathbf{x}$ and $\mathbf{y}$ which differ in exactly two coordinates, we associated four rectangles, two from $\mathbf{x}$ to $\mathbf{y}$ and two from $\mathbf{y}$ to $\mathbf{x}$. Let A be a rectangle from $\mathbf{x}$ to $\mathbf{y}$ and pick a rectangle B from $\mathbf{y}$ to $\mathbf{x}$. The union of A and B will comprise a positive domain D, which geometrically is a strip running around the torus (the direction depending on which rectangle from $\mathbf{y}$ to $\mathbf{x}$ is chosen). See Fig. 6.31. If $\mu(D) = 2$, then the strips will have height or width equal to 1 (otherwise they will contain entire circles, which carry coordinates, driving the Maslov index up).

The composition D of these domains, seen as a domain from $\mathbf{x}$ to $\mathbf{x}$, has a moduli space which is an interval, with one end being the broken flowline given by A (from $\mathbf{x}$ to $\mathbf{y}$) followed by B (from $\mathbf{y}$ to $\mathbf{x}$), and the other end comprising a bubble, see Fig. 6.32.

Remark 6.4.6 In the holomorphic picture the other end of the moduli space, the 'bubble', corresponds to a geometric picture: it is given by a disk which has its entire boundary only on one of the sets of curves (in the grid language either only on the horizontal or on the vertical ones). This explains the pictorial presentation of Fig. 6.32 on the right.

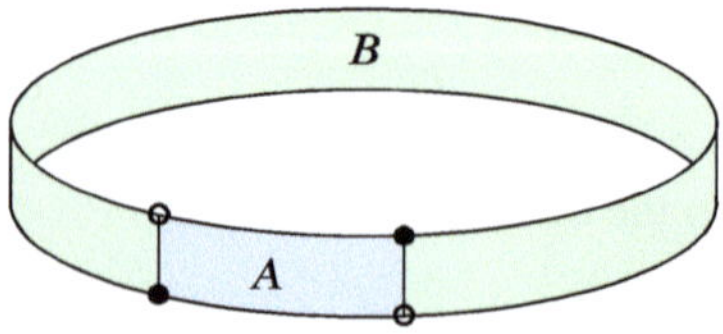

Fig. 6.31 Two rectangles forming a width-one strip around the torus

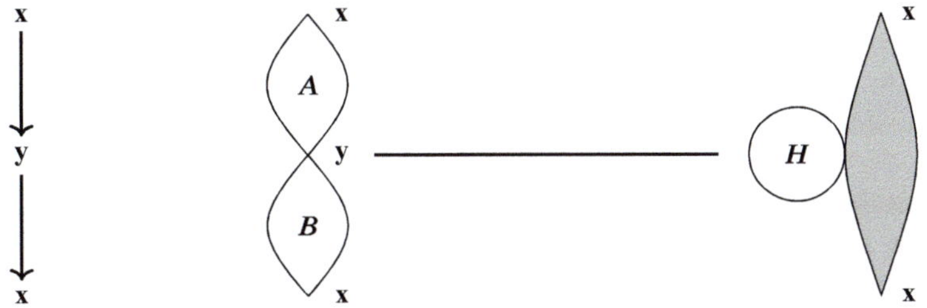

Fig. 6.32 The other end of this moduli space can be viewed as a boundary degeneration. We follow the convention that horizontal bubbles will be drawn on the left, while vertical bubbles on the right

Fig. 6.33 The two thin strips around the torus, passing through the fixed O-marking. Each can be decomposed into two rectangles, as shown (The **x** coordinates are shown as solid dots)

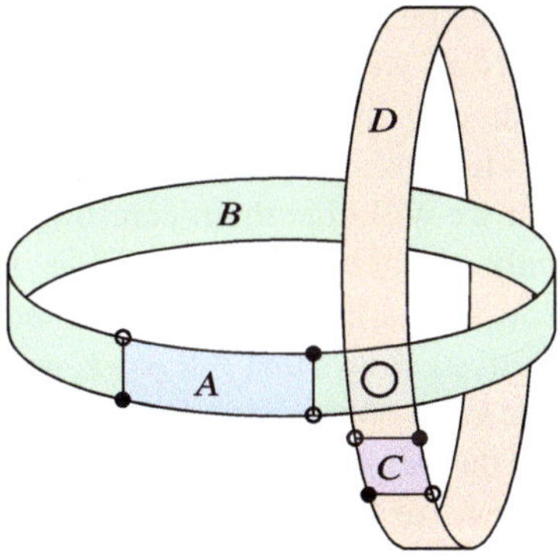

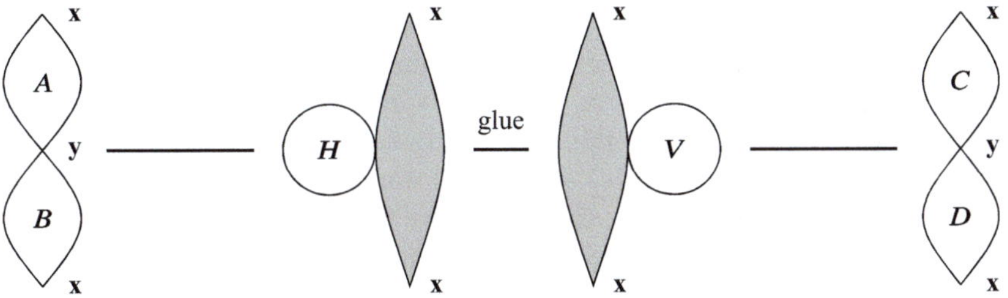

Fig. 6.34 Gluing the two moduli spaces along the two (vertical and horizontal) boundary degenerations

The contribution from the bubbles does not create any confusion in grid *homology*, as these bubbles come in pairs: a horizontal bubble will be cancelled by a vertical one. To exhibit an explicit pairing, we use the O-markings: we take the strips passing through a fixed O-marking to cancel each other, see Fig. 6.33.

In constructing the flow category, we use these pairs to glue the moduli spaces $\mathcal{M}(A + B)$ and $\mathcal{M}(C + D)$ together, along the portion of their 'bubble-type' boundary, as depicted in Fig. 6.34. In this simple example therefore both moduli spaces $\mathcal{M}(A + B)$ and $\mathcal{M}(C + D)$ are intervals, and both admit two different types of boundaries: one corresponding to a broken flowline, and the other to a bubble. The gluing is performed on the bubble-type boundary (where some extra data should be recorded to get a framed manifold with boundary at the end). A crucial feature of the end-result is that the glued-up space no longer has any strata consisting only of bubbles.

The situation encountered above is the first instance when bubbles should be taken into account in our constructions. In grid homology we only need Maslov index 1 domains (to define the boundary map in the chain complex) and Maslov index 2 domains (to prove that the square of the boundary map is zero, i.e. we do have a chain complex). In the spacification process, however, domains with higher Maslov indices should be also taken into account—indeed, we need to work with all positive domains connecting grid states. Therefore higher dimensional bubbles should be examined and handled.

Next we describe a Maslov index 3 example. Continuing from the previous example, consider the domains $2A + B$ and $A + C + D$ from $\mathbf{x}$ to $\mathbf{y}$. Both the moduli spaces $\mathcal{M}(2A+B)$ and $\mathcal{M}(A+C+D)$ should have bubble type boundaries, but we will glue them carefully so that the glued-up space has no strata consisting only of bubbles. For $\mathcal{M}(2A + B)$, the only completely broken flowline is ABA, and after a little analysis, we see that $\mathcal{M}(2A + B)$ should be a triangle. A similar analysis shows that $\mathcal{M}(A+C+D)$ should be a pentagon. Each of these spaces have one edge on the boundary corresponding to bubbling, and the edge is parametrized by the *height* of the bubble. (We will not discuss the concept of the height of a bubble, or the gluing parameters appearing later—these technical data play crucial role in the gluing theory.) We then glue the two edges together—by matching their heights—and the new moduli space is a smooth manifold with corners, with no stratum consisting only of bubbles. This process is illustrated in Fig. 6.35.

Pushing one dimension further, and continuing from the above examples, consider the domains $2A + 2B$, $A + B + C + D$, and $2C + 2D$, all connecting $\mathbf{x}$ to $\mathbf{x}$. As it was noted earlier, this time the three moduli spaces $\mathcal{M}(2A + 2B)$, $\mathcal{M}(A + B + C + D)$, and $\mathcal{M}(2C + 2D)$ will not be manifolds with corners, but will be Whitney stratified spaces. Each will have bubble-type boundaries, but they will be glued together carefully to produce a smooth manifold with corners, with no stratum consisting only of bubbles. We will not describe the full gluing, but rather the gluing near the most complicated portions of their boundaries, namely the portion of the boundary with two bubbles. Each of these three moduli spaces has an interval worth of such 'double' bubbling as shown in Fig. 6.36.

As we got used to gluing horizontal bubbles to vertical bubbles, we might expect that the 'double horizontal bubbling' of I_1 be glued to the 'double vertical bubbling' of I_3, and the double 'mixed' bubbling of I_2 (one horizontal bubble and one vertical bubble) be glued to itself, hence giving rise to a quotient of I_2 in the gluing. What really happens is, in fact, more complicated, because all the above portions of the moduli spaces corresponding to double bubblings are glued together, as we shall see in Fig. 6.37. Nonetheless, I_2 will be glued to itself, and this is why in what follows we consider the quotient $\mathcal{M}(A + B + C + D)/\sim$, obtained by quotienting $I_2 \subset \mathcal{M}(A + B + C + D)$ by $d \sim -d$.

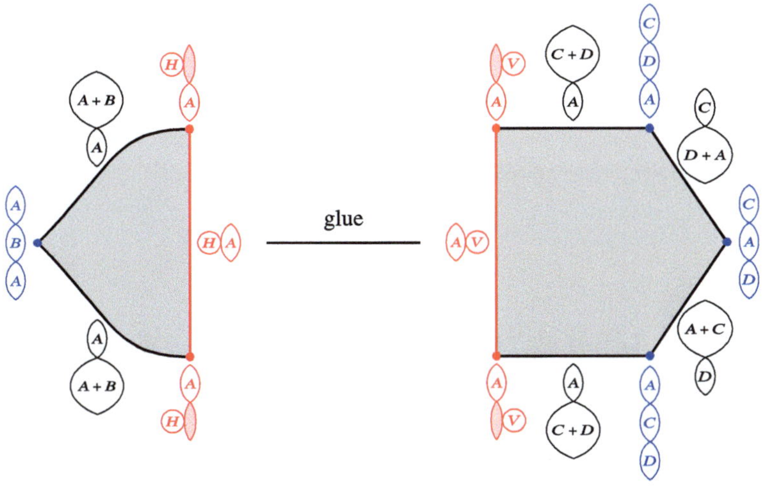

Fig. 6.35 The moduli space $\mathcal{M}(2A + B)$ (the triangle) is glued to the moduli space $\mathcal{M}(A + C + D)$ (the pentagon) along their bubble-type boundaries (Note that the glued-up moduli space is a rectangle, not a hexagon)

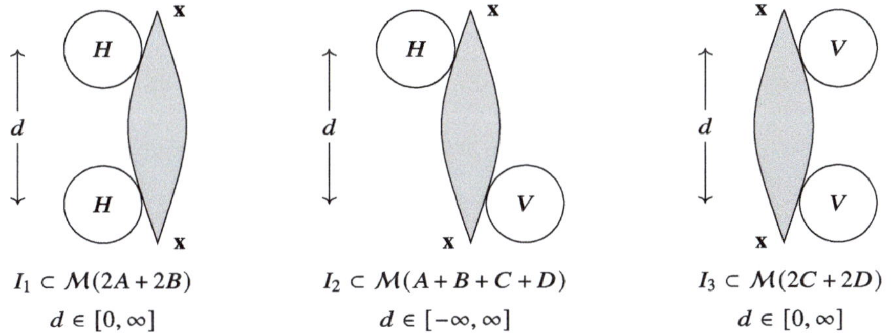

Fig. 6.36 The three intervals, I_1, I_2, I_3, of double-bubbles in $\mathcal{M}(2A+2B)$, $\mathcal{M}(A+B+C+D)$, $\mathcal{M}(2C+2D)$, respectively. The intervals are parametrized by d, the height difference between the bubbles, and we have $I_1 = I_3 = [0, \infty]$ and $I_2 = [-\infty, \infty]$

We now describe how $\mathcal{M}(2A+2B)$, $\mathcal{M}(A+B+C+D)/\sim$, and $\mathcal{M}(2C+2D)$ are glued in a neighborhood of I_1, $I_2/\sim$, and I_3. Each of the intervals I_1, $I_2/\sim$, and I_3 is parametrized by $[0, \infty]$, and they are glued together respecting this parametrization. To extend this gluing to a neighborhood, we first need to understand the neighborhoods of these three intervals in $\mathcal{M}(2A + 2B)$, $\mathcal{M}(A + B + C + D)/\sim$, and $\mathcal{M}(2C + 2D)$.

Towards that end, consider the stratification of $\mathbb{R}^3$ given by the Whitney umbrella: the ambient space is $\mathrm{Sym}^2(\mathbb{C})/\mathbb{R} \cong \mathbb{R}^3$—where the $\mathbb{R}$-action is by translating the real parts—stratified by the signs of the imaginary parts. (There is

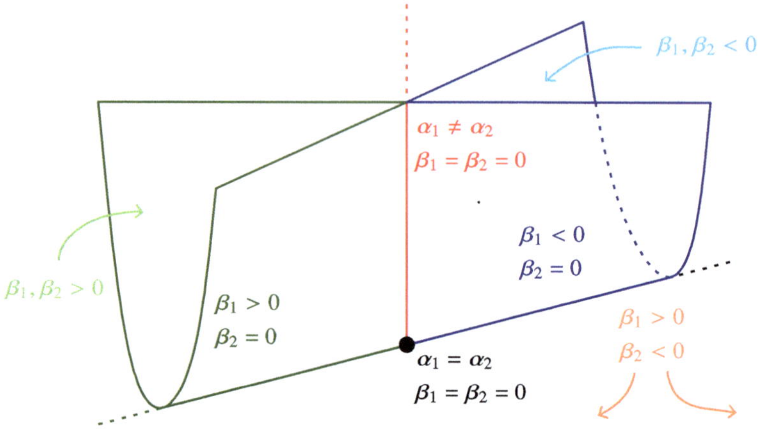

Fig. 6.37 The Whitney umbrella stratification of $\mathbb{R}^3$, seen as $\mathrm{Sym}^2(\mathbb{C})/\mathbb{R}$. Let $\alpha_1+i\beta_1$ and $\alpha_2+i\beta_2$ denote the two complex numbers, which are well-defined up to the $\mathbb{R}$-action $(\alpha_1, \alpha_2) \mapsto (\alpha_1 + t, \alpha_2 + t)$. There is one 0-dimensional stratum, one 1-dimensional stratum, two 2-dimensional strata, and three 3-dimensional strata, which are shown in different colors and are labelled

an extra 0-dimensional stratum when the real parts are equal and both the imaginary parts are 0.) See Fig. 6.37.

Given $(\alpha_1+i\beta_1, \alpha_2+i\beta_2) \in \mathrm{Sym}^2(\mathbb{C})$, the two real parts encode the heights of the two bubbles, while the imaginary parts encode the gluing parameters for gluing the two bubbles, and their signs encode whether the bubbles are glued in as horizontal bubbles or as vertical bubbles.

The closure of the red interval (described by the equations $\beta_1 = \beta_2 = 0$) corresponds to double bubbling, and is parametrized by $d = |\alpha_1 - \alpha_2| \in [0, \infty]$. The interval $I_2 \subset \mathcal{M}(A + B + C + D)$ locally looks like $I_2 \subset I_2 \times \mathbb{R}^2_{\geq 0}$, where the $\mathbb{R}_{\geq 0}$ are gluing parameters for gluing the two bubbles. Therefore, after quotienting, $I_2/\sim\, \subset \mathcal{M}(A + B + C + D)/\sim$ locally looks like the red stratum inside the orange stratum. (The orange stratum has one imaginary part positive and one imaginary part negative, which corresponds to gluing one bubble in as the horizontal bubble H and one bubble in as the vertical bubble V.) Similarly, $I_1 \subset \mathcal{M}(2A + 2B)$ locally looks like the red stratum inside the light green stratum and $I_3 \subset \mathcal{M}(2C + 2D)$ locally looks like the red stratum inside the light blue stratum. The Whitney umbrella witnesses how the light green, orange, and light blue strata glue together to produce the smooth manifold $\mathbb{R}^3$, and that is the local model of how the neighborhoods of I_1, $I_2/\sim$, I_3 are glued together.

Acknowledgments MM acknowledges that: This project has received funding from the European Union's Horizon 2020 research and innovation programme under the Marie Sklodowska-Curie grant agreement No. 893282. AS was partially supported by NKFIH Grant K146401. SS was partially supported by NSF grant DMS-1905717.

References

[Ale28] James W Alexander. Topological invariants of knots and links. *Transactions of the American Mathematical Society*, 30(2):275–306, 1928.

[BN02] Dror Bar-Natan. On Khovanov's categorification of the Jones polynomial. *Algebraic & Geometric Topology*, 2(1):337–370, 2002.

[Car81] Gunnar Carlsson. A counterexample to a conjecture of Steenrod. *Invent. Math.*, 64(1):171–174, 1981.

[CJS95] Ralph L Cohen, John DS Jones, and Graeme B Segal. Floer's infinite dimensional Morse theory and homotopy theory. *The Floer memorial volume*, pages 297–325, 1995.

[Hat02] Allen Hatcher. *Algebraic topology*. Cambridge University Press, 2002.

[Jon85] Vaughan F. R. Jones. A polynomial invariant for knots via von Neumann algebras. *Bull. Amer. Math. Soc. (N.S.)*, 12(1):103–111, 1985.

[Kho00] Mikhail Khovanov. A categorification of the Jones polynomial. *Duke Mathematical Journal*, 101(3):359–426, 2000.

[LS14] Robert Lipshitz and Sucharit Sarkar. A Khovanov stable homotopy type. *Journal of the American Mathematical Society*, 27(4):983–1042, 2014.

[Mil65] John Milnor. *Lectures on the h-cobordism theorem.* Princeton University Press, Princeton, NJ, 1965. Notes by L. Siebenmann and J. Sondow.

[MOS09] Ciprian Manolescu, Peter Ozsváth, and Sucharit Sarkar. A combinatorial description of knot Floer homology. *Ann. of Math. (2)*, 169(2):633–660, 2009.

[MOST07] Ciprian Manolescu, Peter Ozsváth, Zoltán Szabó, and Dylan Thurston. On combinatorial link Floer homology. *Geom. Topol.*, 11:2339–2412, 2007.

[MS21] Ciprian Manolescu and Sucharit Sarkar. A knot Floer stable homotopy type. *arXiv:2108.13566*, 2021.

[OS04] Peter Ozsváth and Zoltán Szabó. Holomorphic disks and knot invariants. *Adv. Math.*, 186(1):58–116, 2004.

[OSS15] Peter S. Ozsváth, András I. Stipsicz, and Zoltán Szabó. *Grid homology for knots and links*, volume 208 of *Mathematical Surveys and Monographs*. American Mathematical Society, Providence, RI, 2015.

[Ras03] Jacob Andrew Rasmussen. *Floer homology and knot complements.* Harvard University, 2003.

[Rei27] Kurt Reidemeister. Elementare Begründung der Knotentheorie. In *Abhandlungen aus dem Mathematischen Seminar der Universität Hamburg*, volume 5, pages 24–32. Springer, 1927.

MIX
Papier aus verantwortungsvollen Quellen
Paper from responsible sources
FSC® C105338

If you have any concerns about our products,
you can contact us on
ProductSafety@springernature.com

In case Publisher is established outside the EU,
the EU authorized representative is:
Springer Nature Customer Service Center GmbH
Europaplatz 3, 69115 Heidelberg, Germany

Printed by Libri Plureos GmbH
in Hamburg, Germany